COMMUNICATION AND
ENGAGEMENT WITH SCIENCE
AND TECHNOLOGY

Science communication seeks to engage individuals and groups with evidence-based information about the nature, outcomes, and social consequences of science and technology. This text provides an overview of this burgeoning field—the issues with which it deals, important influences that affect it, the challenges that it faces. It introduces readers to the research-based literature about science communication and shows how it relates to actual or potential practice. A "Further Exploration" section provides suggestions for activities that readers might do to explore the issues raised. Organized around five themes, each chapter addresses a different aspect of science communication:

- Models of science communication—theory into practice
- Challenges in communicating science
- Major themes in science communication
- Informal learning
- Communication of contemporary issues in science and society

Relevant for all those interested in and concerned about current issues and developments in science communication, this volume is an ideal text for courses and a must-have resource for faculty, students, and professionals in this field.

John Gilbert is Professor Emeritus of The University of Reading, Visiting Professor of King's College London and Editor-in-Chief, *International Journal of Science Education*.

Susan Stocklmayer is Professor of Science Communication and Director of the Australian National Centre for the Public Awareness of Science at the Australian National University.

John Gilbert and Susan Stocklmayer are Joint Editors of *International Journal of Science Education, Part B: Communication and Public Engagement*.

COMMUNICATION AND ENGAGEMENT WITH SCIENCE AND TECHNOLOGY

Issues and Dilemmas

A Reader in Science Communication

Edited by John K. Gilbert and
Susan Stocklmayer

Routledge
Taylor & Francis Group

NEW YORK AND LONDON

First published 2013
by Routledge
605 Third Avenue, New York, NY 10017
4 Park Square, Milton Park, Abingdon, Oxon OX14 4RN

Routledge is an imprint of the Taylor & Francis Group, an informa business

Library of Congress Cataloging in Publication Data
Communication and engagement with science and technology : issues and dilemmas : a reader in science communication / edited by John K. Gilbert and Susan M. Stocklmayer. — 1 [edition].
pages cm
Includes bibliographical references and index.
ISBN 978-0-415-89625-2 (hbk : acid-free paper) — ISBN 978-0-415-89626-9 (pbk : acid-free paper) — ISBN 978-0-203-80752-1 (e-book) (print) 1. Communication in science. I. Gilbert, John (John K.) editor of compilation. II. Stocklmayer, Sue, editor of compilation.
Q223.C6555 2013
501'.4—dc23
2012019156

ISBN: 978-0-415-89625-2 (hbk)
ISBN: 978-0-415-89626-9 (pbk)
ISBN: 978-0-203-80752-1 (ebk)

Typeset in Bembo
by Saxon Graphics Ltd, Derby

CONTENTS

PART III
Major Themes in Science Communication   91

PART IV
Informal Learning   163

PREFACE

The Changing Face of Science Communication

John K. Gilbert and Susan Stocklmayer

'Science communication' is a convenient phrase for 'communications concerned with science, engineering, or technology'. The interplay between these three aspects has increased steadily over the past few hundred years, accompanied by increased industrialization worldwide, but the need to communicate their impact has been recognized only relatively recently. Science communication as a discipline (as distinct from science journalism) has emerged in the past 20 years or so, and has deepened and strengthened as scientific and technological issues of importance to humanity have gripped the public's attention. In this book, science, engineering, and technology—and sometimes mathematics—are often grouped together as 'science'. This is for simplicity of writing, and we acknowledge that it is less than ideal. Other expressions such as STEM (Science, Technology, Engineering and Mathematics) have a real place in educational literature, but we have decided that this acronym, and others like it, have formal overtones that are out of place here.

Increasingly, various interactions of science, engineering, and technology have affected people's personal lives through better health, better living standards, and better nutrition. More efficient transport has altered the social dynamics of villages, towns, and cities. At the same time, the economic impacts of changing agricultural practices and unevenly distributed resources have had an effect on every citizen in the developed and developing world. In sum, our cultural lives have changed to the point where we no longer perceive ourselves or others as we did even fifty years ago. These global changes make a new approach to science communication imperative.

Changes in Who Communicates

When Western science first began to be institutionalized, beginning with the founding of the Royal Society in England in 1660, scientists met face to face in order to share their discoveries. Engineers also began to do so. As the number of practitioners grew, their results were usually published in printed papers intended only for their peers. The specialized technical languages that grew up around this practice made the papers accessible only to the initiated—that is, to other scientists and engineers. Scientists, in particular, began to be seen as an elite group, cleverer than other people, and communicating in a mysterious and confusing secret language. This idea of elitism was reinforced though formal education systems, beginning with the introduction of 'science' as a formal school subject in the late 1800s. From 1860 onwards, in the UK at least, the purpose of the science curriculum was predominantly to screen students and prepare those perceived to be the most able for the study of sciences at university level. The outcome of this approach was that the barrier to access and understanding of science was steadily raised. This was despite some dissenting voices, who from the beginning argued that school science should be included in the education of every citizen.

The practice of science today, however, needs public support and public funding. This need for support has been graphically underlined by crises of science that have affected many nations. Mad cow disease (BSE), bird flu (SARS), genetic modifications of crops and animals, climate disasters, radiation disasters—the list goes on. The public voices from many quarters have been loud in condemnation of poor communication practices attached to many of these crises, and scientists have often been blamed. Those interested in science communication need to hear these varied public concerns and understand their effects on the practices and policies of science. To draw an analogy from chemistry, science communication has become 'multivalent'. That is, it involves a wide range of people in different groups, interacting with each other.

Changes in how Communication can Take Place

The present recognition of science communication as an important aspect of the practice of science itself has come about because ordinary people are now accessing information in an unprecedented way, in both quantity and quality. Information about science, however, is often confusing and still difficult both to access and to understand. It would be easy to say that this is not very important—that scientists will continue to research and publish, no matter what anyone else thinks or does—but this is a naïve view of how science works. The practice of science has changed, and with it, the need for science communication.

Communications are now produced by, and directed at, scientists, mediators, funding agencies, politicians, and many other public groups. There are many players and they communicate through different modes. In addition to a massive growth in the number of scientific journals, 'mediation' is now provided between the communities of scientists and between their many 'publics' in a wide variety of ways including newspapers, magazines, TV, books, museums and science centres and, increasingly, the Internet. Each of these modes has a particular strength and is well received by different audiences. Each also requires to be better understood in terms of its impact, its strengths and its weaknesses.

These changes mean that well-educated scientists, engineers, and technologists must today both know about, and have skills in, science communication. Moreover, there continues to be a steady growth in jobs that entail not only specialist knowledge of science but also these particular science communication understandings and skills. All knowledge that is valued rests on a consensus by some group of people as to what is distinct about it and why it should be valued. Each distinct claim to be a form of knowledge rests on its epistemology, the grounds for making the claim. Science communication, which is only slowly establishing its claim to be a distinct form of knowledge, rests on the epistemologies of science. In addition to those, it rests on the use of social sciences to explore the impact of science on the psychology of individuals, the sociology of groups, and the economic consequences for all. These are complex issues.

Science, Engineering, Technology, and Social Sciences

The aim of science is to understand the world-as-experienced: that is, how it behaves, of what it is composed, and why that behaviour occurs. Despite the fact that there is no one model of how a scientific enquiry should take place or be evaluated, there are some statements about the epistemology of science that can safely be made. For example, science as currently practised assumes realism (the world exists independently of our experience), objectivity (on the part of researchers), methodological pluralism (in the conduct of enquiry), logicality (in the analysis of data), predictability (in respect of future events), and tentativeness (the modification of ideas is always possible). The persistence of scientific ideas over many years is a tribute to their usefulness. Many people also believe that science is socially shaped to some extent, recognising that it is influenced by what seems particularly salient within a given tradition at a particular time.

Engineering, on the other hand, is concerned with design, focused on improving the quality of life or the efficiency of processes through practical applications of science. In the application of engineering principles, the methodology of science is used to make and to test predictions. Technology takes this further: it is the use of techniques at a practical level within an organisational framework, to solve the problems that arise as the products of engineering are developed and implemented. The ideas of science, applied within engineering, are actually implemented in technology and have a direct impact on the lives of humans and other species.

It is concern over the nature and consequences of these impacts that has led to the recent rapid rise of the importance of science communication. Any examination of these effects, however, requires use of the social sciences, which therefore confers on science communication studies a particular multidisciplinary aspect. In this book, the many factors affecting the relationship between these disciplines are explored.

The Aims of this Book

This book is intended to:

- Provide an overview of the field of 'science communication', i.e. what it deals with, important influences that affect it and, in particular, the challenges that it faces;
- Provide an introduction to the researched-based literature about science communication, especially in respect of these challenges;
- Show how this literature relates to the potential and actual practice of science communication;
- Provide suggestions for activities that readers might do to explore the issues raised.

The Themes that are Addressed in this Book

After this introductory Preface, *The changing face of science communication*, the book is organized in five parts:

Part I: Models of Science Communication—*Theory into Practice*

In this section, two theoretical chapters give an overview of the ways in which science is communicated. In Chapter 1, Masakata Ogawa sets out a challenge to communicators, questioning their purpose in communicating and the place of public engagement in science. He introduces the idea of a communication model. In Chapter 2, Sue Stocklmayer develops this theme.

Part II: Challenges in Communicating Science

In Chapter 3, Suzette Searle discusses the engagement between the communities of science and the general public, from the perspective of practicing scientists. Such communication will only be meaningful if there are clear roles for science in personal, social, and public policy. These roles are suggested by Will Grant in Chapter 4. In Chapter 5, Lindy Orthia discusses what is needed if public resistance to science is to be overcome.

Part III: Major Themes in Science Communication

This section addresses important aspects that take centre stage when the themes being communicated are associated with science. In Chapter 6, Craig Trumbo discusses the notion of 'risk'. Maurice Cheng, Arthur M.S. Lee, Ka-Lok Wong and Ida Mok tackle the vexed subject of 'the use of numbers' in science communication' in Chapter 7. The issues of ethics and accountability in both science and in communicating about science are debated by Rod Lamberts in Chapter 8. The nature of and relationships between evidence and belief are discussed by Michael Reiss in Chapter 9.

Part IV: Informal Learning

Part IV discusses learning about science in various settings. In Chapter 10, John Gilbert presents those issues that control the effectiveness of science communication and suggests 'best practice' in them. Whilst 'science education' is established in the formal education sector, the ideas of 'science communication' are still emerging. In Chapter 11, Sean Perera and Sue Stocklmayer discuss the relationships between the two. The great importance of informal environments in science communication are recognised in Chapter 12, where Léonie Rennie discusses 'best practice' in that provision.

Part V: Communication of Contemporary Issues in Science and Society

Many matters of public interest and concern nowadays involve science and their implications for human affairs. This section addresses several important examples. Climate change is an issue that has occupied global attention for more than two decades: in Chapter 14, Justin Dillon and Marie Hobson discuss the evolution of responses to this complex environmental problem. Communication issues become pressing when a general issue turns into a crisis: in Chapter 15, Yeung Chung Lee evaluates the effectiveness of communications to the public using the case study of the outbreak of Severe Acute Respiratory Syndrome (SARS) in Hong Kong. In Chapter 16, Julia Corbett talks about communication of the great challenges associated with on-going attempts to achieve 'sustainability' in respect of the Earth's resources. Indigenous science and local knowledge are increasingly recognized as having important contributions to make to our understanding of science: Yonah Seleti poses the challenges of incorporating these worldviews into a more narrow Western science culture in Chapter 16. Last, this section and the book conclude with an address, by Chris Bryant, to an issue that has only gradually emerged as the study of genetics and behaviour have become sophisticated: What does it mean to be human and what are the implications for the communication of science?

USING THIS BOOK

This reader is intended for all those interested in and concerned about current issues and developments in science communication. It is also intended for the use of university students who are studying for:

- Undergraduate degrees in the Natural Sciences, in Engineering and Technology, and in the Medical Sciences;
- Undergraduate and postgraduate degrees in Science Communication;
- Professional development in Education at pre-service and in-service levels;
- Postgraduate degrees in Education

As you will have seen from the summary above, each chapter in this book addresses a different aspect of science communication. If you are using it on your own, then you will undoubtedly identify themes in it that are of particular interest to you. As is argued in Part IV, the most effective learning takes place if you actively engage with the ideas that are presented in what you see, read, or hear. This process will be helped if you carry out some of the activities that are suggested at the end of the book in *Part VI: Further Exploration.*

You may be using this book as part of an organised course. The chapters could then be used as a preparation for, or follow-up to, any lectures that are part of it. This would enable you to think of important questions to ask about the topic being considered. Specific chapters can readily be used as reading before seminars, and the outcomes of activities used to provide material for discussions in them. Because such discussions will be interactive between you, your fellow students, and your tutor, it is likely that your learning will benefit from a broader range of ideas that you would have considered on your own.

If you can spare the time, the editors would value any suggestions for the further development of the book and the activities suggested. Their email addresses are included in the 'About the Authors' pages at the end of the book.

ACKNOWLEDGMENTS

We are grateful to Professor Mike Gore for critical reading of the chapters in this book. We also acknowledge earlier sources of some of the tables and diagrams—these are identified within the relevant chapters.

PART I

Models of Science Communication—Theory into Practice

1

TOWARDS A 'DESIGN APPROACH' TO SCIENCE COMMUNICATION

Masakata Ogawa

Introduction

Demands for science communication activities from various sectors are increasing worldwide, and progress in science communication research is becoming more visible than ever before. Also visible is a kind of evolution of the emphasis of science communication, from public understanding of science to public engagement with science, and beyond. There has also been a recent overwhelming rush of monographs on science communication with various "flavors" (e.g. Bauer & Bucchi, 2007; Bennett & Jennings, 2011; Cheng et al., 2008; Kahlor & Stout, 2010; Russell, 2010). However, it is difficult to grasp the overall picture of science communication because such prominence has been achieved by many different stakeholders and initiatives, including national and regional governments, universities, museums and science centers, media, citizen groups, non-profit organizations and non-government organizations, as well as attentive individuals. This has not occurred systematically, but rather autonomously and mutually independently.

Trench and Bucchi (2010) have contributed a very short but insightful commentary on the *status quo* of research in science communication and its disciplinary nature. Regarding the current situation of science communication research, they argued that: (1) science communication as a subject of teaching and research has appeared mainly as a response to external needs; (2) it is shaped as much by political and institutional concerns as by intellectual interests; (3) many of the publications in the period 2008-10, including their own, reflected not only the intention to support the formal study of science communication and to promote its further development, but also the difficulty of achieving theoretical unity; and (4) no full length publication, so far as they are aware, proposes a coherent framework for thinking about key issues in science communication. On the disciplinary aspects of science communication, they mentioned that science communication "is not (yet) established as an academic discipline with strong interdisciplinary characteristics or as a sub-discipline in the still-growing field of communication studies." (p.4)

In a sense, therefore, we have not yet seen either the grand picture of the science communication "jigsaw puzzle," or of its missing pieces, if there are any. Theoretical efforts in the area have included explorations of an appropriate definition of science communication itself (e.g., Bryant, 2003; Burns, O'Connor & Stockmayer, 2003) and its disciplinary nature (e.g., Trench & Bucchi, 2010); these have been explored because of a general belief in the need for clear and shared definitions to enable the area to "mature" into a discipline.

Difficulty in definition may come not only from complexity in the area itself but also from a diversity of visions and dreams within the science communication community. Individuals, groups, and organizations within this community have their own ideals of what they see as the "future" in terms of the relationship between sciences and society, but they do not usually explicitly express these ideals and may not even be consciously aware of them. It goes without saying that such images of an ideal "future" emerge from their own *values*. Differences in values or value orientations within the science communication community have not generally been taken into serious consideration thus far.

Current Efforts to Comprehend the Science Communication Arena

For the past several years, science communication researchers have struggled with the issue of how to understand the nature of activities which fall under the name of science communication and how to develop models and frameworks to analyse them. For example, Van Der Auweraert (2005) used a science communication 'escalator' model, identifying four dimensions of science communication—public understanding of science (PUS), public awareness of science (PAS), public engagement with science (PES) and public participation of science (PPS)—each of them with specific characteristics relating to science knowledge and communication. Trench (2008) proposed an analytical framework of science communication models, in which he identified three types of dominant models (deficit, dialogue, and participation) and their associated characteristics. McCallie et al. (2009) developed a framework for discriminating among PUS-PES activities in informal science education, for the purpose of classifying current and proposed activities. Their framework consists of three dimensions: (1) role of the public (audience participation); (2) role of Science, Technology, Engineering, and Mathematics (STEM)-related experts (expert participation), and (3) content focus. Brossard and Lewenstein (2010), however, re-examined to what extent prominent theoretical models accorded with real activities in science communication and concluded that traditional theoretical models often overlap in practice (p.32). They therefore suggested that popular theoretical models were less distinguishable when considering real science communication activities.

The nature of models in science communication is discussed at length in Chapter 2; in this chapter I wish to focus on what is needed from such models in terms of developing a unified framework for science communication.

Reflections: Fundamental Nature of Science Communication

Recent explorations of science communication are mainly based upon analysis of the *status quo* of science communication activities and events (what is), and have not yet provided a satisfying concept of a grand-picture of science communication (what should or might be) in terms of overall goals. So it is a good time for us to do a "health check" (Jenkins, 2000) of the whole arena of science communication. In this chapter, I will try not to stick to the current state, but to consider an ideal state: what the ideal goals of science communication could be. This approach resembles an endeavour of engineering, in the sense of Sparkes' explanation (1993, p.293), "whereas science is concerned with discovering and theorizing about 'what is,' engineering is concerned with creating and theorizing about 'what might be.'"

The current activities of science communication are commonly developed independently and follow their organizers' (scientists, science communicators, etc.) own beliefs, interests and concerns. They are rarely, if ever, developed in a systematically planned manner. Therefore, simple collection and evaluation of the resultant activities and events at a certain point in time may fail to deliver a grand-picture of what science communication should or could be. In order to obtain, or to appropriately *design* such an ideal picture, we must develop alternative analytical approaches. This requires us to look at the whole field but, for the past 20 years, the field of science communication has been differentiated into sub-fields labeled PUS, PAS, PES, PPS, and so on. Because efforts have concentrated on distinguishing one from another, much analysis has focused on their differences. If we are to identify universal grand goals for science communication, then identifying commonality across these differentiated sub-fields becomes important. What is common and what remains unchanged from before the historical differentiation happened? To answer this question, two possible approaches come to mind: (1) exploring the situation before such differentiation emerged (historical reflection); and (2) exploring fundamentally shared presuppositions among the sub-fields. For the former approach, we need to identify a kind of "prototype" or benchmark of science communication activity. For the latter, we may face difficulty: within the sub-fields, their own unique characteristics have, to date, been given greater emphasis, and common characteristics may not be apt to be explicitly and directly expressed.

Historical reflection guides us back to the days of the establishment of the Royal Institution in 1799 and the British Association for the Advancement of Science (BAAS) in 1831. Looking to the original missions or objectives of such institutions may give us a hint of possible overall goals. In the official website of history of the Royal Institution (n.d.), we read that "The Royal Institution was founded in March 1799 with the aim of introducing new technologies and teaching science to the general public." Jones (1871) quoted the founder Count Rumford's proposal for founding the Royal Institution, and it reads:

The two great objects of the Institution being the speedy and general diffusion of the knowledge of all new and useful improvements, in whatever quarter of the world they may originate; and teaching the application of scientific discoveries to the improvement of arts and manufacturers in this country, and to the increase of domestic comfort and convenience.

(*Jones 1871, p. 121*)

In the official website of the current British Science Association (n.d.) we read that the purpose of the organization was as follows:

The original purpose of the organization, expressed through its annual meetings held in different towns and cities throughout the UK was: "to give a stronger impulse and a more systematic direction to scientific inquiry; to promote the intercourse of those who cultivate Science in different parts of the British Empire with one another and with foreign philosophers; to obtain more general attention for the objects of Science and the removal of any disadvantages of a public kind that may impede its progress.

These two examples from earlier days in science history indicate that, from the very beginning, these institutions had the clear intention to diffuse scientific knowledge outside their own circle. This can be viewed as "the science community's intention to intervene in the relationship between science and the public." Emphasizing the point that certain kinds of intervention were intended from the very beginning, these aims could possibly serve as a "prototype" goal of the contemporary science communication movement.

In the second approach—exploring shared presuppositions across present sub-fields of science communication—a promising unifying concept seems to be "the relationship between sciences and society." As is well known, the title of the report on the relationship between the public and science and technology, published by the House of Lords Select Committee on Science and Technology (2000), was "Science and Society," and it has been said that this report was the actual starting point for shifting the current science communication movement beyond "public understanding of science" to "public engagement with science." The report formulated a strong intention to intervene in the relationship between science and society, but in different terms from those implied by The Royal Institution and The BAAS. From 2000 onward, the phrase "science and society" has been used as an umbrella term of inclusiveness for various areas of science communication. For example, the consultation document of the UK Department for Innovation, Universities and Skills (2008) was entitled "A vision for science and society." In its back cover, they explained "science and society" as follows:

We include engagement with society in its broadest sense, from science centres and festivals, through information provision by consultation, active dialogue and other media, to enabling citizen empowerment and decision-making. We include the use of science by society and the provision of scientific advice to policy makers for the benefit of society. We include the range of science skills opportunities, through the education system and beyond, and the importance of diversity in enabling a workforce truly representative of the society which it serves.

The relationship between science and society is, here again, the target of intervention by the authors or by the government.

One of the common features of science communication in its broadest sense could thus be summarized, using either approach, as "the intention to intervene in the relationship between sciences and society (or scientists and the public)." It is no problem to assert that this intention is shared by activities and events under any of the sub-fields. The idea of intervention between science and society would seem, therefore, a useful starting point from which to examine the science communication field.

The critical step, however, is to determine the *intent* behind the intervention. If we allow for different value orientations within the science communication community, we can then explore ways of conceptualizing science communication, emphasizing these differences in value orientations. This approach is different from current perceptions that are simply based upon analyses of science communication activities, but which take no account of value-oriented goals. We must therefore start by making different value orientations within the science communication community much more explicit and visible, using a framework to identify, uncover, and map out unspoken, hidden or unconscious values. This alternative approach addresses the challenge of making the unfinished science communication "jigsaw puzzle" visible and also that of identifying missing (or, not yet found) pieces which should have been included.

A New Framework: Science Communication as Policy

A definition based upon analysis of the *status quo* is problematic, because it gives us neither a grand picture, nor any preferred future state of science communication in a community. An alternative approach should, one would hope, show both. Are there any promising approaches? I propose a "design approach" to conceptualizing the nature of science communication. One of the critical points of such an approach is to focus on the goals, means, and stakeholders (often termed "actors") of each science communication intervention, and also on the stakeholders' intention. Of course, there are a few examples in the literature which do emphasize the importance of goals and outcomes in science communication (e.g., Department of Innovation, Universities & Skills, 2008; Jones, 2011; Powell & Colin, 2008; Stockmayer, 2005), but my present discussion is more concerned with the interactions among various factors.

In a design-centered approach, present, prospective, and potential stakeholders (here I prefer the term "driving actor") set their ideal goals regarding science communication in a particular community *in advance*. They intentionally design the intervention, making use of appropriate means, to alter the current state of science communication within that community. The alteration, if achieved, moves the community towards the science communication goal.

This design approach requires that the driving actors' value orientations and their dreams for their ideal future community be made explicit and visible. This is different from many contemporary practices in science communication which are not framed to express such goals.

In order to achieve such a framework, let me introduce the concept of "policy" from the disciplines of political sciences and policy studies, because it seems to be highly relevant to the relationships between "goals," "means," "actors," "intention," and "future" in science communication. This concept is not an alien one in science communication: a similar idea (school curriculum as policy) was once pursued in the formal education sector, based upon policy deliberation studies (Orpwood, 1981). In this case, "curriculum as policy" was conceptualized as "rules, plans, or guides for the determination of what shall be taught in specific situation" (p.23).

In the political sciences, Brewer and deLeon (1983, p.30) define "policy" as "a broad strategic statement of intent to accomplish aims." Anderson (2000, p.4) defines it as "a relatively stable, purposive course of action followed by an actor or set of actors in dealing with a problem or matter of concern." A similar definition is adopted by Sapru (2004, p.5): policy is "a purposive course of action taken or adopted by those in power in pursuit of certain goals or objectives." In these definitions, emphasis is placed on "purposive," "goals or objectives," and "actor(s)." Another definition by Clark and Keller (1988, p.7) reads: policy is "a specific course of action designed to achieve a desired outcome". Here, "designed" and "desired outcome" are emphasized. May (2003, p.228) argues that policy *intentions* establish the goals and type of policy that are to be put into practice. Landau (1977) explains the nature of policy much more descriptively as follows:

A policy proposes an intervention to alter some existing circumstance or mode of conduct. If it is well formulated, it will contain a description of the desired state condition and the set of means which promise to realize that condition (i.e., to attain its goals). It should be clear, thus, that policy proposals engage the future tense: they fall into that tense. The object of any policy proposal is to control and direct future course of action—which is the only action that is subject to control… the one thing we know about them is that their truth value has not been determined.

(Landau 1977, p.425)

Comparing the nature of policy as described above with the nature of science communication described in the previous section, we realize that the idea of "science communication as policy" seems to be promising. Thus, in a new framework featuring science communication as policy, science communication is defined as "purposive intervention by a driving actor or a group of driving actors to alter the present state of the relationship between sciences and society toward their desired state." Implications of the definition are that the driving-actors need to: (1) express their respective value orientations before proposing any intervention; (2) identify and express their own view of the grand picture or goal; (3) identify the concrete target of their proposed intervention; and (4) design concrete interventions reflecting their value orientations, goal, and target. It is important that the framework strongly reflects that their proposed intervention (whatever it is) is a "means" to achieve their *own* "goal." Driving actors with different value orientations and dreams could design or propose quite different kinds of intervention.

Once various kinds of driving actors (including potential driving actors who are not yet involved in science communication) and their respective visions for the relationship between sciences and society have been identified, they can be mapped out in a grand picture of science communication (assuming we could indeed design it). We may then also find certain missing parts (not yet challenged science communication areas) within the jigsaw puzzle.

Issues Raised by the Framework

As shown above, the framework theoretically requires present, prospective, and potential driving actors in science communication to address their own value orientations, identify their own goals for the community, identify their target groups and then design an intervention. The outcome must reflect all these factors. Issues arising from these points are briefly discussed below.

Driving Actors' Value Orientation

Many surveys on the public's scientific literacy or understanding of science have been reported, and almost all of them have shown that "the public" has always consisted of groups of people with various levels of literacy or understanding of science, as well as attitudes toward science. Thus "target actors' may be very diverse. For example, Miller's (1983) classic work identified "decision makers," "policy leaders," "attentive public," and "non-attentive public," and later he also referred to the "interested public" (Miller, 1986). In the UK, a survey by the Wellcome Trust and Office of Science and Technology (2000) identified six attitude clusters among the British public: Confident Believers (17%), Technophiles (20%), Supporters (17%), Concerned (13%), Not Sure (13%), and Not for Me (15%). While slight drifts among the surveys are visible, such distinct categories among the public still do exist despite various efforts to change them. Thus, it is not appropriate to refer simply to "the public." Many science communication researchers now prefer the term "publics." Also, there is no valid reason why such diversification in terms of knowledge of, and attitudes toward, science and technology issues should not occur among science policy makers, science communicators, and even scientists. For example, in his survey of scientists and science educators, Showers (1993) found that they did not necessarily: (1) hold a high level of content knowledge outside their own professional expertise; (2) vote based on understandings of the scientific dimensions of public issues; (3) have insignificant beliefs in superstition and pseudoscience; and (4) make personal and social choices based upon scientific aspects of decision-making situations. Similar findings about the content knowledge of scientists were described by Stockmayer and Bryant (2012).

Trench and Bucchi (2010) develop this point as follows:

It [science communication] concerns the communication between communities of scientists, interest groups, policy-makers and various publics. But, on further reflection, we have to consider whether science communication also includes communication between and within various scientific institutions and communities of scientists. This has received significantly less attention than the cross-sectoral communication between scientific communities and those of wider society. Even less attention has been given to the communication between various publics—without the involvement of scientists—on scientific issues.

(Trench & Bucchi 2010, p.1)

If such differences are seen in existing scientific literacy and attitudes toward science, why would they not occur in possible goals for a future community, as well as the value orientations leading to those goals? How can we in practice uncover, identify, and visualize such differentiated value orientations among various actors? In terms of the general public, I have previously discussed this issue (Ogawa, 1998, pp.108–110). At this time, I proposed three mutually independent dimensions along which to identify the stances of people toward science: (a) science literacy vs. science illiteracy (knowledge level): (b) pro-science vs. anti-science (attitude level): and (c) pro-scientism vs. anti-scientism (values level). A combination of these dimensions implies a possible six typologies among the general public: (1) Science Believers (science literate, pro-science, pro-scientism); (2) Science Contextualists (science literate, pro-science, pro-scientism); (3) Science Fanaticists (science illiterate, pro-science, pro-scientism); (4) Science Vigilants (science illiterate, pro-science, anti-scientism); (5) Authentic Antiscientists (science literate, anti-science, anti-scientism); and (6) Neo Antiscientists (science illiterate, anti-science, anti-scientism). In addition, I later suggested the inclusion of the category "indifferent" in the attitude level (i.e., pro-science, indifferent, and anti-science) (Ogawa, 2006). A recent empirical study in a Japanese context concerning participants in science cafés (Yoshida, 2011) indicated that a positive orientation toward active engagement with topics in which people are interested is another important trait. If so, a fourth level of commitment to science issues (i.e., active commitment, no commitment, and negative commitment) can be added.

Readers should note particularly that this categorization is also theoretically applicable to scientists, science communicators, science teachers, science policy makers and so on, though the relative ratios of the typologies may be quite different among these groups. Value orientations are not necessarily linked with job orientations.

These categories still require a practical trial to describe various community members with regard to their value orientations, but such categorization might be possible and promising to pursue.

What do we mean by "Science Literacy"?

What is the ideal image of a future community in terms of the "relationship between sciences and society" that various driving actors are eager to achieve or may be dreaming about? While the question is highly relevant to any ultimate goal of science communication, rarely do driving actors (science communicators, for example) refer to this point directly. One of the issues frequently discussed, however, is community members' "scientific literacy levels" in an ideal future community.

Two extreme ideals of scientific literacy are possible: (1) a scientifically literate community, consisting of perfectly scientifically literate individuals, in order for them to serve as scientifically literate decision makers; or (2) a community consisting of individuals with widely diversified levels of scientific literacy. In between, of course, there are many options available.

Usually, the former stance seems to be the majority view (e.g., The American Association for the Advancement of Science's *Science for All Americans: Project 2061*). It is, however, problematic. Should we continue to endeavour to achieve an ideal ultimate goal where all citizens hold a certain level of scientific literacy and/or engagement that is satisfactory to the scientist community—that is, an ideal future community with perfect scientific literacy and perfect engagement? While this indeed may be an ideal state, we unfortunately cannot overlook the fact that the diversity of science literacy levels among the community, described in the previous section, has remained rather stable or unchanged despite various international remedial efforts. This ideal is, therefore, really "ideal." It currently serves as the ultimate goal for certain groups, but these groups also need to accept a reality in which diversity in community levels of "scientific literacy" is not diminishing. These groups may need the wisdom to set up more "practical" goals along the way and to try to design a set of appropriate "means" for different groups of target actors.

The latter position is more reality-oriented, or practice-oriented. If we accept the reality of a diversity of scientific literacy levels, alternative views of what we mean by a "scientifically literate community" are then required. For example, Shamos (1995) did not pursue a vision of a future community filled with perfectly scientifically literate citizens. Instead, he radically proposed that scientifically sound community decisions on science and technology issues only required that 20% of citizens be truly scientific literate. In another example (Ogawa, 2006), I stated that while the community as a whole might demonstrate "perfect" scientific literacy and engagement, individual members are not necessarily required to be perfectly literate and engaged. In both these examples, one of the major tasks of science communication would be to develop and maintain a system of innovative modes of knowledge and information flow, and to set up networks within a community for every citizen in that community to have timely access to appropriate knowledge and information. I have argued (Ogawa 2001) for the need of such a knowledge and information "decipher-ment" system, where every community member can function as a "multiple agency" (each individual serving as a receiver, sender, interpreter, producer etc. in different circumstances) in terms of knowledge and information flows in their own community.

The former approach is idealistic, the latter realistic. The latter position does not necessarily refute the efforts of the former. We can pursue both approaches simultaneously.

Target Actors of Interventions

The target actors of interventions such as science communication activities and events have, to date, been generally vaguely defined. Usually, target actors are named and identified as the "audience" or the "general public." This indicates that driving actors have not addressed the term "target" in its specific meaning. They have been concerned about "performing the activities" themselves, rather than with the question of "what kinds of particular target actors are these activities aiming to reach?" But once we acknowledge the diversity of value orientations among actors (not only publics, but also scientists and science communicators), it is possible to similarly diversify target actors for intervention. For example, I have previously sketched out (Ogawa 2006) various kinds of intervention in terms of "Who drives the particular inter-vention for whom?" (Table 1.1).

TABLE 1.1 Classification of Science Communication Activities

ACTORS		DRIVING ACTORS			
		Professionals, Policy-Makers	Pro-Science Public	Indifferent Public	Negative Public
T A R G E T A C T O R S	Professionals, Policy-Makers	Governmental Councils, Committees	Consensus Conference★, Science Shops, Public Lectures		Negative Campaigns, Public Hearings
	Pro-Science Public	Consensus Conference★, Citizen's Panels★★, Science Café	Science Café, Field Trips		Negative Campaigns
	Indifferent Public				
	Negative Public				
	Public in General	Deliberative Polling★★★, Citizen's Panels★★, Science Shops, Public Lectures, Science Fair, Field Trips, Science Museums, Science Centres	Science Café, Field Trips		Negative Campaigns, Science Cafe

Source: Ogwam, 2006, p.204: Reproduced with Permission:

Notes:
★ Consensus Conference: By convention, a group of 16 lay volunteers is selected for a consensus conference. These people are selected according to socio-economic and demographic characteristics. The members meet first in private, to discuss an issue and to decide the key questions they wish to raise. There is then a public phase, lasting perhaps three days, during which the group hears and interrogates expert witnesses, and draws up a report. ★★ Citizen's Panels: A citizens' jury (or panel) involves a small group of lay participants (maybe 12-20) receiving, questioning, discussing and evaluating presentations by experts on a particular issue, usually over 3-4 days. At the end, the group is invited to make recommendations. ★★★ Deliberative Polling: A large, demographically representative group of perhaps several hundred people conducts a debate on an issue, usually including the opportunity to cross-examine key players. The group is polled on the issue before and after the debate. (Parliamentary Office of Science and Technology, 2001: p.6)

This kind of differentiation of target actors will be necessary if the driving actors are to achieve the goals of the intervention. This idea is enlarged upon in Chapter 2, in which Sue Stocklmayer discusses the intentions of science communication with a greater range of "actors" and "means."

Designing Intervention: Filling the Missing Pieces

Remembering the metaphor of the science communication "jigsaw puzzle," we can identify Table 1 as the beginning of the jigsaw puzzle, where there are many "missing pieces." Why is a particular box still blank? What kinds of new intervention can possibly be developed? If one target actor is changed to another, what modification of means is needed? If this intervention were to be designed by a different driving actor, what differences would we see happening? Such questions would emerge in the minds of science communicators and scientists. Thus, every driving actor (individuals, groups, organization, initiatives, etc.) can develop its own grand picture of a desired future community, and can design various kinds of particular interventions whose goals, means, and target actors are seriously deliberated. Currently, an indifferent public is excluded from serving as driving actors. This is an interesting thought experiment to break our fixed ideas about science communication intervention: What kinds of intervention might be expected to appear if we, as the organizer or producer, could invite indifferent members of the public to be driving actors?

Another interesting aspect of designing interventions, which is not yet fully developed, will be the issue of different actors' roles in a particular setting of an intervention activity. Consider a certain kind of science communication activity, for example, a science café (or *Café Scientifique*). This serves as a communication space for open conversation between a guest scientist and a small number of lay people in casual settings like a bar or cafe (Russell, 2010, pp.92-93). Suppose there exists an individual, group, or organization that wants to develop a café. Usually it will be concretely designed or planned by a team of program designers in order to meet and reflect the needs or intent of the organizer. (In a small-scale science café, it often happens that the organizer and the program designer are the same or come from the same organization or group.) During the preparation process, the planners need to identify and allocate appropriate "actors" (organizer, producer, program designer, scenario writer, presenters, staff, active audience, passive audience etc.). And on the day of the café, the "actors" play their own roles just as was designed or planned. This is the typical scenario of science communication activities. The *issue* here, however, is the value orientations among the driving actors. Often, the actors believe that they share the same value orientation toward the relationship between science and society, but is this the case? What kinds of image of the preferable relationship between science and society in the future do each of the driving actors hold? What is the ultimate goal of this café? Discussion of individual value orientations is rare but such discussions will help the intervention activity to be more goal-oriented and target-oriented.

Moving into the Public Policy Domain (CODA)

Driving-actors' views of a desired future community (their goals) are, in principle, expected to differ widely according to their respective value orientations. While they have freedom to pursue their respective dreams, there are currently no legitimate reasons why the target actors must share those dreams. The target actors (including indifferent publics) have the right to dream their own dreams. The point here is that such dreams, either among target actors or among driving actors, are still private ones: views of a desired future community in this context are private domains. Once various views are openly disseminated within a community where target actors and driving actors live together, there is an immediate need to start negotiation and discussion on developing a shared view of desired future community. At this point, the issue moves from the private domain (private policy decision) to the public domain (public policy decision). How science recommendations translate to policy is further discussed in Chapter 4.

Thus, science communicators can play an important role in the public domain, developing a new type of intervention for consensus-making around their community's desired future. Of course, it will not be easy. Sometimes science communicators may need to withdraw their private dreams, in order to reach a consensus among the community members, which will be painful, indeed. However, in a democratic community, various values among the community members should be negotiated, and compromises must be achieved. Science communicators are obliged to follow these principles.

A prominent public policy researcher, Easton (1971, p.130), once argued that public policy is implicitly defined as " an authoritative value allocation for society." In those days, public policy was believed to be substantially under governmental control. But nowadays, since public participation in the policy domain has become very visible, the definition has shifted. Authority is not necessarily derived from governmental agencies alone, but from public consensus-making processes such as deliberative democracy.

The community has a right to decide its desired future regarding the relationship between science and society. Science communicators are able to contribute to consensus-making processes or value-allocation processes as facilitators, not as authorities. Once the community's image of its desired future is decided, science communicators can serve as designers of appropriate interventions to support the community's dreams and help them come true.

Notes

1   In a complex but nonetheless comprehensive definition, 'value' can be defined as "a conception, explicit or implicit, distinctive of an individual or characteristic of a group, of the desirable [i.e., what I or others ought to want] which influences the selection from available modes, means and ends of action" (Kluckhohn, 1951, p.395).

References

Anderson, J. E. (2000). *Public policymaking: An introduction.* 4th ed. Boston: Houghton-Mifflin Company.

Bauer, M. W., & Bucchi, M. (2007). *Journalism, Science and Society: Science Communication between News and Public Relations.* New York, NY: Routledge.

Bennett, D. J., & Jennings, R. C. (2011). *Successful Science Communication: Telling It Like It Is.* New York, NY: Cambridge University Press.

Brewer, G., & deLeon, P. (1983). *The Foundations of Policy Analysis.* Homewood, Illinois: Dorsey Press.

British Science Association (n.d.). History of the British Science Association. Retrieved from http://www.britishscienceassociation.org/web/AboutUs/OurHistory/ .

Brossard, D., & Lewenstein, B. V. (2010). A critical appraisal of models of public understanding of science: Using practice to inform theory. In L. A. Kahlor and P. A. Stout (eds) *Communicating Science: New Agendas in Communication.* (pp.11-39). New York, NY: Routledge.

Burns, T. W., O'Connor, D. J., & Stocklmayer, S. M. (2003). Science communication: A contemporary definition. *Public Understanding of Science, 12,* 183-202.

Bryant, C. (2003). Does Australia need a more effective policy of science communication? *International Journal for Parasitology, 33,* 357-361.

Cheng, D., Claessens, M., Gascoigne, T., Metcalfe, J., Schiele, B. & Shi, S. (2008). *Communicating Science in Social Contexts: New Models, New Practices.* Springer Science.

Clark, T. W., & Kellert, S. R. (1988). Toward a policy paradigm of the wildlife sciences. *Renewable Resources Journal, 7,* 7-16.

Department of Innovation, Universities & Skills (2008). *A Vision for Science and Society: A Consultation on Developing a New Strategy for the UK.* Retrieved from http://www.bis.gov.uk/assets/biscore/corporate/migratedD/ec_group/49-08-S_b .

Easton, D. (1971). *The Political System.* 2nd ed., New York, NY: Alfred A. Knopf.

House of Lords Select Committee on Science and Technology (2000). *Science and Society.* London: House of Lords.

Jenkins, E. (2000). Research in science education: Time for a health check? *Studies in Science Education, 35,* 1-25.

Jones, B. (1871). *The Royal Institution: Its Founder and its first professors.* London: Longmans, Green, and Co. (Reprinted by Cambridge University Press in 2011.)

Jones, R. A. (2011). Public engagement in an evolving science policy landscape. In D. J. Bennett, & R. C. Jennings (eds) *Successful Science Communication: Telling It Like It Is.* (pp.1-13). New York, NY: Cambridge University Press

Kahlor, L. A., & Stout, P. A. (2010). *Communicating Science:New Agendas in Communication.* New York, NY: Routledge.

Kluckhohn, C. K. (1951). Values and Value Orientations in the Theory of Action. In T. Parsons and E. A. Shils (eds), *Toward a General Theory of Action.* (pp.388-433), Cambridge, MA: Harvard University Press.

Landau, M. (1977). The proper domain of policy analysis. In D. Bobraw, H. Eulau, M. Landau, C. O. Jones and R. Axelrod (eds), The place of policy analysis in political science: Five perspectives. *American Journal of Political Science, 21,* 415-433.

May, P. J. (2003). Policy design and implementation. In B. G. Peters & J. Pierre (eds) *Handbook of Public Administration* (pp.223-233), London, UK: Sage Publications.

McCallie, E., Bell, L., Lohwater, T., Falk, J. H., Lehr, J. L., Lewenstein, B. V., Needham, C., & Wiehe, B. (2009). *Many experts, many audiences: Public engagement with science and informal science education*. A CAISE Inquiry Group Report. Washington, D.C.: Center for Advancement of Informal Science Education (CAISE). Retrieved from http://caise.insci.org/uploads/docs/public_engagement_with_science.pdf .

Miller, D. (1983). Scientific literacy: A conceptual and empirical review. *Daedalus, 112*, 29-48.

——(1986). Reaching the attentive and interested publics for science. In S.M. Friedman, S. Dunwoody and C. L. Rogers (eds), *Scientists and journalists: Reporting science as news,* (pp.55-69). New York: Free Press.

Ogawa, M. (1998). Under the noble flag of 'developing scientific and technological literacy.' *Studies in Science Education, 31*, 102 –111.

——(2001). Kagaku gijutsu kei jinzai ikusei haichi ron: Gendai shakai wo kaidokusuru hohoron to naruka? [Techno-scientific human resource development and allocation: A new research methodology to decipher contemporary society?]. *Journal of Science Education in Japan, 25*, 230-242. (In Japanese with English abstract)

——(2006). Exploring the possibility of developing indifferent public-driven science communication activities. *Journal of Science Education in Japan,* 30, 201-209.

Orpwood, G. W. F. (1981). *The logic of curriculum policy deliberation: An analytic study from science education.* Unpublished doctoral thesis, University of Toronto. (ED 211 372).

Parliamentary Office of Science and Technology (2001). *Open Channels: Public dialogue in science and technology (Report No.153).* Retrieved from http://www.parliament.uk/documents/post/pr153.pdf .

Powell, M. C., & Colin, M. (2008). Meaningful citizen engagement in science and technology: What would it really take? *Science Communication, 30*, 126-136.

Russell, N. J. (2010). *Communicating Science : Professional, Popular, Literacy.* New York, NY: Cambridge University Press.

Sapru, R .K. (2004). *Public policy: Formulation, implementation, and evaluation* (2nd revised edition). New Delhi, India: Sterling Publishers Private Limited,

Shamos, M. H. (1995). *The myth of scientific literacy.* New Brunswick, NJ: Rutgers University Press.

Showers, D. (1993). *An examination of the science literacy of scientists and science educators.* Paper presented at the Annual Meeting of the National Association for Research in Science Teaching, Atlanta, GA, April 15-19. (ERIC ED362393)

Sparkes, J. J. (1993). The nature of engineering and the physics it needs. *Physics Education, 28*, 293-298.

Stocklmayer, S. M. (2005). Public awareness of science and informal learning: A perspective on the role of science museums. *The Informal Learning Review, 72*, 14-19.

Stocklmayer, S. M., & Bryant, C. (2012). Science and the public: What should people know? *International Journal of Science Education: Science Communication and Engagement, 2*, 81-101.

The Royal Institution (n.d.). *R I History.* Retrieved from http://www.rigb.org/content Control?action=displayContent&id=00000002894

Trench, B. (2008). Towards an analytical framework of science communication models. In D. Cheng et al. (eds), *Communicating Science in Social Contexts.* (pp.119-135). Springer Science.

Trench, B., & Bucchi, M. (2010). Science communication, an emerging discipline. *Journal of Science Communication, 9,* 1-5.

Van Der Auweraert, A. (2005). The science communication escalator. *Proceedings of 2nd Living Knowledge Conference.* February 3-5, Seville, Spain, 237-241.

Wellcome Trust and Office of Science and Technology. (2000). *Science and the public: A review of Science Communication and public attitudes to science in Britain.* Retrieved from http://www.wellcome.ac.uk/assets/wtd003419.pdf .

Yoshida, M. (2011). The learning experiences of participants of science cafés in Japan. *Conference Proceedings of EASE International Conference,* October 25-29, Gwangju, Korea, 164-165.

2
ENGAGEMENT WITH SCIENCE
Models of Science Communication

Susan Stocklmayer

Introduction

Until the end of the twentieth century, the dominant mode of communicating science was simple transmission of information. The ideal mode has now shifted, however, from one-way transmission to some form of two-way, participatory practice. The one-way mode originated within the telecommunications industry (Shannon & Weaver, 1949). The communication aim is for a 'source' to transmit a message to a 'receiver' without distortion, but 'noise' is a real and practical problem.

The message must first be encoded into electronic signals, then decoded to make it intelligible to the receiver, with minimum impact from the 'noise' (Figure 2.1).

Quite how this model came to be applied to human communication is not entirely clear. Between humans, however, it is deeply flawed, in that the process of encoding a message about science always requires modification of the science itself in some way. In addition, the encoder can never make the assumption that the message will be decoded by the receiver exactly as it is sent, because the receiver will decode according to their own understandings, thoughts and

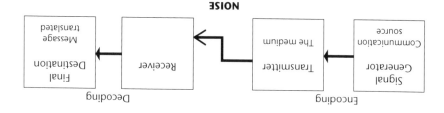

Encoding | Decoding

Signal Generator / Communication source → Transmitter / The medium → Receiver → Final Destination / Message translated

NOISE

FIGURE 2.1 Simple Transmission (*Source:* Shannon & Weaver, 1949)

experiences. The 'noise' here is quite deafening. For example, there may be other messages coming through, such as dissenting voices about current issues like climate change. There may be a strong and confusing media influence. Aspects of culture, environment, and personal circumstances will also affect the process, perhaps inappropriately.

Current rhetoric in science is, therefore, much more about two-way communication, accepting that even with the simplest model, constant feedback is necessary to assist both coding and decoding processes. Attending to the noise created by cultural, environmental and other differences is now considered fundamental to successful communication.

The transmission model assumes the existence only of a single 'source' and 'receiver' but science communication involves many 'stakeholders' who contribute to the message in different ways. Attempts by various authors to represent this complexity are discussed later in this chapter: suffice it to say here that a simple message about environmental sustainability might affect national and local governments, scientists, the worlds of business and industry, farmers, and residents in a particular locality—and all these actors, perhaps thousands, would have to share a vision in order to bring about a positive result.

The one-way model has also been widely criticised for its underlying implication that the transmission of information is from 'expert' to 'layperson'—implying that the public is somehow deficient in their understanding of science. The fundamental assumptions of the so-called 'deficit model' (Layton, Jenkins, McGill & Davey, 1993; Wynne, 1991; Ziman 1991) are that members of the general public know and understand very little science and therefore require further education. This model has now been comprehensively rejected in favour of a style of engagement that respects public knowledge as well as the knowledge of scientists, and regards the public and scientists as equal players in science communication endeavours (e.g. House of Lords Report, 2000; Lewenstein, 2003). Terms used to describe this more equal relationship include 'public engagement', 'dialogue', 'knowledge sharing', and 'knowledge building' (see, for example, Research Councils UK, 2002; Welp, de la Vega-Leinert, Stoll-Kleeman & Jaeger, 2006).

Ideas about uncertainty in science, views of science as an unchallenged authority or as a given body of knowledge, have all shifted in recent times as the concept of authority itself has altered. This was clearly stated in the UK *Science and Society Report* (2000), Section 2.42) that said:

The relationship between experts and the rest of society is changing... In all areas of society it is now normal for assertions of authority to be questioned. When the decision-maker says "Trust me", the response is very often "Show me". Scientific authority is in this respect no different from the authority of parents, teachers, the police or indeed Parliament.

(*Science and Society Report 2000, Section 2.42*)

As far back as 2002, the UK Research Councils published a 'how to' booklet for scientists to facilitate their interaction with the public, which is termed 'dialogue' in the booklet. In this booklet they also stated explicitly that things have changed (p.3):

... we suggest that there are six principal objectives that motivate people and organisations to develop activities to communicate science. These are:

1. To promote an awareness of science as "part of the fabric of society";
2. To promote an individual organisation;
3. Public accountability;
4. To recruit the next generation of scientists and engineers;
5. To gain acceptance of science and new technologies; and
6. To support sound and effective decision-making.

In the old world of public understanding, there was often a view that a monologue approach could achieve the first five of these objectives and that non-expert opinion had no role to play in the last. However, with the increasing recognition that dialogue and multiple inputs are crucial factors in sound decision-making in science, it has become accepted that two-way communication is a more robust way to address all of these objectives.

Have things really changed, however? In other papers (Bauer, Allum & Miller 2007; Trench, 2006; Wynne, 2006), the death of the deficit model has been questioned. Trench (2006, p.1) concluded that 'a deficit model remains the default position of scientists in their public activities and underpins much of what is proposed by public officials in their promotion of science'. Kim (2007) says that 'this is the dominant communication strategy and the reigning behavioural theory' (p.288). In the most extreme cases, the deficit model translates to the 'defence' model (Johnson, 2005 cited in Trench, 2006) in which scientists feel threatened by a perceived devaluing of science and 'advocate even more grimly than before the merits of science, and of careers and studies in science' (p.2). Wynne (2006) reviewed the history of views about the public's trust in science, and concluded that the deficit model is as pervasive as ever:

This monolithic and blackboxed deterministic framework is hardly enlightened and democratic. Yet it is important to note that it is imposed upon society, without deliberate intent, but no less rigidly so, by the prevailing institutional culture in virtually all international innovation and regulation processes. Yet sadly, this occurs in the name of the avowedly post-deficit model, enlightened public engagement with science.

(Wynne 2006, p.218)

With respect to science communication more generally, Trench (2006) concluded: 'Fortified with …notions of plurality and participation, science communication could offer the possibility of a real engagement with and by the publics' (p.8). It is relevant, therefore, to examine these notions of plurality and participation in respect of an emerging framework for the discipline of science communication.

That one-way communication is a true reflection of most current practice is, in my view, uncontroversial. If we regard the activities of science communicators across a continuum, then simple one-way 'promotion of science' is at one end, with many who term themselves 'science communicators' engaged in such activity. Promotion of science underpins a great deal of government thinking (and funding). 'Science literacy' is a term that is part of this rhetoric. Promotion of science also emphasizes the need for more tertiary students to take up science careers, and the need to improve the poor science performance of many school students.

Setting aside these more obvious overt deficit models, however, we can categorise as 'promotion of science' such examples as the inclusion of informative articles in the press, screening a television documentary, placing science on the Internet, or presenting a new exhibition in a science centre. They are certainly overwhelmingly one-way in their design and, therefore, intent. There is clearly no expectation by the writers, designers, and producers that they will engage in two-way communication, but rather that they are 'transmitting' information to whatever audience is willing to listen, play, read, or watch. All these examples nevertheless contribute to a view of scientific knowledge as knowledge worth having, interesting, or important to a variety of people.

Important barriers to effective one-way communication include a lack of identification of clear goals by the communicators themselves (Rennie & Stockmayer, 2003; Stockmayer, 2005). One-way communication becomes less problematic, however, when relevance to those for whom the communication is intended is paramount. Of course it should be emphasized that this knowledge deficit is not restricted to the general public or to non-scientists': it applies to all participatory groups, including 'experts' (Stockmayer & Bryant, 2012).

Trench (2006, p.3) noted that 'Scientists and their professional organizations can call on a range of communication models' which extends from 'imperatives about educating various publics about science and persuading them about its benefits', through statements about a moral responsibility to engage with the public, to propositions about possibly learning about science itself through the insights of others'. He nominates three major categories of communication. These are: 'Deficit' (split into 'marketing' science and 'defending' science); 'Dialogue' (divided into 'duty' to communicate and communicating in a context of public need); and 'Participation' (divided into agendas jointly set by the public and scientists, which he terms 'deliberation', and the negotiation of meanings between these parties, termed 'deference') (p.5). Trench neatly categorises these activities as communicating science *to* them (deficit), communicating science *with* them (dialogue), and communicating about science *among* them (participation). Implicit in these categories for communication is that the chosen method has purposes or outcomes for the communicators themselves. Such outcomes could be public persuasion, engagement, or collaborative learning.

Trench later (2008, p.131) expanded his dialogue category to include 'Engagement'. He also attached 'scientists' orientation to the public' to his three major categories. For the Deficit model, the scientists' orientations to the public are: 'They are hostile', 'They are ignorant', or 'They can be persuaded'. In consultative dialogue: 'We see their diverse needs', and 'We find out their views'. Through engagement, 'They talk back' and 'They take on the issue'. Last, the participatory model has the orientations of 'They and we shape the issue', 'They and we set the agenda' and 'They and we negotiate meanings'.

Thus, a spectrum of models of the purpose of science communication activities may be defined. At one extreme is a one-way transmission which intends only to inform—to put the information 'out there'. Such transmission is not confined to scientists—it can come from any source. At the other end, there is knowledge building, a process that seeks to construct new meaning from many contributions (Figure 2.2).

In 2006, in an address to the American Association for the Advancement of Science, Matthew Nisbet described two 'dominant approaches to science communication': the 'popularization model, personified by Carl Sagan, and the public engagement model, symbolized by the deliberative town meeting' (p.1). In between these extremes, however, there exists a wide range of science communication activities that include 'dialogue events[1]', knowledge sharing exercises, and communication about science for the purposes of research.

This chapter examines science communication activities in terms of purpose, or intended outcome, within which different methods of communication reside. This is not a new way of looking at communication generally: indeed, public relations experts Grunig and Hunt (1984) listed 'purpose' as a major aspect of PR models. Within the domain of communication of science, a joint report by the British Office of Science and Technology and the Wellcome Trust (2000) also used 'purpose' to classify 'the range of science communication' which was then mapped to illustrate 'giving out information about science' (pp 15-17).

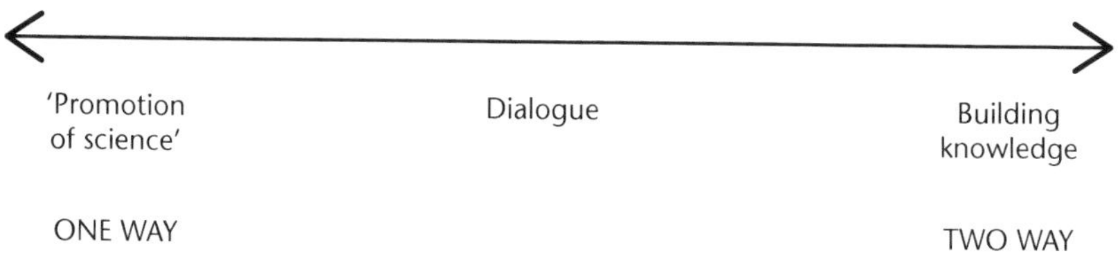

'Promotion of science'	Dialogue	Building knowledge
ONE WAY		TWO WAY

FIGURE 2.2 A Continuum of Science Communication

Models of Science Communication

The style of Shannon and Weaver (1949) is essentially a 'conduit' model. Its simplistic imagery is still reflected in the terms 'getting across' information and 'delivering' curricula. Even the addition of a feedback loop (Schramm, 1954) did not substantially alter the intended outcome of straightforward information output. In acknowledgement of the limitations of this approach, many more complex models have subsequently been described. From the field of communication studies these include, for example, those of Gruing and Hunt (1984) above; helical models (Dance, 1967); and fractal models (Wheatley, 1992). The intent of all these models is to describe the huge complexity of generic communication processes, but it should be noted that they are framed largely from the perspective of the 'source' of the information. Essentially they seek to address the near impossibility of modelling all the influences on communication as a human activity—a process of multiple dimensions and fractal sub-processes.

The communication of science is of course subject to this same underlying complexity and can be similarly described. For example, d'Andrea and Declich (2005) describe 'an elementary model of science communication' from the point of view of researchers in a scientific organization. They describe eight elements contributing to successful communication that include communicating the vision, the politics of communicating, peer and interdisciplinary interactions, networking, and so on. Esteban (1994) has summarised the communication of science between scientists as a complex model that resembles an irrigation scheme, in which the source corresponds to the river and dissemination is achieved by a series of irrigation canals. Both these models are concerned with specific communicators—scientists—and the one-way flow of information from them to other interested parties. They describe how an outward flow of information is disseminated and used.

Models of science communication have also been described, however, from a different perspective which recognizes the role of communicators other than scientists and is more related to 'goals and possibilities' (Lewenstein, 2003). Lewenstein summarized four models that reflect the range described above by Trench (2006). These are the deficit model, the contextual model, the lay expertise model, and the public participation model. The models are increasingly complex in the way that they frame the major communicators. 'Contextual models' place the communication in a real context, 'recognize the presence of social forces but nonetheless focus on the *response* of individuals to information' (p.4, my italics). Lewenstein places health communication in this category. 'Lay expertise' is 'knowledge based on the lives and histories of real communities such as detailed farming and agricultural practices'. The lay expertise model addresses the problem that 'scientists are often unreasonably certain—even arrogant—about their level of knowledge, failing to recognize the contingencies or additional information needed to make real-world personal or policy decisions' (p.4). The model therefore gives recognition to the value of local knowledge, although Lewenstein acknowledged that it is not clear how this model 'provides guidance for practical activities that can enhance public understanding of particular issues' (p.5). Finally, the 'public participation' or 'engagement' model seeks to 'enhance public participation and hence trust in science policy' (p.5), through consensus conferences and similar events.

Greco (2004) introduced a 'Mediterranean model' which is 'slightly different', having four outstanding characteristics: 'interdisciplinary character, acknowledgement of the intrinsic value of knowledge, respect for history, and multimodality' (p.4.). The 'consequences' of Greco's model make this point more clearly:

> The first consequence is that substantial communication does not just imply conveying scientific "ideas" in a linear, top-down approach, from the expert to the layman but it also means conveying scientific "ideals"; a mission which is just as necessary and important.
>
> *(Greco 2004, p.5)*

The second consequence is that there are no guidelines to follow. "The world of science communication is so big and varied, the 'noise' is so frequent and effective that hardly any action can have an effect in a linear way" (p.5).

In many ways the Mediterranean model may be regarded as a subtle mix of the 'British' and 'North American' models described by Lewenstein (2003), who concludes his paper with a plea for more research, urging that we seek to understand which of the models he mentions are 'at work in any public communication project' and 'which public communication activities don't fit any of these models'. He states that such research would contribute to our understanding of how science operates in society (p.7). Considering the potential, however, for the existence of various communication modes, for many different actors, for multiple ways in which 'the public' interacts with 'science', one might conclude that research in science communication may be as big, varied and wicked as the world described by Greco.

A Wicked Problem

Wicked problems are not evil, but complicated. The characteristics of 'wicked problems' (a term coined by Rittel & Webber, 1973) are that they often have many contributing causes, and no clear solutions. They are unique. Rittel and Webber made the point that wicked problems are ill-defined, so solutions are not 'given', they are not right or wrong, and every solution has to be tried experimentally because its progress cannot be predicted. They are also:

> ...ambiguous and associated with strong moral, political and professional issues. Since they are strongly stakeholder dependent, there is often little

consensus about what the problem is, let alone how to resolve it. Furthermore, wicked problems won't keep still: they are sets of complex, interacting issues evolving in a dynamic social context. Often, new forms of wicked problems emerge *as a result* of trying to understand and solve one of them.

(*Swedish Morphological Society, 2011, p.1*)

A typical wicked problem might be 'how do we tackle climate change' or 'how should scientific and technological development be governed?' Sometimes, addressing these problems results in unforeseen outcomes, because they are socially complex and change quite rapidly. Often, they involve people changing their behaviour. Problems such as climate change, public policy, drug abuse, and so on are often termed 'wicked problems' because there are no easy solutions and they are inherently very complex. Thus far, we have little research on which to draw to help us to address these problems.

Wicked problems, however, afflict open, complex and imperfectly understood systems, and are beyond the reach of mere technical knowledge and traditional forms of governance.

(*Hulme, 2009, p.334*)

Communicating science at the right-hand end of the continuum (Figure 2.2) requires a deep understanding of all the players in the communication process. It is characterised by complicated social interactions, which may involve many people and institutions who can affect the outcomes of the science. Mapping these social and governmental networks is, in itself, an enormous task, that is essential for clear formulation of the problem itself. Solutions to the problem are another matter altogether:

The disagreement among stakeholders often reflects the different emphasis that they place on the various causal factors.... Solutions to wicked problems are not verifiably right or wrong but rather better or worse or good enough....

It is clear, for example, that environmental issues... require action at every level—from the international to the local – as well as action by the private and community sectors and individuals.

(*Australian Public Service Commission, 2007, p.1*)

Thus, in the continuum of science communication, the most difficult mode involves highly complex interactions and a degree of time and patience. Each case requires independent analysis and different messages between the actors in the network.

In conclusion, I should like to propose a simple model to embrace the aspects of communication described above. In formulating this model, I have sought to give more coherence to this diversity. If we are to understand which modes of communication are appropriate and which activities 'don't fit' (Lewenstein, 2003), an overview of the domain would be useful. The model proposed here offers a simple framework within which all communication of science may be mapped and analysed for its relevance to the communicators across the whole continuum, including the public.

Construction of the Model

At a meeting in Geneva (Report, 1979) to identify the goals, processes and indicators of scientific development for projects at the United Nations University, 'basic questions' about communicating any scientific material were stated to be: 'from whom?', 'to or with whom?', and 'to what end?' It is these three elements which I have selected to describe science communication as a 'space' in which various actors communicate. If the space can be defined, then activities within the space can be viewed in context.

Who Communicates Science?

A list of categories of the people who communicate science today is given in Table 2.1. The categories are broad, to facilitate construction of a model, and I acknowledge that several of the categories have important sub-divisions which are not defined here (See Chapter 1 for some categorization of the 'general public.')

TABLE 2.1 Categories of Science Communicators

Science communicators
Scientists
Science communication mediators (extension officers, knowledge brokers, etc.)
Historians and philosophers of science and art/science practitioners
Those in social studies of science
Science educators
Popularisers of science e.g. popular authors
Science communicators in science centres, festivals etc.
People in the science media: print and electronic
Science communication researchers
Government, both national and local; administrative organizations, national and international science organizations
Involved public
Disinterested ('lay' or 'general') public

For the purposes of the model, I have arbitrarily placed 'scientists' at one end of the table and 'the lay public' at the other. No scale of size or importance is implied. There is therefore no sense of a 'continuum' down this table. It may be noted that:

- 'Scientists' include industrial and government scientists and research academics;
- 'Mediators' include those employed by scientific organizations with responsibility for communication of the science to involved publics and the lay public. It also includes medical and health professionals;
- 'Social scientists' includes those concerned with the practice of science at the public interface, with matters of risk and ethics, and so on;
- 'Science educators' include primary, secondary, and tertiary teachers;
- 'The science media' include the popular press, magazines, television, film and radio;
- 'Science communication researchers' include independent researchers and those who are commissioned to conduct the research or are given a grant by a concerned organization;
- 'Administrative organizations' are those who communicate science on a regular or intermittent basis. These include such organizations as the World Health Organization, the Wellcome Trust, and so on:
- 'Involved public' includes lobby groups, community organizations, farmers and fishers, the mining and forest industries, all direct 'stakeholders', hobbyists in a scientific domain. NGOs are usually in this group, as are all students studying science because they have a stake in the success of such study;
- 'Disinterested public' includes all who, according to their own perception, have no direct stake or involvement in the practice and knowledge base of science. This includes those who are curious, and those who are not.

All these categories may be communicators of science in different ways and in different contexts. There are thus 12 categories of people who communicate science, and 12 to, with or between whom science is communicated. In some instances, such as one-way media campaigns, the latter group might be described as a 'target audience'. In others, they will be equal partners in participatory processes.

Outcomes of Science Communication

In the simplest view, communicating science varies from one-way information to engagement in wicked problems. As I have stated above, there is clearly a place for information to be simply disseminated, mindful of the intended audience and appropriate to its needs. This style of transmission encompasses both the daily newspaper and the prestigious scientific journal. The purpose of this information output is to inform (not to 'educate') or, possibly, to influence attitudes or behavior (Research Councils UK, 2002, Annex 1, p.42). I therefore term this kind of communication 'one-way information'. Close to this in design and in spirit are other one-way communications such as media campaigns or submissions to Government, which have the overt intentions of affecting attitudes or behavior, influencing policy decisions, and so on. At the opposite end of the continuum are wicked, two-way (one-with-one or one-with-many) participatory processes where the intended outcome (for any of the participants, including experts) is knowledge building and problem solving by mutually respectful sharing of expertise from within and without the formal disciplines of science.

In between lie many variations of intent of the communication process. For example, the Research Councils UK, in their practical guidelines (2002, Annex 1) refer to 'dissemination, consultation, engagement and dialogue'. According to this document, 'dissemination' is 'putting out information with no expectation of a response' (p.42) and 'consultation' is 'a way in which organizations, whether public or private, local or national, can obtain public input to their decisions following an exchange of ideas and information'. (p.43). 'Engagement' is 'stimulating interest in science and generally raising awareness of science and the issues it raises among the public' (p.43).

"Dialogue" is defined by the Research Councils UK as 'generating debate and interaction between individuals and groups... (possibly) with no other end in sight other than that science becomes just another facet of life, rather than something different and difficult' (p.43). The expected result in this last case is, presumably, both attitudinal and awareness-raising. It is difficult to see a great difference between the intentions of 'engagement' and 'dialogue' as defined in these guidelines. In this century, the notion of 'dialogue events' also seems to define dialogue in this way, since such events involve expert scientists presenting a current issue of interest or concern to an interested public for debate and comment (see, for example, Lehr, McCallie, Davies, Caron, Gammon & Duensing, 2007).

It is clear from the literature that terms like 'engagement' and 'dialogue' have many different interpretations. For example Welp et al. (2006, p.172) refer to ambiguity in the literature regarding the terms 'stakeholder dialogues' and 'public participation'. 'A science-based stakeholder dialogue is defined as a structured communicative process of linking scientists with selected actors that are relevant for the research problem at hand... Public participation on the other hand refers to the participation of ordinary citizens in debates on controversial issues'.

For the purposes of a simple model, I have divided the intended outcomes (or purposes) of science communication into three major categories, which are designed to accommodate these different definitions. My categories are shown in Table 2.2.

TABLE 2.2 Intended Outcomes of Science Communication

---

1. One-way information

For example, typically:
To inform the reader, listener or viewer (no other effect)
To inform research in science communication
To inform policy
To affect attitudes (and possibly behaviour)
To inform as an expert witness
To facilitate creation of theoretical models
To 'educate' (as understood by the 'deficit model')

---

2. Knowledge sharing

For example, typically:
To assist in formulation of policy
To mediate diverse perspectives by exchange of knowledge
To facilitate and integrate interdisciplinary approaches

---

3. Knowledge building

For example, typically:
To create new meaning or understanding from different knowledge systems
To enable action in complex environments through integration of knowledge in order to construct new meaning

---

The Research Councils' (2002) definitions of 'dissemination' and 'engagement' both fall into Group 1, despite the usually more equal implications of the word 'engagement'. Trench's (2006) definitions of 'defence', 'marketing', 'duty' to communicate and, possibly, communicating in a "context of public need" are also in this group. So too are Lewenstein's (2003) 'deficit' model and 'contextual' model (which places the communication in a real context but focuses on the response of individuals to information—presumably provided by an expert). This last category might also fit into Group 2, however, depending on the circumstances. Group 2 is about knowledge sharing, sometimes to no other end but achieving insight into other people's views. Lewenstein's 'lay expertise' model, in which scientists gain insight into lay perspectives, and perhaps his 'public participation' model are in Group 2, although this last can also be in Group 3. So also are the Research Councils' definitions of 'consultation' and 'dialogue'. Welp et al's 'public participation' is also classified in Group 2 (the participation of ordinary citizens in debates on controversial issues). Trench's 'dialogue' category would also be in this group. Greco's (2004) 'Mediterranean model' may be placed in either this group or Group 3, depending on the circumstances.

Trench's category of 'participation' is in Group 3, as is Welp et al's 'stakeholder dialogue'. In this category, the implication of equality between the communicating parties is a key to knowledge building. Further, such deference requires not only that other views be heard and noted, as in many public participatory processes, but that cross-disciplinary knowledge be incorporated into the subsequent scientific endeavour. That is, it is fundamental to the scientific outcomes and is a part of these outcomes.

In summary, three aspects must be considered within the model: (1) who communicates the science; (2) to or with whom does the communication occur; and (3) with what purpose? These aspects lead to the need for a three-dimensional model.

A Three-Dimensional Space

To describe the scope and depth of science communication, an appropriate model for these three aspects needs to have the potential to display three non-linear properties. The model can therefore be a simple multi-celled cube (after Zwicky, 1969). According to Richey (2006, p.2) the Zwicky method, which is especially useful for wicked problems, 'encourages the investigation of boundary conditions' of complex systems and provides for 'solution space' (p.2). These spaces enable identification of relationships and, in the case of complex problems, to arrive at possible solutions.

These ideas have inspired my model. A field of activity exists inside each cell of the cube within which we can 'examine all of the configurations… to establish which of them are possible, viable, practical, interesting…' (Richey, 2006, p.5). The three dimensions of the model are shown in Figure 2.3. It may be seen to have 12 categories of communicators on each of two axes, and three major intended outcomes, giving a total of 432 'cells' in the space. This is a minimum number, since each of the intended outcomes may be further subdivided and my categories of communicator may not be all-inclusive. For simplicity, however, I propose the 432-cell model as a useful overview.

Using the Model: Finding Communication Solutions

A model is only useful if it can be used in a practical way. The idea of a Zwicky box is that each cell *contains* information. In this model, the information inside each cell gives the appropriate modes by which the communication might occur. In the case of all cells on the front face, the intended outcomes or purposes of the communication are one-way. These may occur through a variety of *transmission modes* such as television, the Internet, a lecture, or the popular press.

The second layer of cells contains *modes of sharing knowledge*. As an example, the many modes of knowledge sharing for just one group of communicators—scientists—have been classified and are discussed below. The third group contains *modes of knowledge building.*

For example, if popularisers of science wished to disseminate information to the lay public about a recent discovery, they might look inside the relevant cell and find that appropriate modes could be a television documentary, a popular book, and so on. These modes seem quite obvious, but the closer to the back of the cube you go, the more complicated the modes become. Thus the model combines the *goals* of communication (the principal triplet of layers) with the *processes* of communication—the modes within each cell.

A consideration of just one group of communicators (scientists) and one intended outcome (knowledge sharing) focuses on a cross section of my model that involves 12 cells, each cell representing a group of people with whom the scientists might wish to communicate. These people range from other scientists to the lay public. In the Research Councils (2002) informative document for scientists, *Dialogue with the public: practical guidelines*, similar sets of people with whom scientists might wish to communicate were described. These included 'General population (Lay public).'; 'Pressure/interest groups' (Involved public); 'Policy makers' (Government and administration), and Scientists.

For each of these groups, knowledge sharing modes by which scientists might interact with the groups were described. These modes included theatre meetings (large audiences, lecture based), focus groups, citizens' juries, consensus

FIGURE 2.3 The Science Communication field

Who communicates science?

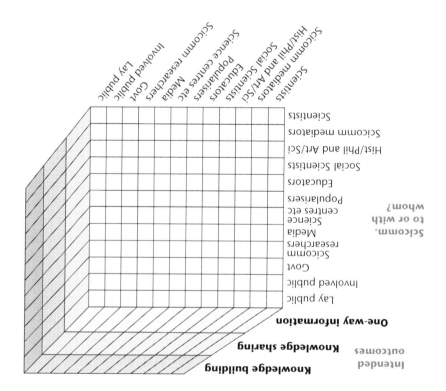

conferences, the Delphi technique (a process of successive consultation), web discussions, and written consultations. All these modes involve sharing knowledge which has originated from the scientist, who is then concerned to obtain other views about the science. In the Research Councils' document, each mode was classified according to its suitability for scientists to use when engaging with different people such as policy makers, other scientists, or pressure groups. For example, theatre meetings were deemed suitable for everyone, whereas a consensus conference applied only to communicating with the general public.

Figure 2.4 shows a selection of these modes mapped *into* the relevant cells (seen as 'exploded' in the diagram). The result is a useful quick reference for scientists who wish to find out about other groups and share knowledge, and who are uncertain of the best technique to use. The relative position of the modes within the cells indicates the degree to which that mode has attributes of one-way transmission or of knowledge building. (To place them in this way, I ranked each mode approximately according to these attributes and a colleague independently ranked them for comparison). Those closest to the front faces in Figure 2.4 have a high level of one-way character even though the intent may be to incorporate some knowledge sharing. To actually achieve knowledge sharing outcomes is likely to be more difficult for modes located closer to the front faces of the cells.

A theatre meeting, for example, offers somewhat limited opportunities for other views and knowledge to be aired. Despite their explicit statement that two-way communication is a preferred outcome of this mode, the Research Councils UK (2002) booklet makes it clear that a straightforward theatre meeting is unlikely to have a strong interactive aspect. As stated in the document, the suggested maximum number of participants is 500 and the following information is given:

> A straight public lecture is probably going to be a monologue event. A lecture with questions however, is starting to build a dialogue. Further down the dialogue path is a talk where not only questions, but comments are invited.
>
> *(Research Councils UK 2002, p.24)*

This is scarcely two-way communication, still less knowledge sharing in a serious sense. The document goes on to describe a more interactive meeting, which also has a suggested limit of 500:

> Moving further still, an event could be constructed where one or more speakers introduce a topic and associated issues or perspectives. The "audience" then breaks up into small groups to discuss the topic and build lists of comments and questions that are shared with everyone else and to which the speaker(s) respond. An event that is basically still a talk is now generating significant amounts of dialogue.

Knowledge sharing

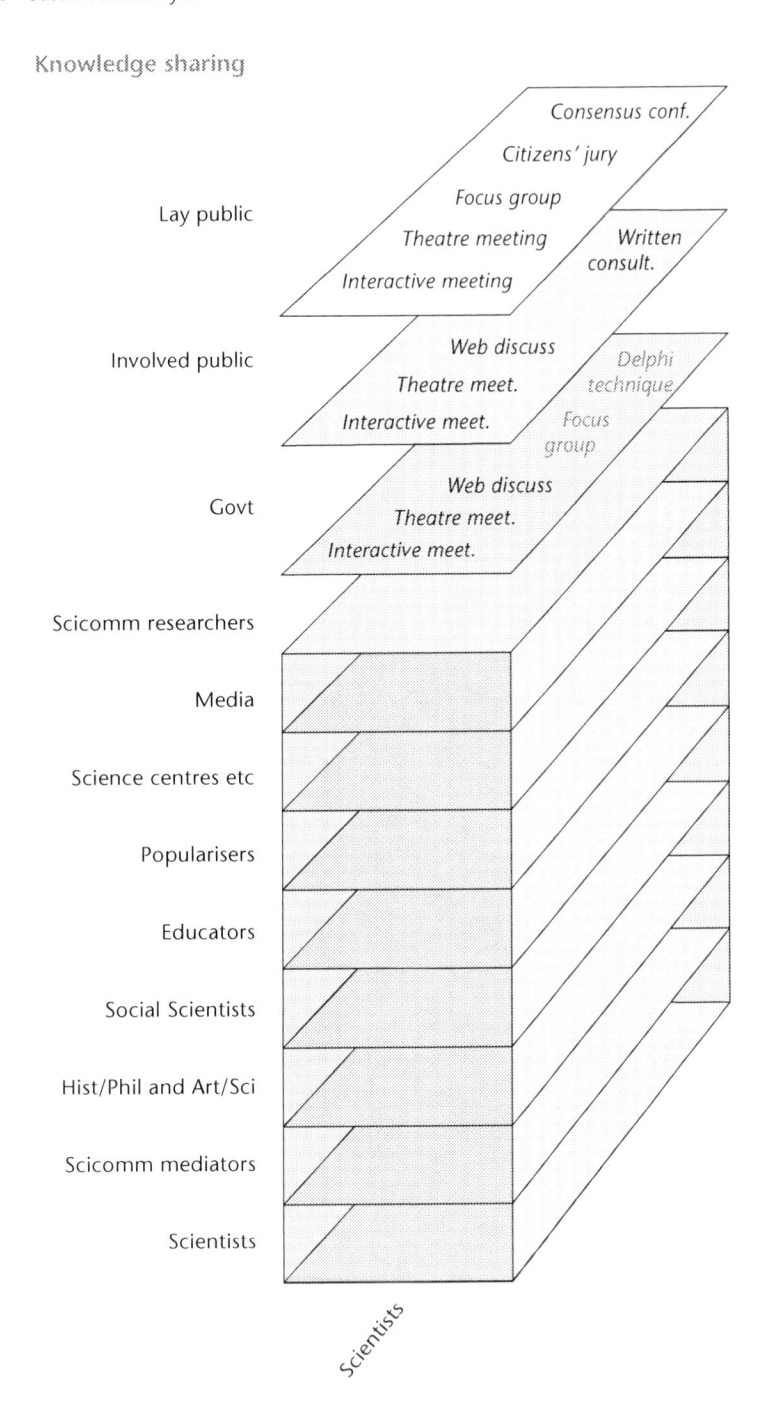

FIGURE 2.4 Cross-Section of the Model Featuring Scientists' Knowledge Sharing with Three Groups

The Delphi technique, on the other hand, is not advocated as a technique for the general public but is useful for interested groups, other scientists and policy makers (pp 32-3). In the Delphi technique, selected participants are

> ... sent a series of questionnaires. The first questionnaire asks each participant to engage in individual brainstorming so as to identify the issues and generate as many ideas as possible for dealing with the issues. The second questionnaire contains all the ideas sent in response to the first questionnaire and provides space for participants to refine each idea, to comment on the strengths and weaknesses of each idea for addressing the issue, and to identify new ideas.

This model therefore provides a quick and simple reference for any group of communicators who can identify their desired outcome for a specific interest group. Clearly it is possible to populate *all* the cells in Figure 2.3 in a similar fashion, indicating, for each different groups of communicators, the variety and appropriateness of techniques for communicating with different people. This cube would then be a quick comparative reference of suitable modes or tools for those intending to communicate science with others.

I should emphasise that these solution spaces offer alternatives. They therefore require further examination to determine the best mode for any particular agenda. To compare the choices of a focus group, a theatre meeting or a citizen's jury means somewhat more than identifying their likely interactive outcomes. The chosen mode will also depend on relative costs, intended participants, time scales, availability of resources, cultural considerations, and so on. These, if separately mapped, would require a multi-dimensional box that would immeasurably add to the complexity of the cube. When it comes to truly wicked problems, which lie right at the back face of the model, the degree of complexity is enormous. Future research may, however, offer some hints about the useful contents of these cells.

One of the purposes of using a Zwicky Box-based model is to identify impossible combinations. In the case of science communication, all cells are possible, but clearly some are less likely to occur.

Conclusion

The model I have proposed describes aspects of science communication: the actors, their communication aims, and the modes by which they communicate. It facilitates the kind of research described by Lewenstein (2003, p.7) when he urged that we seek to understand which models are 'at work in any public communication project and which public communication activities don't fit any of these models'.

The model also recognizes many who communicate science but would not necessarily call themselves science communicators. Included in this group are those who practise what Trench (2006, p.5) describes as 'longer-established disciplines such as history of science, philosophy of science, science education, sociology of science, mass communication, journalism and cultural studies'.

As more information about the processes of knowledge sharing and -building accumulates, through the reporting of participatory practices, cells within the model may continue to be filled. Even in the case of one-way transmission, more information about effective (as opposed to clearly ineffective) modes may assist communicators in deciding which best suits the circumstances. As an example, information about demographics of audiences has revealed that the 'not for me' and 'not sure' groups of the public in Britain are less likely to use the Internet or listen to the radio, and more likely to watch commercial television (Research Councils, 2002). Such information is important for effective communication and may easily be included within the appropriate cell. As information about best practice comes to hand, each group of communicators can draw up a suite of modes to achieve desired outcomes and map them in a simple manner within the cells. Each mode could then be cross referenced with information about participant numbers, costs, personnel required and so on, in order that the somewhat wicked and messy problems of science communication may be clarified.

Notes

1   Dialogue events have been defined as "sites of education for scientifically literate citizenship, in which scientific and technical experts, policy-makers, and the public are understood as citizens both challenged by and contributing to decision-making about complex socioscientific controversies" (quoted in Lehr, McCallie, Davies, Caron, Gammon & Duensing, 2007, which provides an intensive discussion of such events and their purpose.)

References

Australian Public Service Commission (2007). Tackling wicked problems: A public policy perspective. http://www.apsc.gov.au/publications07/wickedproblems2. htm. Retrieved 4.9. 2010.

Bauer, M. W., Allum. N., & Miller. S. (2007). What can we learn from 25 years of PUS survey research? Liberating and expanding the agenda. *Public Understanding of Science, 16,* 79-96.

D'Andrea, L., & Declich, A. (2005). The sociological nature of science communication. *Journal of Science Communication, 4.* http://jcom.sissa.it/. Retrieved 27.1.2007.

Dance, F. E. (1967). *Human Communication Theory: Original Essays.* New York: Holt.

Esteban, L. P. (1994). *Communication among scientists in the center and on the periphery.* www.citeulike.org/user/qaramazov/article/225177. Retrieved 15.2.2007.

Greco, P. (2004). Towards a "Mediterranean model" of science communication. *Journal of Science Communication, 3.* http://jcom.sissa.it/. Retrieved 27.1.2007.

Grunig, J. E., & Hunt, T. (1984). *Managing public relations*. New York: Holt, Reinhart and Winston.

House of Lords (2000). *Report of the Select Committee on Science and Society*. London: House of Lords.

Hulme, M. (2009). *Why we disagree about climate change*. Cambridge: Cambridge University Press.

Kim, H. (2007). PEP/IS: A new model for communicative effectiveness of science. *Science Communication, 28*, 287-313.

Layton, D., Jenkins, E., McGill, S., & Davey, A. (1993). *Inarticulate science? perspectives on the public understanding of science and some implications for science education*. East Yorkshire: Studies in Education Ltd.

Lehr, J. L., McCallie, E., Davies, S. R., Caron, B. R., Gammon, B., & Duensing, S. (2007). The value of "dialogue events" as sites of learning: An exploration of research and evaluation frameworks. *International Journal of Science Education, 29*, 1467-1487.

Lewenstein, B. (2003). *Models of public communication of science and technology*. http://communityrisks.cornell.edu/BackgroundMaterials/Lewenstein2003.pdf. Retrieved 19.02.2007

Nisbet, M. (2006). *Framing science: Understanding the battle over public opinion in policy debates*. Seminar presented to the AAAS, http://www.aaas.org/news/releases/2006/1018framing.shtml. Retrieved 22.1.2007

Office of Science and Technology & The Wellcome Trust (2000) *Science and the Public: A review of science communication and public attitudes to science in Britain*. London: The Wellcome Trust

Research Councils UK. (2002). *Dialogue with the public: Practical guidelines*. London: Research Councils UK.

Report of a meeting (Geneva, 1979) of the sub-project on Forms of Presentation (FoP) of the Goals, Processes and Indicators of Development (GPID) project of the United Nations University. http://www.laetusinpraesens.org/docs70s/79forms.php. Retrieved 15.2.2007.

Rennie, L. J., & Stocklmayer, S. M. (2003). The communication of science and technology; past, present and future agendas. *International Journal of Science Education, 25*, 759-773.

Richey, T. (2006). *General morphological analysis. A general method for non-quantified modelling*. Adapted from the paper: 'Fritz Zwicky, morphologie and policy analysis', presented at the 16th EURO Conference on Operational Analysis, Brussels, 1998. ritchey@swemorph.com. Retrieved 5.1.2007.

Rittel, H. W. J., & Webber, M. M. (1973). Dilemmas in a general theory of planning . *Policy Sciences, 4*, 155-69.

Shannon, C. E., & Weaver, W. (1949). *A Mathematical Model of Communication*. Urbana, IL: University of Illinois Press

Schramm, W. (1954). How communication works. In W. Schramm, (ed.), *The Process and Effects of Communication*. Urbana, IL: University of Illinois Press, pp. 3-26.

Stocklmayer, S. M. (2005). Public awareness of science and informal learning— a perspective on the role of science museums. *The Informal Learning Review, 72*, 14-19.

Stocklmayer, S. M., & Bryant, C. (2012). Science and the public: What should people know? *International Journal of Science Education Part B: Communication and Public Engagement, 2*, 81-101.

Swedish Morphological Society (2011). *Wicked problems.* http://www.swemorph.com/wp.html. Retrieved 22.6.2011.

Trench, B. (2006). *Science communication and citizen science : How dead is the deficit model?* Unpublished manuscript, based on a paper presented to Scientific Culture and Global Citizenship, Ninth International Conference on Public Communication of Science and Technology (PCST-9), Seoul, Korea, May 17-19, 2006.

——(2008). Towards an analytical framework of science communication models. In D. Cheng, M. Claessens, T. Gascoigne, J. Matcalfe, B. Schiele and S. Shi (eds). *Communicating science in social contexts.* European Commisssion: Springer, pp.119-133.

The Danish Board of Technology (2011). *Methods.* http://www.tekno.dk/subpage.php3?survey=16&language=uk. Retrieved 21.6.2011

Welp, M., de la Vega-Leinert, A., Stoll-Kleeman, S., & Jaeger, C.C. (2006). Science-based stakeholder dialogues: Theories and tools. *Global Environmental Change, 16,* 170-181.

Wheatley, M. J. (1992). *Leadership and the new science: Learning about organization from an orderly universe.* San Francisco: Berrett-Kohler.

Wynne, B. (1991). Knowledges in context. *Science, Technology and Human Values, 16,* 111-121.

——(2006). Public engagement as a means of restoring public trust in science—hitting the notes, but missing the music? *Community Genetics, 9,* 211-220

Ziman, J. (1991). Public understanding of science. *Science, Technology and Human Values, 16* 99-105.

Zwicky, F. (1969). Discovery, invention, research—through the morphological approach. Toronto: Macmillan.

PART II

Challenges in Communicating Science

3

SCIENTISTS' ENGAGEMENT WITH THE PUBLIC

Suzette D. Searle

The Rhetoric

Since the 1970s, commentators such as Goodell (1977) have observed that science was 'being pressured to update its antiquated concepts of how much to tell the public, when, and how' (p.6). This update has been very slow to occur: in 1993, Neidhardt stated that that communication between scientists and the public was poorly done (1993, p. 340), and, in 2005, it was reported that '[t]here is a widespread feeling that the communication of science needs to be improved and that the public would like more information on science and science issues' (MORI, 2005, p. 12).

Many social reformers, political leaders, presidents of professional scientific associations, philosophers, and senior scientists have recognised the importance of scientific knowledge and the power of scientific and technological developments to create, save, and destroy life as we know it (Obama, 2008; Pai, 1999). People's concerns, curiosity, and their right to know about scientific knowledge and the effects of its applications upon their daily lives, and on the lives of their children's children, have prompted scientific and political leaders to call for scientists to communicate more effectively with the general public beyond their traditional method of peer-reviewed scientific journals. By explicitly demanding changes in the practice of communication by scientists, these leaders implied that scientists should re-examine their assumptions about themselves and the general public (Searle, 2011). Previously, as Yankelovich observed, 'In scientific circles, it is always assumed that the public and society at large must catch up with science and technology…Little is said about what science must learn about the public' (Yankelovich, 1984). In recent years however, leaders have begun to acknowledge that scientists need to listen to the views and knowledge of non-scientists. Words such as 'dialogue' and 'engagement' have often been used, interchangeably and confusingly, to generally describe a two-way communication that is much more than the scientists' traditional one-way delivery of information to an unquestioning audience. But what do scientists think and do about these public exhortations? As pithily observed by Jensen (2011, p. 26): 'Officially, researchers and academic institutions alike have accepted the importance of public engagement. It is not clear whether these generous intentions translate into effective popularisation actions from individual scientists or career recognition from the institutions for these actions'.

To help improve communication with the public about science and technology, various social science researchers have conducted surveys, interviews, and focus groups to explore and describe people's views and beliefs about science, scientists, scientific issues, and how they like to receive or find scientific information. As a result, descriptions of the public's perceptions and expectations of the direction, conduct, regulation, and communication of science research are being refined. The public's opinions about particular issues, such as climate change, and about emerging scientific applications such as genetically modified organisms (GMOs), cloning, stem cell research, and nanotechnology, are often sought.

At the same time, much research has identified effective ways or processes that bring scientists and non-scientists together to discuss science and technology issues of common interest or concern, often in the name of dialogue, public engagement, or public participation in science and technology. Many of these participatory processes are initiated by governments to consult the public about their ideas and concerns, and to give the public a voice. Few of these processes, however, take the step of promising to implement the public's views (Wilsdon & Willis, 2004).

Comparatively less research has been conducted to understand the views of scientists about their communication with the public, although scientists are integral to communication between science and society. Some researchers believe that scientists choose whether they communicate with the public. For example, Trench and Junker (2001) stated that 'the decision whether or not to engage in public communication and the manner and content of such communication were matters for the individual scientist, and thus wholly distinct from professional communication' (p. 1). This may be true to some degree, but the relatively few studies of scientists that have been done this century indicate that scientists are not as free to communicate as they wish (MORI, 2001a; People Science and Policy Ltd., 2006b; Searle, 2011).

Responsibility

The social responsibility of any profession, according to Frankel (1989), is part of a negotiation process that is based on the tension 'between the professions' pursuit of autonomy and the public's demand for accountability' (p. 110). Over the last four decades, those who have written about the specific responsibility of scientists to communicate with society have included scientists themselves. For example, in 1971, Brown described the 'new' relationship between government and the scientific estate and stated that scientists had a responsibility to inform the public: 'The scientific estate must become public informer and educator by initiating the objective dissemination of its specialised knowledge and information to a deserving, interested and sometimes confused public' (Brown, 1971, p. 228). Goodell (1977) was much more cynical about the sincerity of the science community's increasing sense of social responsibility, linking this with 'public anti-science feeling, job shortages, funding restrictions, and technological dilemmas...Sceptics question whether the new sense of social responsibility runs very deep in the scientific community' (pp. 96-7). Asimov warned of the consequences of the science community's not taking this responsibility seriously: 'Without an informed public, scientists will not only be no longer supported financially, they will be actively persecuted' (Asimov, 1983, p. 119).

In the United Kingdom, the word 'duty' has often been used, rather than 'responsibility' regarding scientists' communication with the public. For example, the Bodmer Report's directive to the scientific community stated emphatically that communicating with all segments of society was a scientist's personal duty: '...learn to communicate with the public, be willing to do so, indeed consider it your duty to do so' (Bodmer, 1985, p. 24). This duty to communicate also incorporates the need to secure public support, which no doubt includes public funding. Mayor, Director-General of UNESCO, did not mince words about the importance of public support and strong research policy for science when he addressed the many government representatives present at the World Conference on Science in 1999. He said, 'direct, public support is the lifeblood of basic research and of all levels of science education. Make no mistake: science needs political will'. In return for 'funding and structured support' he said that science 'must respond to the needs of society' (Mayor, 1999, p. 26).

Research shows that most scientists do indeed feel that that they have a responsibility or duty to communicate their research findings with the general public (MORI, 2001b; People Science and Policy Ltd., 2006b; Searle, 2011). Such responsibilities and duties are being articulated and incorporated in recent national and international guidelines for science communication. These include the International Council for Science Committee on Freedom and Responsibility in the Conduct of Science (2010), the US Committee on Science Engineering and Public Policy et al. (2009), the UK Government Office for Science (2007), and the European Commission (2005).

Scientists are also motivated to communicate with the general public by feelings of accountability to the taxpayer; or a need to promote their area of research to maintain or increase public funding. Other motivations include a desire to educate and inform the public or share knowledge and learn, as well as to gain public approval or recruit new scientists and science students (European

Communities, 2007; Gascoigne & Metcalfe, 1997; Gregory & Miller, 1998; Martin-Sempere, Garzon-Garcia, & Rey-Rocha, 2008; Pearson, Pringle, & Thomas, 1997). Yet there continue to be numerous statements that the public does not understand science and does not get the information it wants and needs. Further, scientists do not understand the public. Why this failure to communicate when, according to Merton (1973, p. 33), science is inherently a communicative culture?

The Rules

Scientists communicate very effectively with other scientists within their own fields. They have to, because communication, especially but not exclusively through research publications, is the basis for an academic career in science. 'Paper publication in peer-reviewed journals is the only accepted form of communication of results that the scientific community engages in en masse' (Suleski & Ibaraki, 2010, p. 117). Scientists' communication with peers is essential for professional recognition and rewards and the advancement of scientific methods and knowledge.

According to a recent review by the InterAcademy Council of the Processes and Procedures of the IPCC, 'Scientists have long struggled to effectively communicate their findings to wider audiences (InterAcademy Council, 2010, p. 47). Metcalfe and Gascoigne (2009) observed that, 'To many scientists the opportunity to discuss their work publicly is more a threat than an opportunity…. But pressure on scientists to communicate…is mounting' (p. 41).

This problem can be partially attributed to the science community itself, which for many years has had firmly entrenched rules for scientists' communication with the public. As stated by Trench and Junker (2001), 'Communication beyond the boundaries of the scientific system was governed by implicit or explicit rules of professional conduct on speaking only about a recognised specialism and only after formal scientific publication' (p. 1).

These implicit and explicit rules of professional conduct evolved during the 20th century, when public funding for science was more assured, to protect and enhance the status and reputation of scientists, their employers, their scientific disciplines, the profession and scientific knowledge as a whole (Marburger, 2005, p. 96). As part of the culture of science, these rules are taught to, or absorbed by, scientists during their training as students or as apprentices in the workplace, and they are enforced by critical peers within the scientific community.

Goodell (1977) listed six ways in which she believed the scientific community handled the conflict between long-range ambitions to inform the public and more pressing needs to concentrate on basic research. Although written more than four decades ago, these six rules still have an influence upon scientists' communication today (See Figure 3.1), although scientists these days are just as likely to be 'she'.

Rule 1: He should confine his activities to the government advisory system if at all possible. The scientist who perhaps once a month evaluates new research proposals for technological programs as part of a respected, selective elite, is fulfilling scientists' public obligation in the approved way.

Rule 2: The scientist would limit his public activities to a small percentage of his time. After all, research is his goal, and the rest is mostly distraction.

Rule 3: The scientist should try to postpone most of his public efforts until after his most productive years are over. Since it is also an adage of the scientific community that 'science is a young man's game', an older scientist, especially a successful one, is allowed more time for public activities. Whether or not it is true, the adage becomes a self-fulfilling prophecy: during the life course of the typical scientist, the amount of time spent on research steadily decreases, and the amount of time on public activities, administrative duties, and 'gate keeping' functions (like refereeing scientific papers and distributing research funds) increases.

Rule 4: The scientist should restrict his public communications to those that can be considered in his 'area of expertise'. He is only an expert on subjects related to his PhD, no matter how much he may have studied other areas.

Rule 5: The scientist should confine his remarks and activities to those that will enhance the public image of science and its propensity to provide funding. Popularisation is better than politics, because it is safer and extols the virtues of science. As a corollary, the scientist, by all means, should not dredge up and expose controversies that are raging in the scientific community behind closed doors.... Whether political or scientific (if there is a distinction), controversies will detract from the image of science and scientist as objective and rational.

Rule 6: If the scientist feels he must express political opinions, he should keep them in the moderate range of the political spectrum, avoiding extremes....
In general, political activities that protect and enhance the scientific establishment are acceptable; those that threaten it, not surprisingly, are not. Since the scientific establishment is vitally connected to government and industry, activities that question the overall established structure of the nation in the long run are rejected.

FIGURE 3.1 The Science Community's Rules for Communicating with the General Public (*Source:* from Goodell, 1977, pp. 91-92)

Three decades later Leshner (2006) wrote his 'lessons for the scientific community', and while he repeated some elements of Goodell's observed rules, he went further to state that scientists should 'never insert their personal values into discussions with the public about scientific issues', and that they should help the public to understand the nature of science. He also, however, encouraged scientists to meet with the community and to listen. 'The most important—and most difficult—lesson to learn is that public engagement involves genuine dialogue, which means both parties must listen and be willing to modify their own positions. Studies conducted for the Department of Trade and Industry in Britain, have suggested that the public is very sceptical of so-called public-engagement events. We have to mean it when we do it' (Leshner, 2006, p. B20).

How scientists have communicated their knowledge in the past was memorably described by Ziman (1998): 'In pursuit of complete "objectivity"— admittedly a major virtue—the norm rules that all research results should be conducted, presented, and discussed quite impersonally, as if produced by androids or angels' (p. 1813). For fear of criticism from their peers, perhaps, many scientists do not tailor their presentations of research findings to appeal to and engage with the general public—a very different but perhaps no less critical audience. Perhaps scientists also fear that they will lose their credibility with a general audience (or their scientific peers) if they present their results with the same passion with which they pursue their research.

This dichotomy of the detached public pretence and the private passions within scientists has long been recognised, and psychological studies were published on the subject in the 1960s. For example, Roe (1961) states that the public image of scientists that includes 'coldness, remoteness, and objectivity' as a reinforcement of impersonal character and emotional neutrality, another of Merton's norms of science, 'could hardly be further from the truth' (p. 456).

Scientists themselves have been shown to believe that this 'naïve' view of 'purely objective, emotionally disinterested scientists' was only taken 'literally and seriously by the general public or beginning science students' (Mitroff, 1974, p. 588). Gieryn (1983) pointed out, however, that it is scientists who continue to present ideologies of science as 'distinctively truthful, useful, objective or rational' because they are 'useful for scientists' pursuit of authority and material resources' (p. 793). 'Especially when scientists confront the public or its politicians, they endow science with characteristics selected for an ability to advance professional interests' (ibid., p. 783).

Contradictions and tensions remain to this day in the scientific community's attitudes to popularisation or making scientific knowledge accessible and comprehensible to a wide audience. For example, Lewenstein was quoted as stating that 'the rules of appropriate behaviour with regard to popularisation are used self-servingly—they are stressed by scientists who want to criticise or limit other scientists' behaviour but are ignored by the same scientists with regard to their own behaviour'. Lewenstein also wrote about the contradiction in the views of those scientists who do not popularise who 'see popularisation as something that would damage to their own career' but 'also think that other scientists use popularisation to advance their career' (quoted in Gregory & Miller, 1998, pp. 82-83).

These rules of communication have arguably restrained or prohibited scientists from being more effective communicators. Inadvertently they have also set scientists up for criticism as poor communicators from the public, the media, and their own political and scientific leaders. The scientific community now recognises that these rules are losing their usefulness, and in response to scientists' need for more public support (including funding), their failure to communicate effectively, and the publics' increasing desire to know and influence what is happening in science, they are changing. Nevertheless, these rules present dilemmas for scientists who, on the one hand, are advised to speak only to their area of specialist knowledge, after they have published in scientific journals and preferably when it does not distract them from their research or applying for funds. On the other hand, they are increasingly expected to seek opportunities to communicate and to present their findings in a broader and more relevant context for public audiences (Leshner, 2005, 2006, 2007). To do this means they have to first understand the broader context and the implications of their research themselves—something they often lack the time, confidence and expertise to do.

The reality for most scientists is that communicating with the general public is not formally part of their job and they are not trained to do so. Research has shown, however, that employers expect them to communicate when required (Searle, 2011). It has also been found that most scientists communicate at least once a year or more with the general public, but that this varies with their age and seniority. Those who are paid to communicate with the public as part of their job are more likely to be older scientists in more senior positions (Jensen, 2011; Kreimer, Levin, & Jensen, 2011; People Science and Policy Ltd., 2006b; Searle, 2011).

The Risks

Scientists risk more than missed opportunities or criticism from their peers, their employer, and the public when they communicate with the general public. Edmeades (2009, p. 36) stated that scientists risk losing their funding, their jobs, or both, by speaking out publicly against the wishes of their employer or funder. There are other personal risks, such as lost promotions for scientists who hold views or conduct research with which others disagree (Gascoigne & Metcalfe, 1997). Media reports of scientists who have been abused and threatened for their views in recent years concerning climate change, animal testing, genetically modified organisms, stem cell research, or the Large Hadron Collider, are evidence of these risks.

The risk, however, that most frequently influences the public communication of most scientists, irrespective of their field of study or employer, is the professional risk of disapproval or other negative reactions that affect their reputation for reliability or their credibility with their scientific peers. Within the last decade, quantitative data has also been collected regarding scientists' views on peer criticism of communication with the public. The 2006 United Kingdom study found that: 'A fifth of respondents said that taking part in public engagement activities was perceived as a barrier to career progression by their peers' (The Royal Society, 2006, p. 32). A study in Spain of scientists participating in PCST[1]-Madrid Fairs found that, 'According to some respondents, certain colleagues consider that those who participated in this type of PCST event "have nothing better to do" or "aren't good enough for more important activities'". The authors reported that, 'This is an opinion that extends to any activity other than carrying out funded research and the subsequent publication of results in prestigious international journals' (Martin-Sempere et al. 2008, p. 357).

According to a survey of media contacts of scientists in top R&D countries (United States, Germany, France, United Kingdom and Japan) in 2008, 'possible critical reactions from peers' were considered important concerns for 42% of the respondents. A similar proportion (39%), however, found 'enhanced personal reputation among peers' to be an important outcome of media contacts (Peters, et al., 2008, p.204).

In 2009 Burchell, Franklin and Holden reported that a number of scientists in their study 'rejected the notion that scientists' participation in public engagement brings with it the risk of professional stigma or opprobrium (p. 61). Despite this, the authors concluded that there was a 'professional anomaly' because 'although [public engagement] is increasingly recognised as valuable to science in general, and as individually rewarding', they wrote that, 'public engagement activity is also seen to be potentially detrimental to a professional scientific career' (Burchell, et al., 2009, p. 7). Criticism by conservative and jealous peers has been shown to arise from a number of cultural beliefs such as scientists should not 'air their dirty laundry', blow their own trumpets, or trust the media to report their science accurately. Serious researchers, apparently, do not spend their time communicating with the public nor risk their professional reputation by being seen as a 'media tart'.

The Restraints

Scientists' are restrained by more than just their slowly changing professional rules of conduct. Their employers and funders, and a more questioning and a less deferential public, also play their part. In a recent Australian study, however, I found that most examples of hindrances to scientists' communication occurred within their workplace. These included a lack of time to organise, prepare, and communicate. A typical comment was:

> Time—it is difficult to just publish work and complete research projects let alone make time to improve public communication or actually do some communication. If it is or was part of my job I think I would improve and obviously do it more.

Heavy workloads and finding time to apply for funding grants and actually do research were necessarily a higher priority for scientists than finding time or opportunities to communicate with the general public. Employers' public comment policies, approval processes or protocol requirements and lack of opportunity to communicate also hindered scientists from communicating. These influences are shown in Figure 3.2.

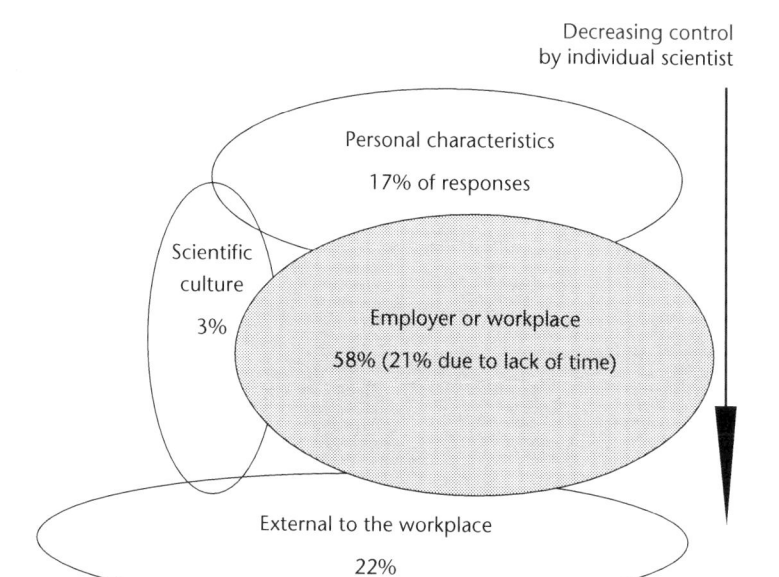

FIGURE 3.2 Hindrances—Broad Emergent Themes from Scientists' Comments (*Source:* Searle, 2011)

Factors beyond workplace control included commercial considerations, government politics, media, national security, or the lack of interest or misconceptions of the public. Scientists' personal characteristics, such as a 'certain outgoing personality', a lack of confidence or knowledge or motivation, and fear, were also mentioned. Only a small minority described aspects of the scientific culture, such as the difficulty of communicating to those without scientific knowledge, as a hindrance.

These restraints also reflect that, for most scientists, communicating with the general public is not recognised or rewarded. These sentiments have been recorded in other studies of scientists' communication in Australia (Gascoigne & Metcalfe, 1997), the United Kingdom (People Science and Policy Ltd., 2006b) and New Zealand (Edmeades, 2009) in recent years. Even in workplaces where communication with the public is expected, lack of opportunities, especially for the younger scientists, and time-consuming approval processes hinder scientists from communicating.

Research has shown that scientists' communication is also inhibited by secrecy imposed by collaboration, commercialisation, and competition; an increase in short-term employment contracts; and scientists' beliefs that the public is ignorant, uninterested, distrustful, or critical. There is also an often uneasy alliance between scientists and journalists, whom scientists fear, or expect, will include inaccuracies and misquotes in the presentation of their research, or over-sensationalise it.

The Reality

Despite the oft-expressed belief that scientists' communication must improve, relatively few studies describe scientists' actual communication practices, although national studies have been conducted in the UK (MORI, 2001a; People Science and Policy Ltd., 2006b; Poliakoff & Webb, 2007), Sweden (Vetenskap & Allmanhet, 2003), Denmark (Nielsen, Kjaer, & Dahlgaard, 2007), France (Jensen, 2011), and Argentina (Kreimer, Levin, & Jensen, 2011). Together with my own Australian study, these generally show that most scientists communicate seldom and those who communicate more, do so as a result of their job, seniority, or age.

It would also seem that most scientists communicate with the general public beyond the range of microphones or cameras (The Pew Research Center for the People & the Press, 2009; The Royal Society, 2006). For example, Australian scientists said that they communicated most frequently through a mixture of informal and formal situations and none of these included the mainstream media. They answered inquiries from the public via email, letter, or telephone, for example; or spoke with students or teachers or spoke at events specifically intended to inform the general public. Face-to-face communication was much more frequent than communication via the media or the Internet. Of those who had communicated via the media, most had communicated only once or twice within a year, and that was more likely to be through an interview for a national or local newspaper.

The Rewards

Beyond the motivations already described, which may or may not be realised, why do scientists communicate? Relatively few are paid to communicate with the general public as a formal part of their job, and fewer are recognised or rewarded for it in other ways, as was found by Jensen (2011). The latter concluded from his studies of thousands of French CNRS researchers that included statistical analyses of the influence of dissemination activities on their promotions, that, 'dissemination activities are *not* [original emphasis] bad for scientists' careers' and 'They are not very good either' (Jensen, 2011, p. 34).

Nevertheless, many scientists say that they do benefit both professionally and personally from communicating with the general public. For example, communication with the public has been found to contribute to scientists' professional success through direct public participation or co-operation in research, and networking with potential employers and funders (Searle, 2011). This study, of 1,521 Australian scientists, also found that nearly one in five of the scientists described positive feelings as a benefit from communicating with the general public. Many scientists valued connections with people who were interested in and could use their knowledge and help with their research, because of how it made them feel about themselves and their work. Given that research has shown that many scientists—in particular women, applied scientists, younger scientists, and those who choose to work in the public sector—are motivated to become a scientist by their desire to contribute to the public good, these feelings can not be ignored as they strike at the very heart of why many people choose to become, and remain, scientists.

Emotional benefits were also described in a survey conducted by UK researchers Burchell, et al. (2009, p.52). These authors wrote about the 'unexpected enthusiasm' described by scientists for the benefits of public engagement such as identifying scientific priorities, positively refocusing scientific objectives and improving clinical practice. Their interviews with 30 UK scientists revealed that, 'Many of the interviewees' perceptions of the value of public engagement were based on first-hand experience of it, often accompanied by a "conversion narrative" of sorts, in which interviewees described unexpected enthusiasm for this type of activity in spite of its potential limitations, time-consuming nature and unconventional demands'. 'Enjoyment' associated with communicating with the public via the media was also a finding in Australia. 'Some scientists enjoyed interacting with the media, and also enjoyed the public exposure it brought their work' (Metcalfe & Gascoigne, 2009, p. 42).

In my own study, I found that scientists identified many intrinsic benefits from communicating about their work, including feelings of satisfaction, self-esteem and pride from making a contribution to society, and enjoyment from sharing what they knew and helping others.

According to Yankelovich (2003), however, 'Emotions and beliefs linked with society's changing issues …are not often part of scientists' public image'. Is this because scientists are confusing the need for a disinterested and emotionally neutral conduct of research, as described by Merton and others, with the objective, unemotional delivery of results to maintain the credibility of their knowledge and themselves? Is this style of delivery reinforced by critical peers? Are scientists deliberately undermining the effectiveness of their communication with the general public by limiting the affectiveness of their delivery? When in doubt, perhaps scientists moderate their style of presentation to ward off critical peers, rather than appeal to a more engaged public. If so, they may be unaware of the cost, because, to paraphrase John Maxwell's oft-quoted advice about good leaders, '…good [scientists] understand that people do not care how much you know until they know how much you care' (Maxwell, 2007, p. 304).

Few studies have explored the personal importance to scientists of their communication with the general public, but one of them, the UK study *Survey of factors affecting science communication by scientists and engineers*, (People Science and Policy Ltd., 2006a, p. 22) revealed interesting trends amongst academic scientists and engineers. Participants were asked how important they felt it was that they personally, in their current post, engage directly with groups such as journalists, policy-makers, schools and school teachers and the non-specialist public. A key finding of the study was that, 'There was a strong positive relationship between the number of activities undertaken by a scientist and their perceived importance of public engagement' (The Royal Society, 2006, p. 10). The results also showed that those who rated direct engagement as more important than their counterparts were male researchers and more senior and older researchers. Those with teaching as well as research responsibilities, and who thought their work had social implications, were also more positive, as were researchers trained in communication skills or who felt it was easy to get involved in engagement activities (The Royal Society, 2006, p. 16). These results also indicate the positive influences upon scientists' direct engagement with the public.

The Remedy

Many scientists feel a responsibility to communicate with the general public and it is personally important to them (People Science and Policy Ltd., 2006b: Searle, 2011; The Royal Society, 2006). It is logical therefore, that research has also shown that scientists want more time, more opportunities, and more training to help them communicate. The training they want is to help them improve their communication of science, their communication with the media, and their general presentation skills. They also want more encouragement and recognition for their communication activities. A number want help from professional science communicators, too.

With regard to training, since 1997 in Australia funding has been provided by The Australian National University for graduate scientists to complete a university award in science communication. Other organisations such as the UK Royal Society, The Wellcome Trust, Research Councils UK, the British Science Association, and, in the United States of America, the American Association for the Advancement of Science, and the National Academies of Science, have contributed significant funding for training as well as incentives and rewards for scientists to communicate with the general public. All have aimed to improve the relationship between those who create and govern the science, and those who use it.

It is instructive to see how much funding has been allocated to what purpose. In the UK, public understanding of science activities, for example, received £4.5 million annually from the budget of the Parliamentary Office of Science and Technology (House of Lords, 2000) and more than £100 million up to 2006 from the private charity The Wellcome Trust (Turney, 2006, p. 3). Further resources from several of the funding streams created by the National Lottery were also used to support grants for these activities (House of Lords, 2000 3.3). Funding for public engagement since 2006 has included, for example, an £8 million initiative for four years by the Higher Education Funding Councils for England and Research Councils UK, in association with the Wellcome Trust. It was launched in 2006 to 'promote excellence in public engagement and effect a culture change in UK universities towards engaging with the public'. It aimed to co-ordinate efforts to recognise, reward and build capacity in public engagement. Higher education institutions (universities and research institutes) were invited to bid to become 'Beacons for Public Engagement' to change the culture in universities, assisting staff and students to engage with the public and removing any barriers' (Higher Education Funding Council for England & Research Councils UK, 2006).

Significantly, these universities and research institutes focused on training young scientists to be better equipped for communication activities. Miller stated that this emphasis represented a change, 'in contrast to the pre-CoPUS view that only very senior and (probably) research inactive scientists had earned the right to talk to the general populace…. In particular, Royal Society university fellows and postdoctoral researchers at the start of their academic careers were encouraged to go on media training courses and even to take a month or two out to work as journalists and broadcasters' (Miller, 2001, p. 116). 'From being an activity carried out by superannuated boffins or second-ranked minds, popularising science was legitimised by Bodmer' (Miller, 2001, p. 115)

In the USA, commitment to scientists' communication training has progressed relatively quickly within the last decade, considering that it was only in 1997 that Neal Lane, former Director of the National Science Foundation (NSF) wrote 'An Open Letter to Scientists and Engineers. Let's Get the Word Out Together about Why Science Matters'. Its purpose was to promote the concepts of 'civil science' and the 'civic scientist', and yet Lane only 'ventured' to suggest that outreach become a professional responsibility for scientists and engineers and that they should be taught communication skills at universities (Lane, 1997).

In 1998, the National Science Board of the United States National Science Foundation, referred to Lane's call for 'civil science' stating:

> The public has a reasonable expectation that scientists will contribute to demystifying for others what is so personally and professionally engaging to them. The challenge to do so is the essence of what former NSF Director Neal Lane has called 'civic science.' Scientists and engineers need to be more accessible and more accountable; they need to be articulate and clear about their work and the good it is doing for society; and they need to be willing to lead or participate in public information efforts in a wide variety of public forums, from schools to the media.
>
> *(National Science Foundation Board 1998, pp. 15-16)*

In 2006, the President of the National Academy of Sciences, Cicierone, opined that it was getting harder for scientists to communicate (which he seems to have defined as educating and influencing) with the general public through the media. He suggested that scientists would have to 'do a better job of communicating directly to the public' and 'Our goal will be to communicate the valuable role science plays in the world and to reinforce and enhance positive attitudes toward science and the scientific process' (Cicerone, 2006, p. 3).

The Centre for Public Engagement with Science and Technology, launched in 2004 by the American Association for the Advancement of Science (AAAS), published online in 2009 its 'Communicating Tools for Scientists and Engineers' to provide training and advice. This included a statement that emphasised that although scientists' traditional training did not prepare them to be effective communicators outside of academia, funding agencies were increasingly encouraging researchers to communicate their results directly to the greater public (AAAS, 2009). Scientists were being asked to change their ways and in the USA and the UK funds were made available to help them.

The 20th century culture of science, defined by masculine norms that arguably encouraged and maintained a disconnect between scientists and a respectful society to maintain its authority, status, and independence, is perhaps giving way to a younger and more gender-balanced culture that recognises the mutual benefits of communication between scientists and society, beyond the production of academic papers. Scientists are being urged 'to move beyond simply stating the facts and technical complexities to focus on ways to make complex topics personally relevant' to different audiences (Nisbet & Mooney, 2007, p. 56). As stated by Cribb and Sari (2010, p. 12), 'true communication is not just about sharing information, but more about sharing meaning and achieving a common understanding'.

Within the culture of science, there is evidence that its rules for communication with the general public are gradually changing as the scientific community realises that it can no longer take a supportive public for granted. Research has shown that many scientists communicate for a range of intrinsic and altruistic reasons, and most feel a sense of responsibility to communicate to the public that funds them. They have also identified many personal and professional benefits from doing so.

It is imperative, therefore, that the employers and funders of scientists enable them to respond to the international calls from scientific and political leaders, and the demands of the public, for more effective communication between science and society. To do so, scientists need their communication with the public to be judged as equally important for performance assessment and promotion as conducting and publishing research and securing funds. This would lead to the provision of what many scientists want: more time, more opportunities, and more communication training. From this, both scientists and the public would all benefit.

Notes

References

AAAS. (2009). Communicating science tools for scientists and engineers. Centre for Public Engagement with Science and Technology. Retrieved 17 Jan 2009 from http://communicatingscience.aaas.org/Pages/newmain.aspx

Asimov, I. (1983). Popularising science. *Nature, 306*(5939), 119.

Bodmer, W. (1985). *The Public Understanding of Science.* Retrieved from http://royalsociety.org/displaypagedoc.asp?id=26406 (accessed 14/7/2009).

Brown, W. T. (1971). The scientist's responsibility to the public. *Psychiatric Quarterly, 45*(2), 227-233.

Burchell, K., Franklin, S., & Holden, K. (2009). Public culture as professional science: final report of the ScoPE project—Scientists on public engagement: from communication to deliberation? London: London School of Economics and Political Science.

Cicerone, R. J. (2006). Celebrating and Rethinking Science Communication. *The National Academies IN FOCUS magazine, 2009.*

Committee on Science Engineering and Public Policy, National Academy of Sciences, National Academy of Engineering, & Institute of Medicine. (2009). *On being a scientist. A guide to responsible conduct in research* (Third ed.). Washington, D.C.: The National Academies Press.

Cribb, J., & Sari, T. (2010). *Open science: sharing knowledge in the global century.* Melbourne: CSIRO Publishing.

Edmeades, D. C. (2009). Science is under threat. [Opinion]. *Australasian Science Magazine, 30* (No. 8 September 2009).

European Commission (2005). The European Charter for Researchers and the Code of Conduct for the Recruitment of Researchers. from http://ec.europa.eu/euraxess/index.cfm/rights/index

European Communities. (2007). European research in the media: the researchers point of view. Report December 2007.

Frankel, M. S. (1989). Professional codes: Why, how, and with what impact? *Journal of Business Ethics, 8*(2), 109-115.

Gascoigne, T., & Metcalfe, J. (1997). Incentives and Impediments to Scientists Communicating Through the Media. *Science Communication, 18*(3), 265-282.

Gieryn, T. F. (1983). Boundary-Work and the Demarcation of Science from Non-Science: Strains and Interests in Professional Ideologies of Scientists. *American Sociological Review, 48*(6), 781-795.

Goodell, R. (1977). *The Visible Scientists.* Boston, United States of America: Little, Brown and Company.

Government Office for Science United Kingdom. (2007). *A Universal Ethical Code for Scientists.*

Gregory, J., & Miller, S. (1998). *Science in public: Communication, culture, and credibility.* New York: Plenum Press.

Higher Education Funding Council for England & Research Councils UK. (2006). £8M for new initiative to boost public engagement. Retrieved 9 February 2009, from http://www.hefce.ac.uk/NEWS/HEFCE/2006/beacons.htm

House of Lords. (2000). *Science and Society—Third Report of the Science and Technology Committee, Session 1999–2000.* Retrieved from http://www.publications.parliament.uk/pa/ld199900/ldselect/ldsctech/38/3805.htm#a26.

InterAcademy Council. (2010). InterAcademy Council (IAC) Review of the processes and procedures of the IPCC—Full report.

International Council for Science Committee on Freedom and Responsibility in the Conduct of Science. (2010). Advisory note on science communication. Retrieved 28 January 2011, from http://www.icsu.org/publications/cfrs-statements/science-communication/

Jensen, P. (2011). A statistical picture of popularization activities and their evolutions in France. *Public Understanding of Science, 20*(1), 26-36.

Kreimer, P., Levin, L., & Jensen, P. (2011). Popularization by Argentine researchers: the activities and motivations of CONICET scientists. *Public Understanding of Science, 20*(1), 37-47.

Lane, N. (1997). An open letter to scientists and engineers: 'Let's get the word out together about why science matters'. Retrieved 12 March 2009, from http://www.nsf.gov/od/lpa/news/media/nlaaultr.htm

Leshner, A. I. (2005). Where Science Meets Society. *Science, 307,* 815.

——(2006). Science and public engagement. [Opinion]. *The Chronicle Review The Chronicle of Higher Education, 53*(8), B 20.

——(2007). Outreach Training Needed. *Science, 315*(5809), 161-.

Marburger, J. (2005). Science and Technology Policy in the Real World. *Distinguished Series on Science Policy*, from http://www.ostp.gov./cs/issues/education

Martin-Sempere, M. J., Garzon-Garcia, B., & Rey-Rocha, J. (2008). Scientists' motivation to communicate science and technology to the public: surveying participants at the Madrid Science Fair. *Public Understanding of Science, 17*(3), 349-367.

Maxwell, J. C. (2007). *The 21 irrefutable laws of leadership* (10 , revised, annotated (1st ed. was 1998) ed.): Thomas Nelson Inc, 2007.

Mayor, F. (1999). *Opening address.* Paper presented at the World Conference on Science. Science for the 21st century. A new commitment, Budapest, Hungary.

Merton, R. K. (1973). *The Sociology of Science: Theoretical and Empirical Investigations* (Illustrated ed.): University of Chicago Press.

Metcalfe, J., & Gascoigne, T. (2009). Teaching scientists to interact with the media. *Issues, 87* 41-44.

Miller, S. (2001). Public understanding of science at the crossroads. [CRITIQUES AND CONTENTIONS]. *Public understanding of science, 10,* 115–120.

Mitroff, I. I. (1974). Norms and Counter-Norms in a Select Group of the Apollo Moon Scientists: A Case Study of the Ambivalence of Scientists. *American Sociological Review, 39*(4), 579-595.

MORI. (2001a). The Role of Scientists in Public Debate. Executive Summary. Retrieved 23 January 2011, 2011, from http://www.wellcome.ac.uk/About-us/Publications/Reports/Public-engagement/wtd003429.htm

——(2001b). The Role of Scientists in Public Debate. Retrieved 9 April 2009, from http://www.wellcome.ac.uk/About-us/Publications/Reports/Public-engagement/wtd003429.htm

——(2005). *Science in Society. Findings from qualitative and quantitative research.* Retrieved from http://www.ipsos-mori.com/content/uk-public-is-largely-positive-about-science.ashx.

National Science Foundation Board. (1998). Toward the 21st Century: The Age of Science and Engineering, National Science Board, Strategic Plan NSB-98-215, November 19, 1998. from http://www.nsf.gov/publications/pub_summ.jsp?ods_key=nsb98215

Neidhardt, F. (1993). The public as a communication system. *Public Understanding of Science, 2*(4), 339-350.

Nielsen, K., Kjaer, C. R., & Dahlgaard, J. (2007). Scientists and science communication: A Danish survey. *Journal of Science Communication, 6*(1), 1-12.

Nisbet, M. C., & Mooney, C. (2007). Framing Science. *Science, 316*, 56.

Obama, B. (2008). Search for knowledge. *Barack Obama Weekly Address – 20 December 2008* Retrieved 01:46, January 26, 2009, from http://en.wikisource.org/w/index.php?title=Barack_Obama_Weekly_Address_-_20_December_2008&oldid=951158

Pai, P. (1999). *Opening address.* Paper presented at the World Conference on Science. Science for the 21st century. A new commitment., Budapest Hungary.

Pearson, G., Pringle, S. M., & Thomas, J. N. (1997). Scientists and the public understanding of science. *Public Understanding of Science, 6*(3), 279-289.

People Science and Policy Ltd. (2006a). Factors Affecting Science Communication: A survey of scientists and engineers. Report on quantitative research prepared for The Royal Society. London: People Science and Policy Ltd.

——(2006b). Factors Affecting Science Communication: A survey of scientists and engineers. Retrieved 12 February 2011, from http://www.peoplescienceandpolicy.com/projects/survey_scientists.php

Peters, H. P., Brossard, D., De Cheveigne, S., Dunwoody, S., Kallfass, M., Miller, S., et al. (2008). Interactions with the mass media. [policy forum]. *Science, 321*, 204-205.

Poliakoff, E., & Webb, T. L. (2007). What Factors Predict Scientists' Intentions to Participate in Public Engagement of Science Activities? *Science Communication, 29*(2), 242-263.

Roe, A. (1961). The Psychology of the Scientist. *Science, 134*(18 August 1961), 456-459.

Searle, S. D. (2011). *Scientists' communication with the general public - an Australian survey.* Canberra: The Australian National University.

Suleski, J., & Ibaraki, M. (2010). Scientists are talking, but mostly to each other: a quantitative analysis of research represented in mass media. *Public Understanding of Science, 19*(1), 115-125.

The Pew Research Center for the People & the Press. (2009). Scientific Achievements Less Prominent Than a Decade Ago. Public praises science; scientists fault public, media.

The Royal Society. (2006). Survey of factors affecting science communication by scientists and engineers. Final report on The Royal Society website.

Trench, B., & Junker, K. (2001). *How Scientists View Their Public Communication.* Paper presented at the Sixth International Conference on Public Communication of Science and Technology. Trends in Science Communication today: Bridging the Gap between Theory and Practice. Proceedings of the PCST2001.1-3 February 2001, CERN Geneva.

Turney, J. (2006). Engaging science. Thoughts, deeds, analysis and action. In J. Turney (ed.) Available from http://www.wellcome.ac.uk/About-us/Publications/Books/WTX032706.htm

Vetenskap & Allmanhet. (2003). VA Report 2003:4 How researchers view public and science. Retrieved 25 March 2009, from http://www.v-a.se/downloads/varapport 2003_4_eng.pdf

Wilsdon, J., & Willis, R. (2004). See-through Science: Why public engagement needs to move upstream. In DEMOS (Ed.). London: DEMOS.

Yankelovich, D. (1984). Science and the public process: Why the Gap Must Close. *Issues in Science and Technology* Jul 9, 2003 Summer. from http://www.issues.org/19.4/updated/yankelovich.html

——(2003). Winning Greater Influence for Science. *Issues in Science and Technology*, (July 9 Summer 2003). Retrieved from http://www.issues.org/19.4/yankelovich.html

Ziman, J. (1998). Essays on science and society: Why must scientists become more ethically sensitive than they used to be? *Science, 282*(5395), 1813-1814.

4

THE ROLE OF SCIENCE AND TECHNOLOGY IN PUBLIC POLICY

What is Knowledge for?

Will J. Grant

Introduction

Modern political discussion is heavily influenced by the idea of rational decision-making. This is the idea that the actions of society—the decisions made by governments, organisations, and institutions—should be guided by reason and logic, using the best available scientific knowledge.

In large part, this is a compelling and sensible idea. Just as we all hope that our bridges are made using the best possible mathematics, so too we hope that our hospitals are organised on the basis of the best available evidence. Just as we might hope that our governments consider the scientific evidence on the impact of pollution when developing environmental laws, so too might we hope that they consider psychological evidence when deciding on the best means to deal with violent offenders.

Indeed, politicians and scientists around the world have long called for a relationship as close as possible between science and our policy-making processes (Banks, 2009; McNie, 2007, p.17). In 19th century England, Florence Nightingale admonished the House of Commons for "chang[ing] your laws so fast and without inquiring after results past or present that it is all experiment, seesaw, doctrinaire" (cited in Banks, 2009, p.2). As candidate for the US presidency in 2008, Barack Obama promised voters that his policies would be based on "evidence and facts" (Bhattacharjee, 2008).

Yet as compelling as the idea of rational decision-making may be, it is far from reflecting how things really are, or indeed how they can ever be. We live in a world in which the best scientific knowledge rarely answers all the questions asked by policy-makers, nor answers them in ways that provide the certainty they crave. While it may seem intuitively sensible to base all our public policy decisions on the best possible science, there is much that stops this from happening. Indeed, a significant gap exists between the worlds of science and policy. This gap—and how scientists, science communicators, and policy-makers might work to rectify it—is the focus of this chapter.

What is Policy-Making?

What is policy-making? What do policy-makers need? In the discussion that follows, 'policy' can be taken as the formal designation of governmental will for the society and material world that are governed.[1] A policy on pollution, for example, might define what pollution is permissible and where, and punishments for breaching that limit. 'Policy-makers' can be taken as those who are able to define and set such policy—the elected officials, bureaucrats, and public servants who make up the organs of government (see Garvin, 2001, p.444).

Policy-making, as defined above, is inherently normative (Payne, 2007). Bound up in all policies, whether they are made explicit or remain implied, are assumptions and aspirations about what the particular governed society and material world *should be*. To set a policy is to argue that 'our society should be like X'. To set a policy is to argue that pollution is a thing that should be monitored and limited by governments, that a world with limits on pollution is better than one without.

Because there will always be contestation regarding what any particular society and material world should be, this means that policy-making is inherently political. Where one person may argue that a world with limits on pollution is better than one without, another may argue the converse. This contestation over what any particular society and material world should be is a key part of democratic society.

Yet if we agree on our normative aspirations, we find another crucial question in the relationship between science and policy-making: how do we achieve our goals? Even when we agree on ends, how do we ensure that the policy we set achieves them? To make *good policy*, what do policy-makers need? To achieve their particular normative ends, policy-makers need two things: the ability to act; and useful information about the problem at hand and the effects of their actions on that problem. The ability to act is largely a question of capacity, outside of the focus of this discussion. Of interest here is the question of useful information.

What then is useful information? Elizabeth McNie contends that policy-makers require information that is salient, credible, and legitimate (McNie, 2007, pp.19-20; see also Jones, Fischhoff and Lach, 1999, p.583). To be useful, information must be *salient* in that it must be relevant to the specific context in which it will be used. It must respond to the specific needs of the policy-maker regarding the problem at hand, and it must be relevant to the context in which the problem is situated. To draw again on the example of pollution, a policy-maker might require information about the substances currently being added to the environment, about how those substances affect the society and environment in question, and about the effects of limiting the release of those substances on that society and environment.

To be useful, information must be *credible* in that it must be perceived by the policy-maker to be accurate, valid, and of high quality. It must provide the policy-maker with as true a picture of the problem at hand as possible. What exactly are the substances being released into the environment? What are their effects? Do we know all that is necessary about the pollution in question? Is the information presenting a true representation of the situation?

Finally, to be useful, information must be *legitimate*: it must be produced in a way that is seen to be free from political bias. As noted above, policy-making is inherently normative. We cannot make a policy limiting pollution without assuming that the effects of that pollution are *bad*. Yet as McNie argues, for information to be legitimate it must be produced without bias towards the normative goal of the policy-making. In our pollution example, this means that the relevant information—the substances being added to the environment and their effects—must be produced without prejudging whether we should have a limit on pollution, whether those effects are bad or good.

On the face of it, it would seem that science would play an excellent partner to the policy-making process. Scientific knowledge—knowledge produced through the disciplined interaction between systematically developed theory and rigorous observation of the physical universe—is knowledge produced with the express goal of providing accounts of phenomena that are as clear and accurate as possible. Whether it is about physical, biological, chemical, or social phenomena, the scientific method has done much to accumulate a body of information that is of use to many.

Yet policy-makers throughout the world have long called for information that better serves their needs (see Garvin, 2001; Jacobs, 2002; McNie, 2007). Scientists have made reciprocal calls for policy-makers to pay greater attention to their insights (see for example Clement, 2011; Risbey, 2011). Despite all the science that is done, policy-makers still have questions that remain unanswered, and still craft policy without a complete understanding of the problem at hand and the effects of their solutions. Despite the power and size of our policy-making apparatus, there remain scientists who are dismayed at governmental responses to the problems uncovered through their work, labelling policy-making "irrational and politically motivated", "based more on expediency than on scientific evidence" (Garvin, 2001, p. 445).

Why is this? Why doesn't science instantly give us good policy? Why—to turn things around—don't policy-makers just do what scientists suggest?

Useful information? The Science-Policy Gap

Policy-makers have long called for useful information: information about their problems and the effects of their solutions that is salient, credible, and legitimate.

Yet despite such calls, a significant gap exists between our key processes of knowledge generation and decision-making. For a number of reasons, our scientists and policy-makers can't always talk easily to one another: our science doesn't turn instantly into perfect policy.

This section discusses some of the key issues that contribute to this gap, looking at the ways the worlds of science and policy differ according to timeframe, language, scales of focus, understandings of evidence, and goals. Underpinning all of these differences is the fact that policy-makers and scientists possess radically different worldviews: radically different understandings of how knowledge works.

Timeframes

Scientists and policy-makers work within and according to dramatically different timeframes. These differences are not only evident in the time it takes each to do their work, but also the timeframes each think about.

Scientists typically take many years to do their work. It has taken, to use climate science as an example, many decades to progress from the 19th century insights of Joseph Fourier, John Tyndall, and Svante Arrhenius into the role played by atmospheric carbon dioxide in the warmth of our planet (Hulme, 2009, chapter 2; Jones, 2011) to our current understanding of the complexities of the global climate system. Even now, many questions remain regarding the dynamics of this system (see Steffen, cited in Grant, 2011). Some of this long intellectual gestation in science is due to the complexities of the issues under examination. Continuing with the climate science example: in its Fourth Assessment Report, the Intergovernmental Panel on Climate Change discusses the roles of various human and natural climate-forcing constituents (including greenhouse gases, aerosols, surface albedo changes, aircraft contrails, solar variability, and volcanic activity), before turning to their non-linear interplay in the complicated global climate system (IPCC, 2007, WG-I Chapter 2). To build a robust understanding of these various systems and their interplay is complicated work that simply takes time. Beyond the complexity of many scientific questions, some scientific research simply cannot be done at a faster rate. We cannot fully understand the lifetime effect of any particular childhood intervention until the children in question live out their lives. Finally, as taking many years to develop an adequate understanding of their topic, many scientists often focus on timescales of thousands, millions, or billions of years.

Yet while scientists typically consider many years a reasonable timeframe for the gradual accumulation of knowledge, most policy-makers focus on much closer horizons. Some will be focused on the 24-hour news cycle; others will be focused on this year's budget; others will be focused on elections that come every few years (Jacobs, 2002, p.8). Even though many will seek to incorporate longer-term strategic thinking into their policy development—perhaps paying attention to where they want their country to be in a decade—the pressures of the policy-making environment are overwhelmingly on 'as soon as possible'. Many do not have sufficient time to search for information that might be considered scientifically best (Tribbia & Moser, 2008, p.317). Typically, 'adequate now' is far more useful to policy-makers than 'perfect later'.

This significant difference in timeframe means that for many questions, a full scientific assessment is just too slow for policy-makers. We can imagine a possible problem—perhaps a doctor notices a few children being diagnosed with higher levels of lead in their blood than the national average. To turn from merely a hunch into a well-understood problem would require examining more children (across the street, the suburb, the town, or the region?) to see if the problem is more widespread. If it is more widespread, what might be the cause? Is it natural geography, decomposing house paint from a generation earlier, inadequate cleaning practices, or the factories and mines of the town? What are the pathways from these possible sources to the children? What solutions might reduce the poisoning? Yet while full investigations along these lines might take many months or years, policy-makers feel the constraints of more immediate timeframes. What action are they taking that addresses the pressing needs of their citizens, or the call of the 24-hour news cycle?

Language

All fields of expert work have their own specific terms, phrases, and ways of saying things. Such jargon is useful for communicating within the field, yet it can be a serious barrier to communication beyond. This is particularly clear in the gap between science and policy-making. On the one hand, these differences can manifest as an inability to understand what the other is actually talking about (Tribbia & Moser, 2008, p.317). We can draw an illustrative example from the IPCC's Fourth Assessment Report discussion of sea level rise:

> Sea level is expected to continue to rise over the next several decades. During 2000 to 2020 under the SRES A1B scenario in the ensemble of AOGCMs, the rate of thermal expansion is projected to be 1.3 ± 0.7 mm yr.$^{-1}$, and is not significantly different under the A2 or B1 scenarios. These projected rates are within the uncertainty of the observed contribution of thermal expansion for 1993 to 2003 of 1.6 ± 0.6 mm yr.$^{-1}$
> *(IPCC, 2007, WG–I TS.5.1)*

This scientific report is, more than most, a public-focused document—yet it still contains much that might be difficult for policy-makers to understand.

The converse is also true. Much governmental language is managerial and legalistic, or imprecise and vague compared to the standards set by scientific writing. Much can be unclear to outsiders. Consider this example from the Office of Fair Trading (OFT) UK:

OFT has many strengths and has been steadily addressing its capability needs. It still has some significant skills gaps in current and potential leadership and managerial skills which will inhibit its effectiveness to meet the challenges identified by the leadership itself. Much of the progress has been driven by individuals responding to specific circumstances including, for example, encouragement from CEO, NAO report, IIP review. There is a need to develop more collective capability.

(http://www.weaselwords.com.au/Government1.htm)

Beyond this, differences in language can manifest in more subtle ways, with differing understandings of seemingly shared words. Many of us use the word 'uncertain': for many policy-makers it can suggest 'something about which little is known'. Yet for scientists it suggests a reasonable—and very different—recognition that knowledge can never be truly certain.

Scale

Policy-makers and scientists work, almost inherently, at scales that rarely line up simply. Policy-makers typically have their scale of operation defined for them by their jurisdiction: by their local government area, their city, their country, or their portfolio of concern. Scientists, meanwhile, may focus on any scale possible, from the sub-atomic to the universal. Indeed, the many natural systems of scientific concern—cell functions, river systems, ecosystems, tectonic plate movements, oceanic circulation systems—rarely match up with any particular governmental lens (Cash, et al. 2006).

This difference in scalar focus can lead to a poor fit between the scientific picture and the area of concern for the policy-maker. Climate scientists may provide an understanding of the global climate system, yet a local government official may see little of use for them in managing their stretch of coastline. Is the '1.3 ± 0.7 mm yr.$^{-1}$' sea rise noted above in the IPCC AR4 (IPCC, 2007, WG-1 TS.5.1) useful information? How does this information map onto the various concerns of local government officials charged with coastline management (see Tribbia & Moser, 2008, p.319), including inland water quality, coastal water quality, species/habitat protection, inland flooding, coastal erosion, public access, salt water intrusion, cliff failure, and wetland loss?

Beyond the inadequate fit between most scientific and policy-making scales, it is also important to note that policy-makers will often hear contrasting scientific advice from different scientific disciplines and different scalar focuses (McNie, 2007, p.24). A compelling example of this can be seen in bicycle helmet legislation. For the neuroscientist and the trauma specialist, focused as they are on individuals, policies requiring helmet use are a clear benefit, reducing dramatically the severity of bicycle accidents (Olivier, 2011). Yet scientists focused on the larger scale of population health might advise policy-makers differently, arguing that such requirements might negatively affect population health by making people less likely to ride in favour of less healthy transport options (Rissel, 2011).

Evidence

Scientists and policy-makers both routinely deal with evidence, yet their understandings of the concept differ significantly. For scientists, the purpose of evidence is to provide as clear an understanding as possible of the phenomenon under examination. Importantly, their methods of collecting that evidence are integral to that understanding. Here we can consider the so-called 'gold standard' of evidence in medical research: systematic reviews of randomised control trials. Building on the work that randomisation (to remove any participant selection bias) and comparison with a control (to remove the placebo effect of any intervention) do to provide evidentiary robustness, systematic meta-reviews allow researchers to remove even the particularities of each trial. With a suitable number of participants in a suitable number of studies, researchers can gain a clear picture of the efficacy of their treatment against other possible interventions. This provides an abstracted view of their intervention in relation to their target problem. Such standards are paralleled (or sought) in other scientific fields of enquiry. Critically, as scientific work is cumulative—because uncorrected errors can waste years of subsequent work—scientists seek the highest quality evidence possible.

In contrast, policy-makers must work to a very different standard; they simply don't have the ability to access similar evidentiary tools. The situatedness of the problems they face within a highly complex social world means that interventions cannot be assessed against an equivalent control group. Consider, for example, a possible trial of a public bicycle share program. Though some outcomes could certainly be assessed, comparing for example the situation in the trial town before and after on a number of criteria (see for example Pucher, Dill & Handy, 2010), it would be impossible to compare the implementation of the trial with an equivalent control. The necessarily different human and infrastructural dynamics—the messy 'stuff' of networks, leadership, social capital, built environment, and other infrastructure—of the trial town and any possible control would make any comparison anecdotal at best. This has been described as 'the counterfactual problem': the inability to know "what would have happened in the absence of the policy" (Leigh, 2009, p.3). It is also important to remember that even when enacted, trials of policy interventions are difficult to run, given that the objects of enquiry (social groups of people) are self aware and affected by the enquiry process itself. Finally, there remains an ethical question: if a policy-maker believes with some justification that a policy will be successful, who should receive the benefits of a trial? Who should bear the risks? (Leigh, 2009, p.5).

So what evidence do policy-makers use? This question cannot, of course, be answered exhaustively. What can be said is that in general, policy-makers use the best evidence they have to hand. This will rarely (if ever) meet the 'gold standard' of a systematic series of randomised control trials. Rather, policy-makers will typically seek to weave together whatever evidence does exist to develop as clear a picture of the problem as possible (Payne, 2007). Depending on the issue, this may involve evidence collected following the methodological principles of the physical sciences, the natural sciences, the social sciences, or the humanities. It may involve anecdotal evidence, such as the stories of those closest to the problem at hand. It may involve the social evidence that emerges from media coverage. Critical to this will always be a strong attempt by policy-makers to understand the implications and acceptability of any action—or lack of action—on the affected society or environment, including both the direct effects on the problem and indirect effects on the economic, social, and environmental milieu in which the problem is situated.

Both policy-makers and scientists seek the best possible evidence, depending on their timeframes and the particularities of the problem at hand. Yet the fact that scientists and policy-makers seek to understand their problems in radically different ways—that scientists seek to abstract their problem from the world, and policy-makers seek to see that problem in its context—means that what matters in each sphere is a radically different thing. It is this difference in worldview to which we now turn.

Orientation: One Worldview of Abstraction, One Worldview of Situatedness

The gap between the worlds of science and policy can be summed up as one of worldview (Garvin, 2001, p.446; Jones, Fischhoff & Lach, 1999, p.582; McNie, 2007, p.24). Towards what goals do the inhabitants of these two different worlds orient themselves? At what do they look? What do they seek to do?

The scientific worldview can be described as essentially about abstraction—the isolation and measurement of particular reactions, events, molecules, particles, cells, and networks from their normal worldly existence in order to better understand their nature. In essence, all scientific experimentation and examination is about looking at these objects of enquiry without the clutter and confusion of everything else. Our laboratories and tools, from surveys and focus groups to the Large Hadron Collider, are all designed to work towards this abstraction. Ever better evidence, in the world of the scientist, can be described as an ever better abstraction of the phenomenon in question.

Yet the worldview of policy formation is essentially one defined by messy, complicated situatedness, by the groundedness of the problem in the governed society and material world. In its essence, this means that the problems and issues on which policy-makers focus—problems and issues which may in themselves not be radically different from those observed by scientists—cannot be abstracted from their normal worldly existence. As an example, malaria can be examined both as a eukaryotic parasite and as a public health problem. The crucial point to make here is that while it can be examined by scientists in an abstract manner as a eukaryotic parasite (or indeed as a public health problem), public health policy-makers must recognise the problem as inherently situated in the living world. We cannot expect an adequate policy response to malaria without knowledge of its functioning as a eukaryotic parasite, its behaviour in the body, its behaviour in mosquitoes, the behaviour of mosquitoes in human environments, and the behaviour of humans. Each of these factors is crucial, yet each in turn depends in complex ways on the variety of other things in which it is situated. Indeed, the complexity of interaction between the different factors points to the fact that many (or perhaps most) social problems are in some senses 'wicked problems' that resist agreed definition (Rittel & Webber, 1973). Where scientists seek always to abstract and clarify their problems, policy-makers at all times deal with problems situated in a messy and complex world. It is this fundamental difference in approach that most contributes to the science-policy gap.

These worldviews are, of course, essential types, somewhat removed from the actual day-to-day practice of policy-makers and scientists. Many scientists have a robust appreciation of the demands on policy-makers; many policy-makers have a thorough grasp of how knowledge is produced by science. Many scientists are themselves policy-makers, and in many areas (as will be discussed below) science has been brought into the policy-making tent. Yet these broad concepts do speak to the overriding dispositions of the two groups, and they do point to why the gap between science and policy worlds can be so great. At heart, if one group seeks to see an issue with all other distractions removed—and the other to understand that issue within those very distractions—then they can very easily talk right past each other.

Solutions: Bridging the Gap

How do we close the gap between science and policy-making? How do we ensure that the policies that shape our lives are based on the best possible evidence, and that the problems discovered by scientists are adequately incorporated into policy decision-making?

In recognising the problems inherent in the science-policy gap, various voices have begun to suggest ways to facilitate a closer relationship between the worlds of science and policy-making. The remainder of this chapter describes some of the more interesting possibilities for scientists, policy-makers, and those in between.

Scientists Bridging the Gap

If scientists are to engage with policy-making, what should shape their action? What role should they play? Scientists, Roger A. Pielke Jr has argued, can play one of four roles when interacting with the policy-making landscape: 'Pure

Scientist', 'Science Arbiter', 'Issue Advocate', or 'Honest Broker of Policy Alternatives' (Pielke, 2007, chapter 1). Consider, he suggests, the simple problem of a visitor from out of town looking for a place to go for dinner. How might one provide relevant information to assist their decision? With no interest in the visitor's decision-making process or goals, the 'Pure Scientist' simply gives basic information, perhaps about nutrition. They might give them a scientific report on dietary guidelines and leave it at that. In contrast, the 'Science Arbiter' would act as a factual resource for the visitor, providing scientific information on the questions they might have. They would be able to provide factual answers to such questions as 'What is the closest Thai restaurant?' or 'Which Indian restaurants accept American Express?' An 'Issue Advocate' might argue for a particular outcome—a particular restaurant or area—perhaps thinking that that outcome is actually really good, or that they know the visitor's interests well enough to make decisions for them. Finally, the 'Honest Broker of Policy Alternatives' would seek to provide relevant information for the visitor to clarify the choices available to them. In the case of the restaurant this might mean incorporating a range of views, experiences, and knowledge about food in general and places to eat locally—so that the visitor can then make a choice based on their own goals or values.

Though ideal types, these roles echo the common ways scientists engage with the policy-making process, and are useful as types for scientists and science organisations to think about their engagement behaviour. What Pielke argues is that Issue Advocates (explicitly) and Pure Scientists and Science Arbiters (by stealth or by accident) tend to compel particular decision outcomes, encroaching on the legitimate role of the decision-maker. Though Pure Scientists and Science Arbiters might not seek to do so, by unconscious decision-making about what information is relevant—by deciding, for example, that nutritional information is more important than information about taste—they become advocates for a particular way of understanding the problem, and in turn particular solutions. It is only the Honest Broker of Policy Alternatives who consciously defends the right of the decision-maker to make the decision, opening up and clarifying their options via the provision of relevant, useful information (Pielke, 2007, p.3). It is only the Honest Broker of Policy Alternatives who truly seeks to address all aspects of McNie's argument that useful information is information that is salient, credible, and legitimate.

This argument echoes the suggestion of Sarewitz and Pielke that what is needed is a reconciliation of the supply of scientific knowledge with users' demands (Sarewitz & Pielke, 2004; see also Reid, 2007). At the heart of this is an argument that the scientific supply of knowledge should be in constant communication with policy-making demands, working to understand what information would be useful to decision-making. This means—drawing on the concept of the Honest Broker—paying close attention to the specific needs and context in which the policy-maker is situated, engaging in 'strategic listening' to the needs of policy-makers (Pidgeon & Fischhoff, 2011). What is the issue the policy-makers are facing? When do they need to make a decision? What is the geographical area in which the problem sits? What other information might they consider relevant? What other stakeholders and stakeholder interests might be relevant? What wider policy-making goals has the policy-maker previously articulated? If information is provided that takes account of these factors, then it is much more likely to be useful.

The core caveat to this argument is that there remains a critical role for scientists to play in demanding that society pay attention to specific problems. There are occasions in which scientists must play a role in talking about problems and goals. It was scientists, after all, who fought long and hard to put climate change on the international political agenda. Yet importantly, this is a qualification, not a rejection of Pielke's Honest Broker argument. Whether they are listening to policy-makers or attempting to get an issue on the agenda, scientists must seek to make their information salient, credible, and legitimate—and this will only happen in dialogue with policy-makers and society more broadly.

Policy-Makers Embracing Science

Policy-makers want, on the whole, to make the best possible public policy. They may argue with each other about normative goals—and even about the nature or existence of the problem at hand—but once these have been agreed to, the great bulk want the best possible solutions. For this they need *useful information* about the problem at hand. In times past a call for more useful information may have focused on the words 'more' and 'information', with policy-makers simply funding an increase in the supply of scientific information. This strategy is doomed to fail. Increasing supply without heed to what is useful is likely to be wasteful and counterproductive (Sarewitz & Pielke, 2007).

To address this problem on the policy-making side, commentators have advocated interventions centred on enhancing the communication between scientists and policy-makers, and enhancing the understanding of the processes, capabilities, and limits of science within the policy-making world.

To enhance the communication between science and policy, commentators have suggested a number of interventions. Firstly, an intervention widely advocated by the open access community (see for example Willinsky, 2006) is to ensure that all policy-makers have unhindered access to published science. At present much is obscured behind publisher pay-walls. Though costs per journal article may be small, this may preclude the curious policy-maker from perusing the literature. A similar intervention has been suggested by the so-called Gov2 or WikiGovernment movements, with suggestions for policy-makers to be more open in online modes during the policy-formation process (see for example Dunleavy, Margetts, Bastow & Tinkler, 2008; Noveck, 2009). This can provide scientists a clearer understanding of what policy-makers are looking for. Finally, organised social networking, such as the 'Science meets Parliament' event run in Australia, can provide greater communication connections between the science and policy spheres. In this yearly event, scientists are brought to meet members of parliament to discuss their research and learn about the policy-making process.[2] Though the event cannot address all of the issues policy-makers might be facing, it works chiefly to provide each side a greater understanding of the other.

To enhance understanding of the processes of science within the policy-making orbit, researchers have suggested physically embedding scientists within policy-making circles, giving other policy-makers an instant touchstone on how scientific evidence works (Pouyat, et al., 2010).[3] Others have suggested randomised policy trials, whereby small interventions (perhaps funding for school lunches) are trialled in random areas before a full roll out (Leigh, 2009). Such trials are unlikely to reach the 'gold standard' of scientific work described above (and indeed, many areas should not be subject to such trials (Smith & Pell, 2003)), but they can give policy-makers a more robust appreciation of whether an intervention is working or not, and how to test it. Others have suggested collaborative work between policy-makers and scientists, whereby policy-makers actively participate in the data collection work of science.

Though many of the interventions in the scientific and policy-making worlds show much promise, some of the most interesting movements to enhance the relationship between science and policy have focused on the space in between: on movements to open up the policy formation landscape so that all parts of society—including science—can usefully contribute.

Between Science and Policy

There is much that scientists and policy-makers can do on their own to ensure a closer relationship, yet it is important to recognise finally that there are many other actors in society who can also facilitate a better relationship between science and policy-making. After all, if we in society are to ask for better policy outcomes, should we leave it to scientists and policy-makers alone to achieve this? Or should we force our policy-makers to craft policy based on the best available evidence?

Indeed, many have argued that it is civil society—the broad and complex collection of individuals and organisations outside the state or the market[4]—that is best able to force a closer relationship between science and policy. Here movements in the last few decades to democratise science (Gallopin, Funtowicz, O'Connor & Ravetz, 1999; World Conference on Science, 2001; and to open the discussion of public policy to actors outside the traditional policy-making worlds (Walzer, 1998) have legitimised a role for civil society actors in the policy formation process. Environmental and social justice groups, for example, have begun suggesting policy to government. Science, properly democratised and embraced by such organisations, can play a critical role in this process.

Science communicators and other boundary organisations are also crucial, communicating the needs of policy-makers to scientists, and the latest science to policy-makers. Bringing policy communication capabilities inside the peak science bodies, whether academies, learned societies, or disciplinary bodies can also assist this communication. For many this has meant employing specific policy or lobbying officers.

What these movements suggest is that though many problems remain, we sit at the cusp of a revolution in the connections between science and policy. The democratisation of science, properly supported by science communicators, civil society and more open forms of government, promises to bring the most useful information ever closer to our policy-making process. This will involve recognition of the many other important voices and stakeholders of society— and other forms of knowledge—alongside the processes of science. Yet this is not a detraction from science. Having fully engaged, context aware—*useful*— science will lead to better policy outcomes. This points, finally, to a crucial role for science communication and science communicators: listening and talking to both sides of the science-policy divide, facilitating a closer connection between science and policy-making to better solve our key social problems. It is this vitally important role that places science communication as one of the critical social activities of the coming century.

Notes

1 Organisations other than governments (and individuals) can, of course, also set policies. An institution might, for example, formally designate a policy for its members to follow. Given its importance in society, however, much of the discussion in this chapter focuses on governmental designations of policy.

2 See http://scienceandtechnologyaustralia.org.au/what-we-are-doing/science-meets-parliament/

3 See for example the New Science Policy Fellowships of Health Canada, at http://www.youtube.com/watch?v=jX6UiI_6hg4

4 'Civil society' is a term that, more than most, resists easy capture. Michael Walzer provides a useful definition by arguing that "the words *civil society* name the space of uncoerced human association and also the set of relational networks—formed for the sake of family, faith, interest, and ideology—that fill this space" (Walzer, 1998: 291-2).

References

Banks, G. (2009). *Challenges of Evidence-Based Policy-Making. Challenges.* Canberra: Australian Government Productivity Commission, Australian Public Service Commission. Retrieved from http://www.apsc.gov.au/publications09/evidence basedpolicy.pdf

Bhattacharjee, Y. (2008). Barack Obama. *Science, 319*(5859), 28-29. doi:10.1126/science.319.5859.28a

Campbell, S., Benita, S., Coates, E., Davies, P., & Penn, G. (2007). *Analysis for policy: Evidence-based policy in practice*. London: Government Social Research Unit. Retrieved from http://www.civilservice.gov.uk/wp-content/uploads/2011/09/Analysis-for-Policy-report_tcm6-4148.pdf

Cash, D. W., Adger, W. N., Berkes, F., Garden, P., Lebel, L., Olsson, P., & Young, O. (2006). Scale and Cross-Scale Dynamics: Governance and Information in a Multilevel World. *Ecology and Society*, *11*(2). Retrieved from http://www.ecologyandsociety.org/vol11/iss2/art8/

Clement, M. (2011). Climate change is real: an open letter from the scientific community. *The Conversation*. Retrieved from https://theconversation.edu.au/climate-change-is-real-an-open-letter-from-the-scientific-community-1808

Dunleavy, P., Margetts, H., Bastow, S., & Tinkler, J. (2008). Australian e-Government in comparative perspective. *Australian Journal of Political Science*, *43*(1), 13-26. doi:10.1080/10361140701842540

Gallopin, G. C., Funtowicz, S., O'Connor, M., & Ravetz, J. (2001). Science for the Twenty-First Century: From Social Contract to the Scientific Core. *International Social Science Journal*, *53*(168), 219-229. doi:10.1111/1468-2451.00311

Garvin, T. (2001). Analytical paradigms: the epistemological distances between scientists, policy makers, and the public. *Risk analysis: an official publication of the Society for Risk Analysis*, *21*(3), 443-55. Retrieved from http://www.ncbi.nlm.nih.gov/pubmed/11572425

Grant, W. J. (2011). Will Steffen: phoney debate is over, now for the carbon policy. *The Conversation*. Retrieved from https://theconversation.edu.au/will-steffen-phoney-debate-is-over-now-for-the-carbon-policy-2015

Hulme, M. (2009). *Why we disagree about climate change*. Cambridge: Cambridge University Press.

IPCC. (2007). *Intergovernmental Panel on Climate Change Fourth Assessment Report: Climate Change 2007*. Retrieved from http://www.ipcc.ch/publications_and_data/publications_ipcc_fourth_assessment_report_synthesis_report.htm

Jacobs, K. L. (2002). *Connecting science, policy, and decision-making: a handbook for researchers and science agencies*. *Water Resources*. NOAA Office of Global Programs. Retrieved from http://www.isse.ucar.edu/water_conference/CD_files/Additional_Materials/Science%20and%20Decision%20Making,%20Jacobs.pdf

Jones, B. (2011). In climate change, everything old is new again. *The Conversation*. Retrieved from https://theconversation.edu.au/barry-jones-in-climate-change-everything-old-is-new-again-1914

Jones, S. A., Fischhoff, B., & Lach, D. (1999). Evaluating the science-policy interface for climate change research. *Climatic Change*, *43*(3), 581–599. Retrieved from http://www.springerlink.com/index/ng60835839642511.pdf

Leigh, A. (2009). Evidence-Based Policy: Summon the Randomistas. *Submission to the Productivity Commission, Strengthening Evidence Based Policy in the Australian Federation*, (August), 1-12. Retrieved from http://andrewleigh.org/pdf/PC_Randomistas.pdf

McNie, E. (2007). Reconciling the supply of scientific information with user demands: an analysis of the problem and review of the literature. *Environmental Science & Policy*, *10*(1), 17-38. doi:10.1016/j.envsci.2006.10.004

Noveck, B. S. (2009). *Wiki Government: How Technology Can Make Government Better, Democracy Stronger, and Citizens More Powerful.* Washington DC: Brookings Institution Press.

Olivier, J. (2011). Putting a lid on the debate: Mandatory helmet laws reduce head injuries. *The Conversation.* Retrieved from https://theconversation.edu.au/putting-a-lid-on-the-debate-mandatory-helmet-laws-reduce-head-injuries-1979

Payne, J. (2007). The Function of Public Policy. *International Journal of Diversity, 6*(6).

Pidgeon, N., & Fischhoff, B. (2011). The role of social and decision sciences in communicating uncertain climate risks. *Nature Climate Change, 1*(April). doi:10.1038/NCLIMATE1080

Pielke, R. A. Jr. (2007). *The Honest Broker: Making Sense of Science in Policy and Politics.* Cambridge: Cambridge University Press.

Pouyat, R. V., Weathers, K. C., Hauber, R., Lovett, G. M., Bartuska, A., Christenson, L., Davis, J. L., et al. (2010). The role of federal agencies in the application of scientific knowledge. *Frontiers in Ecology and the Environment, 8*(6), 322-328. doi:10.1890/090180

Pucher, J., Dill, J., & Handy, S. (2010). Infrastructure, Programs, and Policies to Increase Bicycling: An International Review. *Preventative Medicine, 48*(2).

Reid, W. V. (2004). Bridging the science-policy divide. *PLoS biology, 2*(2), E27. doi:10.1371/journal.pbio.0020027

Risbey, J. (2011). Speaking Science to Climate Policy. *The Conversation.* Retrieved from http://theconversation.edu.au/speaking-science-to-climate-policy-1548

Rissel, C. (2011). Ditching bike helmet laws better for health. *The Conversation.* Retrieved from https://theconversation.edu.au/ditching-bike-helmets-laws-better-for-health-42

Rittel, H. W. J., & Webber, M. M. (1973). Dilemmas in a general theory of planning. *Policy Sciences, 4*(2), 155-169. doi:10.1007/BF01405730

Sarewitz, D., & Pielke, R. A. J. (2007). The neglected heart of science policy: reconciling supply of and demand for science. *Environmental Science & Policy, 10*(1), 5-16. doi:10.1016/j.envsci.2006.10.001

Smith, G., & Pell, J. P. (2003). Parachute use to prevent death and major trauma related to gravitational challenge: systematic review of randomised controlled trials. *British Medical Journal, 327*(7429), 1459. Retrieved from http://www.bmj.com/content/327/7429/1459.short

Tribbia, J., & Moser, S. C. (2008). More than information: what coastal managers need to plan for climate change. *Environmental Science & Policy, 11*(4), 315-328. doi:10.1016/j.envsci.2008.01.003

Walzer, M. (1998). The Civil Society Argument. In G. Shafir (Ed.), *The Citizenship Debates.* University of Minnesota Press.

Willinsky, J. (2006). The Access Principle. *Bulletin of the World Health Organization 85.* Retrieved from http://mitpress.mit.edu/catalog/item/default.asp?tid=10611&ttype=2

World Conference on Science. (1999). Declaration on Science and the Use of Scientific Knowledge. *World Conference on Science: Budapest 1999.* Retrieved from http://www.unesco.org/science/wcs/eng/declaration_e.htm

5

NEGOTIATING PUBLIC RESISTANCE TO ENGAGEMENT IN SCIENCE AND TECHNOLOGY

Lindy A. Orthia

Introduction

Scientists and science communicators often wish to communicate science with the public because they believe that science is exciting, important, and useful. Not everyone, however, feels wholly positive about science. Some members of the public may be offended by certain scientific theories, or feel that particular technologies are frightening and dangerous. It has become commonplace for people from all walks of life to protest against scientific pursuits and technological developments, in the deeply held conviction that there is something wrong with the way that science is done or the way it is used. Meanwhile, other people ignore both those protesting and those defending science, bored and unable to engage with scientific ideas and issues.

Whether someone identifies as a defender of science, a protestor against it, or an uninterested bystander, too often people with different perspectives on science do not understand each other. This can lead to a communication stalemate, with opposing sides stating their views but no genuine conversation taking place, and uninterested bystanders becoming increasingly alienated from science.

This chapter deals with this stalemate situation. Section 1 explores reasons why people might be hostile to science. Section 2 illustrates a subset of these reasons using examples of recent science controversies that have divided the community. Section 3 summarises current thinking on including people in societal debates about science controversies. Section 4 discusses practical ways in which opposing parties may be brought closer together through sound science communication practice.

Attacking or Defending Science: Are We Talking About the Same Thing?

Science can mean different things in different contexts. Part of understanding conflicts about science involves teasing apart those different meanings. Science can be:

- *A secular, rational belief system.* Science is a way of seeing the world which prizes measurable, physical explanations for phenomena instead of supernatural, mystical, or religious explanations. This way of seeing is often called *secular*, meaning non-religious, and *rational*, meaning based on logic and reason, not feelings and beliefs.
- *An empirical, statistics-based method of inquiry.* The scientific method of answering questions involves particular, rigorous routines of hypothesis testing. It is *empirical*, which means it involves testing theoretical ideas in the real world. It is also based on *statistics*, which means that conclusions may only be drawn after testing ideas many times—a single test is never enough.
- *A cultural institution like art or music.* Science offers us insight into the wonders of the universe from the farthest galaxy to the smallest subatomic particle, from the bizarre lives of deep sea creatures to the intriguing workings of the human brain. Science also contributes to the creation of technologies that make our lives interesting—from computer games and sunglasses to chewing gum and jet skis.
- *A source of medical, environmental, agricultural and industrial technologies.* Scientific research can lead to technologies that have an impact in remarkable, life-changing ways on our health and safety, as well as our living and working environments. Such impacts can be very positive or disastrously negative.

These four categories of meaning are not the same. It is possible to defend science for its method of inquiry, but attack it for its impacts on the environment, or vice versa. How one feels about each category of meaning depends on personal experiences with science and technology and the social and cultural contexts in which these experiences took place.

There are no right and wrong ways to feel about science. Science has its strengths and weaknesses, its advantages and disadvantages. In a world filled with competing ideas and obligations, it is not surprising that some people are not interested in it at all. Table 5.1 outlines some of the main kinds of reasons why people feel the way they do about science.

TABLE 5.1 Positive and Negative Aspects of the Four Categories of Meaning for Science

Category of meaning for science	Reasons to like science	Reasons to dislike science	Reasons to be indifferent to science
A secular rational belief system	• It strives to explain the world without resorting to higher powers or magic. • It aims to remove the uncertainty and powerlessness of relying on deities to shape the world.	• It denies the significance and value of deeply held religious or spiritual beliefs. • It takes the mystery out of life and reduces life's wonder to mechanistic explanations.	• It doesn't affect personal beliefs.
An empirical, statistics-based method of inquiry	• It is objective, because its conclusions are driven by the real world, not ideas. • Its conclusions are generalisable into laws and principles, because they are based on extensive testing, large sample sizes, and average out the exceptions to show major trends.	• It has pretensions to objectivity which are unsuccessful because human ideas and ideologies always shape conclusions of research. • It is an estimation which ignores important exceptions, outliers, and the influence of subjective experience.	• It is just one of many valid ways of seeing the world.
A cultural institution like art or music	• It enriches life, adding interesting insights that help us contemplate our place in the universe. • Its fun gadgets and inventions add enjoyment to our day to day lives.	• The money that goes into science's interesting insights could be better spent on something else. • Its technological gadgets and inventions are distractions from more meaningful pursuits.	• People do not find it personally interesting or enjoyable.

Category of meaning for science	Reasons to like science	Reasons to dislike science	Reasons to be indifferent to science
A source of medical, environmental, agricultural or industrial technologies	• It has enhanced our quality of life, health, longevity, safety, efficiency and environmental sustainability.	• It has produced dangerous drugs and surgical methods that have caused deaths, illness and mutilation, as well as polluting industries and technologies of destruction.	• It does not have an impact on ordinary everyday life.
	• It has enhanced our understanding of social problems and how to manage them.	• It has been used to justify genocidal and otherwise destructive social policies.	

Disadvantages and Dangers of Science and Technology

Each category of meaning carries its controversies. But the controversies in the fourth category are often the most emotionally loaded, because for better or worse, science has an impact directly on our health, environment, and even whether we live or die. This section presents three examples of controversies that have raised passionate responses to science and technology in recent years.

Example 1 – The Deaf Community and the Cochlear Implant

In societies dominated by hearing people, deaf people can find life isolating and communication difficult. Technologies such as captions for popular television programs have enabled deaf people to enjoy them more fully. In the 1980s, scientists produced the cochlear implant, a technological tool that enables deaf people to hear, thus helping deaf members of hearing families and communities to participate in speech-based communication. Some deaf people have greatly valued what cochlear implants have given them.

The cochlear implant has, however, also met significant opposition from within deaf communities (Gonsoulin, 2001; Seelman, 2001; Stern, 2004). While hearing people may consider deafness a disability to be cured, generally this is not how deaf people see themselves. People who have grown up communicating through sign language often identify strongly with the community of signers and its culture. They consider themselves to be simply different, not disabled, and certainly not in need of "fixing". For such people, the cochlear implant is equivalent to cultural genocide. It is the hearing community—and importantly, *scientists* from the hearing community—imposing their values and language on the signing community and thus eliminating an entire culture. As one American deaf man put it, the cochlear implant "is like making black people white or white people black" (Stern, 2004, online).

Example 2 – Nuclear Power, Nuclear Disasters, and Climate Change

Debates over the safety and desirability of nuclear power have raged for many decades. Opponents of nuclear power are concerned about the risks of a meltdown or accident, the problem of how to dispose of nuclear waste, and the environmental and social damage caused by uranium mining to create the nuclear fuels. Concerns have also been raised about the relationship between nuclear power and the increasing availability of nuclear weapons. For people who remember the bombings of Hiroshima and Nagasaki, or nuclear testing in Australia and the Pacific in the 20th century, all of which resulted in deaths, severe injuries, and illnesses for people in these areas, the risks of tampering with nuclear power may be too great because the consequences are too traumatic (Brown & Sowerwine, 2004; Clague, 2009; Prosise, 1998).

The Kupa Piti Kungka Tjuta, a group of senior Aboriginal women from central Australia, have spent years opposing new uranium mines and nuclear waste dumps that the Australian government has proposed for their lands. Their opposition is influenced by memories of British atomic tests on their lands in the 1950s, as shown in this quote:

> All of us were living when the Government used the country for the Bomb. Some were living at Twelve Mile, just out of Coober Pedy. The smoke was funny and everything looked hazy. Everybody got sick ... The Government thought they knew what they were doing then. Now, again they are coming along and telling us poor blackfellas "Oh, there's nothing that's going to happen, nothing is going to kill you."
>
> *(Brown & Sowerwine, 2004, online)*

The same may be true for people who remember the nuclear accidents at the Three Mile Island reactor in the USA in 1979, or the Chernobyl reactor in Ukraine in 1986.

Recently the threat of climate change has renewed public discussion about the risks and benefits of nuclear power, since nuclear power itself does not generate greenhouse gases. For people who have grown up fearing climate change more than they fear nuclear disaster, nuclear power can seem like a very desirable option (Bickerstaff, Lorenzoni, Pidgeon, Poortinga & Simmons, 2008). Yet with nuclear technology still posing a potential threat, as became apparent in 2011 with the Fukushima nuclear reactor disaster in Japan, science communicators will need to monitor how attitudes to nuclear power evolve in the coming decades.

Example 3 – Animal Homosexual Behaviour and Anti-Gay Science

In 2004-5, scientists in the USA began a research program into modifying the sexual behaviour of rams in agricultural sheep flocks (Roselli & Stormshak, 2009). Many rams have a preference for sexually engaging with other rams rather than ewes. The aim of the research was to re-orient the rams towards sexual engagement with ewes, to increase the likelihood of lambs being produced from the rams' sexual activity.

The research faced strong community opposition from lesbian and gay rights activists and animal activists, including people from outside the USA. In part, this opposition was a defence of the animals' rights to express themselves freely in the context of their flocks, and a condemnation of science's maltreatment of animals in the service of human need.

The outrage was also related to science's maltreatment of same-sex-attracted people. For example, psychologists have employed many cruel techniques over the years to modify lesbian and gay people's behaviours and attractions (Grace, 2008). These have ranged from giving people electric shocks every time they feel attracted to someone of the same sex, to lobotomisation. Such "treatments" for homosexuality were relatively common in Western countries in the 1950s, '60s and '70s, but also during the German Nazi regime in the 1930s and '40s. A British gay activist described the ram experiments in these words:

> These experiments echo Nazi research in the early 1940s which aimed at eradicating homosexuality. They stink of eugenics. There is a danger that extreme homophobic regimes may try to use these experimental results to change the orientation of gay people.
>
> *(Oakeshott & Gourlay, 2006, online)*

In more recent times, the field of psychology has officially recognised homosexuality as part of the normal expression of human sexuality. But the treatment of homosexuality as an "unnatural" aberration in need of fixing continues today, both in the clinical practice of some individual psychologists (Bartlett, Smith & King, 2009), and also within parts of evolutionary and behavioural biology (Bagemihl, 1999). Cultural beliefs about homosexuality have unduly influenced the way that scientists have seen homosexual behaviour amongst humans and animals, despite science's ideal of objectivity.

In all these examples, it is people's personal experience, needs, and contexts that shape their attitudes to particular technologies and scientific methods. This can turn public attitudes sharply against scientific research. It can also turn them against scientists in general, if scientists do not share people's values and beliefs, seem to be acting against their best interests, or ignore their concerns.

Given the complexity of our lives and how much science and technology we encounter each day, these effects can be magnified into general suspicion and mistrust of scientists, the scientific method, and the scientific worldview. If scientists or science-defenders make a decision that directly puts people in danger, our trust in science can plummet dramatically. Regardless of the benefits of science and technology, it only takes one frightening risk to make people wary.

In the United Kingdom, this occurred in the late 1980s and early 1990s during the controversy over Bovine Spongiform Encephalopathy (BSE), a lethal syndrome found in cattle, commonly known as "mad cow disease." British scientists and politicians foolishly reassured the public that eating British beef was not dangerous, despite the prevalence of BSE amongst British cattle. This turned out to be incorrect advice, with over 150 people dying in the UK of a disease caused by eating BSE-contaminated meat. Millions of cattle have since been slaughtered. This incident has been linked to record low levels of trust in science among the British public (House of Lords Select Committee on Science and Technology, 2000; Jasanoff, 1997).

The question arising from these examples is: how might these controversies have been avoided or hostilities minimised? Science communicators often believe the problem lies with the public, for being irrational and ignorant and thus drawing paranoid conclusions about the risks of scientific developments. This is not founded in truth, since people may become *more opposed* to controversial scientific developments such as biotechnology as they learn more about the science behind them (Purdue, 1999; Sjöberg, 2004). In addition, the attitude of blaming public ignorance and irrationality does not facilitate effective communication. It is more helpful to see the situation as a set of problems for communicators, scientists and publics to manage together.

These problems include:

- *Frustration and powerlessness.* In modern Western society, people's concerns are usually not heard by those in power. In addition, most people have no control over the way that scientific research is practised, or which technologies are funded for development. Ironically, individual scientists may also feel this way. Nonetheless, science is a powerful social institution because of the money it generates for governments and corporations. This means that when something science-related causes disgruntlement or anger, people can feel very powerless to fix the situation.

- *Being made to feel stupid and ignorant.* Scientific knowledge is inherently complex, and many people feel that they could never understand it. When scientists talk in complicated jargon terms, it excludes people, reinforcing a power hierarchy between people who understand and people who do not. Sometimes defenders of science abuse this situation to win arguments, stopping non-scientists from participating in debates about controversial issues by telling them "you don't know the science and I do." But this approach does not recognise what people *do* know, including why the issue is relevant to their own lives. Making people feel stupid and ignorant is bad for the communication process, because it leaves them dependent on "experts" and also resentful of them. This can reinforce a situation of mistrust or make them lose interest in the issue altogether (McKechnie, 1996).

- *Suspicion of scientists' motives.* If people feel attacked by science, it can call into question everything about science. Although scientists may believe they are working altruistically to benefit society, members of the public do not always see it that way. Science is often associated with social institutions people do not trust at all. Historically, Western science developed in Europe via the Middle East, and in recent centuries became part of the toolkit used by European nations to colonise the planet and enslave the peoples they met (Harding, 1993; Smedley, 1999). For example, biologists helped justify slavery by classifying human beings into racial categories and stating that some races were superior to others. Although this theory of race has been shown to be false, including by more recent science, biological science as an institution still carries the taint of its genocidally racist past (Lee, 2009; Morning, 2008; Relethford, 2009). This kind of problem understandably leads to suspicion about scientists and science the world over. Similarly, if scientists work for unpopular corporations, such as tobacco companies, or for corrupt governments or think-tanks that people do not like, this calls into question what their agenda really might be. It is difficult to know who to trust in any walk of life, and in science it is no different. Studies have shown that we tend to trust people to whom we feel similar (Cvetkovich & Nakayachi, 2007; Street, O'Malley, Cooper & Haidet, 2008; Twyman, Harvey & Harries, 2008). This means that if I agree with you on one issue, I am more likely to trust your opinions on an unrelated matter. If scientists seem alien to their audience because they have completely different social or political values, they will find acceptance of their science is difficult.

All three of these problems are applicable to many areas of society. People who are not specialists in an area—be it science, economics, the law, or even the arts—often feel unheard, uninformed, and uncertain of what other people want from them. This is the case despite the fact that all these areas have an enormous impact on our daily lives, and despite the fact that we all have opinions and feelings about these things. We generally want to earn more money, we want to avoid gaol, and we may not know much about art, but we know what we like. But we generally do not want to be lectured at by economists, lawyers, and art curators who believe that we should learn more about their specialised area, that their area is more important than any other, or that their way of thinking is best. If we think we are paying too much interest on a loan, or that the law has treated us unfairly, or that a mural in our neighbourhood is ugly, we do not want these "experts", telling us that the decisions they made are good for us and that we should simply accept them.

The same is true for science. In all areas of life, we need to feel that our concerns will be heard and taken into account in decision-making. We need to know our opinions are valued and our intelligence respected. We also need to know where we stand with respect to the motives of the people with whom we are communicating.

Recognising this broader context of human need—and therefore the importance of involving people in making decisions on issues that affect their lives—is a critical step towards making the science communication process much smoother and more agreeable for all concerned.

The Importance of Public Involvement in Decision-Making

For decades, science communicators have debated the importance of public involvement in decisions about science controversies, and the most appropriate ways to involve people. Three main approaches to public involvement in science have been proposed:

- *Excluding the public.* Some science communicators believe that members of the public have little to contribute to engaging in debates about science controversies, aside from "making the process inefficient and causing incoherent and irrational decisions" (Fischer, 2008, section 2.3). This approach has led to poor communication decisions in the past because it has left people feeling disempowered and has disintegrated the trust between scientists, communicators, and the public.

- *Inviting both scientists and the public to express their views on science developments.* Some science communicators have taken into consideration the problems of ignoring the public, and advocate granting a legitimate voice to members of the public affected by science-based risks and controversies (Collins & Evans, 2002, 2007; Funtowicz & Ravetz, 1993). According to this view, the public voice should be heard alongside the technical assessments provided by scientists estimating the level of risk the development entails. This will ensure that there is a balance between the scientific view and the public view when decisions are made, thereby fulfilling the demand for democratic involvement while avoiding the possibility that the public will make foolish decisions based on ignorance or irrational beliefs.

- *Ensuring that the public frames the issue from the beginning.* Other science communicators think that merely giving people the opportunity to express their views on the specific scientific developments that affect them is not a sufficient level of public involvement (Jasanoff, 2003; Wynne, 2003, 2008). In some cases, inviting public opinion on a scientific controversy has resulted in tokenism, because decision makers have ignored what people have said (Irwin, 2006). In addition, commenting on an issue *after the terms* of the debate have already been set by decision-makers is not particularly empowering for members of the public. It is much more profoundly empowering for people to set the terms of debate themselves: in other words, to *frame* the issue. To again draw a parallel between science and art, this means that people will not merely be invited to express their views on an ugly neighbourhood mural, but will be involved in the neighbourhood's early planning stages, contributing to decisions on how streets should be designed, what kinds of artistic expression might appropriate for the neighbourhood, and whether there should even be buildings on which to paint murals at all.

To illustrate the problems with the second option, which the third option seeks to correct, consider the controversy over genetically modified (GM) crops. The reasons GM crops are controversial are numerous and highly complex, and are not merely restricted to technical scientific considerations. They include: sustainable resource use; threats to the environment; consumers' health; farmers' lifestyle, culture, and heritage; money, profits and economics; international politics and globalisation; the accountability of large corporations; the freedom of small farms to be self reliant; the right of consumers to know what they eat; and ethical objections to manipulating genes. If we as science communicators were to seek consumers' opinions on a government proposal to let local supermarkets stock GM products, we would not be giving people in the community an opportunity to fully express their concerns on all of the above issues—on whether GM technology, or indeed billion-dollar transnational agribusiness corporations, should be developed in the first place. Hence, we would deny them the opportunity to frame the issue in the terms that are of concern to them.

It is common for scientific controversies to be framed in narrow terms that artificially separate the scientific aspects from other aspects such as economics and ethics. The result is that people in the community often feel frustrated and powerless, and see scientists as out of touch with the rest of society. This is *exactly* the kind of situation that leads to open conflict between members of the public and scientists, manifesting as public protests, hostility to science, and mistrust of scientists. Seeking public involvement in decision making processes *from the beginning* of a potentially controversial science development makes people feel empowered, avoids *unnecessary* conflict, and builds trust between scientists and other members of society.

Success Stories in Improving Communication with Audiences Hostile to Science

Experienced science communicators have found that there is a number of strategies likely to open the communication gates and foster a meaningful exchange of knowledge, ideas and concerns between scientists and publics.

way (adapted from Chilvers, 2008):

1. Widespread involvement of affected people from the beginning of the process;
2. Transparency in explaining the relevant science;
3. Discussions over a long time period that include diverse alternative perspectives.

Involvement from the Beginning

We have already seen the importance of involving members of the public early in potential science controversies. This is easier said than done, but there are some basic steps science communicators can take to make the communication process go smoothly.

The first step is to stop thinking of members of the public as a "target audience", and to start thinking of them as communication partners (Kirk, 2009). Whether communicating with the general public or bringing scientists, governments and publics together, thinking about these groups as communication partners resets our mindset towards inviting two-way dialogue and away from a lecture approach. It also shifts the communication dynamic from a competitive "us and them" approach to being more cooperative, friendly, and inclusive. Whereas a "target audience" sounds like a group with whom one can choose to speak or choose to ignore, a communication partner is someone who is entitled to an equal place at the communication table, and who has valuable things to say.

A second step is to challenge our own assumptions about science and open our minds to the possibility that people hostile to science have many good reasons for their views. This does not mean we have to agree with everyone or accept ideas we disagree with. It does mean we must take other people's views seriously. Our core beliefs can be the thing we hold most precious. An important part of establishing trust with communication partners is letting them know we will not trample on what they hold dear. For this reason, listening to a person is not just a matter of letting them talk and moving on. We all need to know that we have been *heard*, that our communication efforts have been taken seriously. When interacting with communication partners, listen respectfully to ensure people have time to express their views. Also listen actively: process what others are saying, repeat it back to them to ensure that it has been heard correctly, and incorporate their views into future communication efforts.

An example that illustrates the value of being inclusive and non-judgemental from the beginning is the contentious issue of discussing creationism or intelligent design in the biology classroom. One biology teacher found that by respectfully making room for creationism in his classroom at the beginning of a biology course on evolution, he avoided setting up an "us and them" dynamic.

Instead he invited all students to participate openly in discussions and express their views, whether they were creationists or evolutionists or unsure of their beliefs. By the end of the course, most creationists had shifted their beliefs towards the evolutionist perspective advocated by the teacher, because everyone involved was able to find common ground, having avoided a confrontational standoff (Verhey, 2005).

Of course, this kind of shift towards a scientific perspective will not always occur. In addition, aiming for such a shift would undo our attempts to treat our communication partners' views with respect and open-mindedness. But if all of us are granted the opportunity to speak our minds, be heard, and be respected right from the beginning of a controversial discussion, we open up communication, and are more likely to come together and find common ground.

Transparently Explained Science

If scientific research contributes to the understanding of a controversial issue, the science must be made accessible to everyone involved, in ways that are relevant to people's points of interest, and in ways that respond to people's ongoing questions. The scientific research process should also be transparent, with underlying uncertainties and assumptions made explicit, so that everyone can see them.

Communication partners may or may not want to spend time discussing the technical details of the scientific perspective on a controversy since there are often many other aspects of the issue to discuss. Forcing them to learn scientific details will only lead to frustration. But if people ask for clarification of the science, we as science communicators must be prepared to give it to them using everyday language but without oversimplifying. It may take considerable time and effort to do this, but it is worth mastering the art of explaining things clearly for different audiences. People always appreciate having the *opportunity* to be informed when they feel it would benefit them.

Non-scientists can find scientists and scientific language intimidating. This is partly because of science's perceived social power. It is also because people communicating science so often claim that science is simply the truth, and then go on to explain why it is the truth in a technical language that most people cannot understand. This gives people who understand the science a lot of persuasive power in conversations, because others simply cannot argue back on the same terms. While this may win arguments by forcing others to back down and bow to an "expert," it is not effective or helpful. It is an abuse of the technical and complex nature of science, because despite the communicator's best intentions, *it does not argue the issues*. It uses the *idea of science as truth* as a shortcut to force people to accept anything that scientists say. To make communication more open and productive, it is important that we get away from using these tricks and shortcuts. Using language that everyone can understand and making time to discuss all angles on the issue, valuing the scientific and non-scientific alike, is much more effective.

Scientists themselves often feel frustrated that non-scientists do not listen to their advice and knowledge. This can make them feel powerless. In order to avoid this feeling, scientists may leave out the uncertainties inherent in science, worrying that expressing uncertainty will weaken their argument. Research has shown, however, that scientists who "hedge" their communication by admitting the uncertainties of the science—for example, when talking to the media—are seen to be more trustworthy and believable by members of the public. Science communicators who *report* this uncertainty—for example journalists—are also seen to be more credible (Jensen, 2008). In a world that is full of uncertainty, people who claim certainty and brush over potential risks are not perceived to be trustworthy; on the contrary, they are perceived to be patronising. This is counterproductive for communication, especially with communication partners who already have little trust in science.

Long-Term, Inclusive Discussions

Communication about controversial science issues should ideally be conducted in an equal way, so that everyone's views are heard and alternative viewpoints are meaningfully explored. When trying to facilitate genuinely open communication, in order to reach a mutual understanding, communication must continue to take place over a long period of time. Time gives people an opportunity to become fully informed and to develop complex understandings of the issues, because they have time to think about them, talk about them, and digest the information. Even more importantly, time grants people an opportunity to forge relationships of trust.

Trust is easily lost and difficult to gain. Scientists and science defenders lament the lack of public trust in science, yet fail to recognise the power of science's mistakes in this regard (Wynne, 2006). As the BSE crisis demonstrated, trust must be matched by trustworthiness. If science defenders prove themselves untrustworthy by withholding critical information and uncertainties, such as those surrounding the potential risks of mad cow disease, then they do not deserve public trust. In addition, it is unhelpful to see trust as equivalent to faith, expecting people to believe blindly that science is good for them. Trust must be earned, not expected (Barnes, 2005).

Exploring alternative views means ensuring that communication is set up to accommodate diversity. Science communication is often characterised as communication between "scientists" and "the public," and this division can reinforce the "us and them" approach. Just as scientists are a diverse bunch of people, so too "the public" is made up of of billions of individuals, all with different knowledge and experiences and values and priorities. In any communication situation, we should try to get to know the people with whom we are communicating. We need to understand the views they hold as a group and also value their individuality and differentness. This means researching potential communication partners when first embarking on a communication activity. We are all experts in something: find out what they *do* know, rather than assuming they know nothing. Not all publics are the same. Remember that all of us—even scientists—are lay people when discussing areas in which we are not specialists.

If the communication situation is very volatile, with passions running hot, the first aim should be to create a safe space for dialogue. This does not necessarily mean calming people's strong feelings, because this may never occur if the controversy conflicts with deeply held values. It does mean opening the communication channels for people to express their views freely without fear of retribution. This can be difficult to achieve when people are very angry. In such situations it is important to put a willingness to listen over a desire to talk. It may be preferable to forget about talking altogether, at least for a while. Together, communication partners should set up an environment with which everyone feels comfortable, particularly including those people who are most likely to feel disempowered and excluded from decision making.

Complex and controversial matters can rarely be resolved in one communication event. It may be necessary to set up further opportunities for all parties to continue information sharing, discussions, debates, and decision making in the future. It may be important to separate these elements of the process to allow time for trust to grow. When making decisions it is critical that people trust each others' willingness to cooperate, even if their views are diametrically opposed.

The importance of taking time with communication is illustrated by the example of natural resource management committees peopled by Indigenous and non-Indigenous Australians, who together must determine how an environment is to be managed. A long social and political history of animosity underlies a great deal of communication between Indigenous people and the non-Indigenous colonisers who make up the majority of the Australian population, even when individuals involved have the best of intentions. Therefore trust must often be built from scratch if there is to be any productive communication. Such committees succeed best when:

- they are given a long time frame in which to make decisions, allowing trust to be built.
- they take place in a space where all parties feel comfortable. Seemingly "neutral" meeting places such as office board rooms can be intimidating to people with less social power if they are not familiar with them.
- there is enough time within each meeting to talk about broader contextual issues that have an impact on people's lives, not just the issue "at hand". People's lives are complicated and often difficult, and science issues cannot be separated from general life issues such as cultural expectations or being able to put food on the table. (Smyth, Szabo & George, 2004; see also Woo, et al., 2007)

Social and political power dynamics affect all of us in different ways, so this not a scenario unique to Australia nor to natural resource management. Although it is easy to idealise science as non-political, ignoring the social and political context of a communication situation simply sweeps problems under the carpet. Instead, by acknowledging the politics and inequalities of power, by taking time—even years—to build trust one conversation at a time, and by being open to alternative perspectives and outcomes, communication about contentious scientific issues can take place in a way that builds bridges between science and people who are hostile to it, rather than building walls between us.

Finally, the most important strategy for dealing with hostility to science is being prepared to accept that we as scientists or science communicators might be wrong. If we expect others to change their views when they hear our arguments, why should the reverse not be the case? Because science is a product of society that affects the lives of real people, in a democratic society it is ultimately up to the citizenry to determine how it is managed. If there is a big groundswell of opposition to some scientific development, perhaps it should not proceed.

Science has made mistakes in the past when dealing with public opposition to its developments. It is very important for everyone's sake—scientists, publics, governments—that such mistakes are not made again.

References

Bagemihl, B. (1999). *Biological Exuberance: Animal Homosexuality and Natural Diversity*. New York, USA: St Martin's Press.

Barnes, B. (2005). The credibility of scientific expertise in a culture of suspicion. *Interdisciplinary Science Reviews, 30*(1), 11–18.

Bartlett, A., Smith, G., & King, M. (2009). The response of mental health professionals to clients seeking help to change or redirect same-sex sexual orientation. *BMC Psychiatry, 9*(11), online.

Bickerstaff, K., Lorenzoni, I., Pidgeon, N. F., Poortinga, W., & Simmons, P. (2008). Reframing nuclear power in the UK energy debate: nuclear power, climate change mitigation and radioactive waste. *Public Understanding of Science, 17,* 145-169.

Brown, N., & Sowerwine, S. (2004). Irati Wanti: senior Aboriginal women fight a nuclear waste dump. *Indigenous Law Bulletin, 23*. Retrieved from http://www.austlii.edu.au/au/journals/ILB/2004/23.html

Chivers, J. (2008). Deliberating competence: theoretical and practitioner perspectives on effective participatory appraisal practice. *Science, Technology, and Human Values, 33*(2), 155-185.

Clague, P. (Producer). (2009). Maralinga: The Anangu Story [Television program] In M. Carey (Series Producer) *Message Stick.* Australia: Australian Broadcasting Corporation.

Collins, H., & Evans, R. (2007). *Rethinking Expertise.* Chicago, USA: The University of Chicago Press.

Collins, H. M., & Evans, R. (2002). The third wave of science studies: studies of expertise and experience. *Social Studies of Science, 32*(2), 235-296.

Cvetkovich, G., & Nakayachi, K. (2007). Trust in a high-concern risk controversy: a comparison of three concepts. *Journal of Risk Research, 10,* 223-237.

Fischer, R. (2008). European governance still technocratic? New modes of governance for food safety regulation in the European Union. *European Integration Online Papers, 12*(6).

Funtowicz, S. O., & Ravetz, J. R. (1993). Science for the post-normal age. *Futures, 25*(7), 739-755.

Gonsoulin, T. P. (2001). Cochlear implant/deaf world dispute: different bottom elephants. *Otolaryngology - Head and Neck Surgery, 125*(5), 552-556.

Grace, A. P. (2008). The charisma and deception of reparative therapies: when medical science beds religion. *Journal of Homosexuality, 55*(4), 545-580.

Harding, S. (Ed.). (1993). *The "Racial" Economy of Science: Toward a Democratic Future.* Bloomington, USA: Indiana University Press.

House of Lords Select Committee on Science and Technology. (2000). *Science and Society.* London, UK: The Stationery Office.

Irwin, A. (2006). The politics of talk: coming to terms with the 'new' scientific governance. *Social Studies of Science, 36*(2), 299-320.

Jasanoff, S. (1997). Civilization and madness: the great BSE scare of 1996. *Public Understanding of Science, 6,* 221-232.

——(2003). Breaking the waves in science studies: comment on H.M. Collins and Robert Evans, 'The third wave of science studies'. *Social Studies of Science, 33,* 389-400.

Jensen, J. D. (2008). Scientific uncertainty in news coverage of cancer research: effects of hedging on scientists' and journalists' credibility. *Human Communication Research, 34,* 347-369.

Kirk, L. (2009). *Taking a more strategic approach to science communication.* Paper presented at the The 11th Pacific Science Inter-Congress, Tahiti, 2-6 March 2009.

Lee, C. (2009). "Race" and "ethnicity" in biomedical research: how do scientists construct and explain differences in health? *Social Science and Medicine, 68*(6), 1183-1190.

McKechnie, R. (1996). Insiders and outsiders: identifying experts on home ground. In A. Irwin & B. Wynne (Eds.), *Misunderstanding Science? The Public Reconstruction of Science and Technology* (pp.126-151). Cambridge, UK: Cambridge University Press.

Morning, A. (2008). Reconstructing race in science and society: biology textbooks, 1952-2002. *American Journal of Sociology, 114*(S106-S137).

Oakeshott, I., & Gourlay, C. (2006). Science told: hands off gay sheep. *The Sunday Times.* Retrieved from http://theratiocinator.blogspot.com/2006_12_01_archive.html

Prosise, T. O. (1998). The collective memory of the atomic bombings misrecognizes as objective history: the case of the public opposition to the National Air and Space Museum's atom bomb exhibit. *Western Journal of Communication, 62*(3), 316-347.

Purdue, D. (1999). Experiments in the governance of biotechnology: a case study of the UK National Consensus Conference. *New Genetics and Society, 18*(1), 79-99.

Relethford, J. H. (2009). Race and global patterns of phenotypic variation. *American Journal of Physical Anthropology, 139*(1), 16-22.

Roselli, C. E., & Stormshak, F. (2009). The neurobiology of sexual partner preferences in rams. *Hormones and Behavior, 55*, 611-620.

Seelman, K. D. (2001). Science and technology policy: Is disability a missing factor? In G. L. Albrecht, K. D. Seelman & M. Bury (Eds.), *Handbook of Disability Studies* (pp. 663-692). Thousand Oaks, USA: Sage Publications.

Sjöberg, L. (2004). Principles of risk perception applied to gene technology. *EMBO Reports, 5* (Special issue), S47-S51.

Smedley, A. (1999). *Race in North America: Origin and Evolution of a Worldview.* Boulder, USA: Westview.

Smyth, D., Szabo, S., & George, M. (2004). *Case Studies in Indigenous Engagement in Natural Resource Management in Australia*: Report prepared for the Australian Government: Department of Environment and Heritage.

Stern, J.-M. (2004). The cochlear implant - rejection of culture, or aid to improve hearing. *Reporter Magazine On-Line.* Retrieved from http://www.deaftoday.com/news/2004/01/the_cochlear_im.html

Street, R. L., O'Malley, K. J., Cooper, L. A., & Haidet, P. (2008). Understanding concordance in patient-physician relationhips: personal and ethnic dimensions of shared identity. *Annals of Family Medicine, 6*(3), 198-205.

Twyman, M., Harvey, N., & Harries, C. (2008). Trust in motives, trust in competence: separate factors determining the effectiveness of risk communication. *Judgment and Decision Making, 3*(1), 111-120.

Verhey, S. D. (2005). The effect of engaging prior learning on student attitudes toward creationism and evolution. *BioScience, 55*(11), 996-1003.

Woo, M.-K., Modeste, P., Martz, L., Blondin, J., Kochtubajda, B., Tutcho, D., et al. (2007). Science meets traditional knowledge: water and climate in the Sahtu (Great Bear Lake) region, Northwest Territories, Canada. *Arctic, 60*, 37-46.

Wynne, B. (2003). Seasick on the third wave? Subverting the hegemony of propositionalism. *Social Studies of Science, 33*, 401-417.

——(2006). Public engagement as a means of restoring public trust in science - hitting the notes, but missing the music? *Community Genetics, 9*(3), 211-220.

——(2008). Elephants in the rooms where publics encounter "science"?: a response to Darrin Durant, "Accounting for expertise: Wynne and the autonomy of the lay public". *Public Understanding of Science, 17*, 21-33.

PART III

Major Themes in Science Communication

COMMUNICATING THE SIGNIFICANCE OF RISK

Craig Trumbo

Introduction

Risk is one of the most dynamic and potentially charged aspects of scientific and technical communication. Issues of risk arising from science and technology capture the attention of the public, of policymakers, and of the media in such a manner as to often take on a life of their own. As a result, scientists and professional science communicators are sometimes stymied in their efforts to appropriately communicate risk. It is therefore important that science communicators have a good understanding of how people perceive risk, as well as how best to communicate it.

In perhaps its simplest definition, risk can be seen as a tripartite concept: the *probability* of *harm* occurring due to some *hazard*. The components of this definition bear additional consideration.

Probability is perhaps the most difficult element of risk for many people to understand. In risk circumstances the expression of probability often becomes statistical because certain parameters (e.g., actual disease rates) are not known, making inference and estimation necessary. The probability of being affected by a risk is therefore some estimated value within some estimated range of error. In addition to this complication, estimated probabilities are somewhat interchangeably expressed in terms of *absolute* and *relative values*. This is especially acute in medical studies when a report states that there is some percentage increase due to an exposure. The percentage may be large, say 15% but this is the increase over what may actually be a very small absolute chance, say 0.01%. Small absolute risk values are also difficult, with the estimated probability of harm from exposures often being in the neighbourhood of one in a million, for example.

Likewise, the idea of harm is complicated. To a large extent this is also due to the fact that actual harm must be estimated. On one end of the spectrum, of course, there is the harm of immediate death, say, in the catastrophic failure of a space vehicle launch. In such extreme circumstances death is assured. In the case of the US Space Shuttle program, some estimates of catastrophic failure on launch were 1 in 60 (Martz & Zimmer, 1992). So that was the chance of death. But rarely is it so cut and dried. Harm is much more typically something that does not occur immediately. For example, a good many risks involve the possible development of some form of cancer. Cancer can take many years to appear, even in cases of acute exposures. Further, concrete proof that some exposure eventually leads to development of cancer in a specific individual is very rare, owing to our incomplete understanding of the disease and its multiple causes.

Finally, the dizzying array of hazards in the modern world hardly simplifies things. Early work on risk perception by Paul Slovic and colleagues (discussed in more detail below) delineated the way in which people differentiated a variety of risks, ranging from the commonplace (e.g., swimming pools) to the highly unlikely (e.g., a nuclear exchange). In this series of studies, nearly 100 specific hazards were compared. Yet 100 hazards is just the tip of the iceberg, and new items are added daily. Hazards may be grossly categorized by ideas such as natural vs. human-made, those affecting individuals vs. large populations, those with immediate effects vs. those with long-term consequences, and so forth. In many cases, however, the true nature of a hazard can be obscured by such classification. For example, the Hurricane Katrina disaster in New Orleans is variously understood as a natural disaster (the hurricane) or as a human-caused disaster (the engineering failure of the levees). In fact, a good many large-scale disasters are such combinations (the Fukushima Daiichi nuclear disaster certainly qualifies), and are coming to be known as "natech" disasters (Picou, 2009). The time-horizon problem with hazards is also a significantly complicating problem. At least in the US, while the political controversy involving climate change is shaped by many factors, the slow-motion nature of clearly identifiable effects thwarts any sense of urgency on the part of the public or elected officials. Such hazards are coming to be called slow-motion disasters (Cline, et al., 2010).

So while the initial definition of risk offered here is at least superficially straightforward, any casual scan of the current landscape makes it immediately obvious that there are complicated processes involved. One process that significantly complicates the manner in which we face risk is the nature of human perception and judgment.

Perspectives on Risk Perception

Risk has for some time been approached from one of three competing scholarly perspectives: cultural, sociological, and psychological. The main thrust of this chapter is to explore the psychological perspective because this perspective has been to a fair degree most influential in shaping both the research and the practice of risk communication.

Before considering the more psychologically-based aspects of risk, it is important to also emphasize that risk exists in cultural and social dimensions (Kahan, Jenkins-Smith & Braman, 2010; Zinn, 2008). In terms of culture, consider that all three components of the risk definition above are present in some form regardless of, for example, a nation's technological development (even on the end of the spectrum represented by contemporary agrarian or nomadic peoples). The earliest work by Douglas and Wildavsky on risk and culture was based on the idea that the perception that people have of risks is shaped, or actually constrained, by such factors as the degree to which a culture is collective versus individualistic in its outlook, or strongly stratified socially versus more egalitarian (Douglas & Wildavsky, 1980). Daniel Kahan and colleagues, who are working with the concept of Cultural Cognition, are undertaking some of the most current work on culture and risk. Their work is built upon the work of Douglas, asserting that the core values held by individuals interact with the various mechanisms of psychological risk perception (more below). Policy preferences to address risk are shaped in the light of their cultural orientation toward the specific issue (Kahan, 2011; Kahan & Braman, 2006; Kahan, Jenkins-Smith & Brahman, 2010).

The Sociological perspective on risk has also offered important insights. Regardless of its positive or negative connotations, risk can be seen as a condition of modern society (or modernity). Giddens (1998) describes modernity as "a society—more technically, a complex of institutions—which, unlike any preceding culture, lives in the future, rather than the past" (p. 94). The very notion that we can consider what is and is not risky, and make individual and societal choices, is certainly not new. The array of factors that must be considered today is, however, greater by orders of magnitude than in most of human history.

In his work *Against the Gods*, Peter Bernstein describes the mastery of risk as the vehicle that propelled society into modernity:

> The ability to define what may happen in the future and to choose among alternatives lies at the heart of contemporary societies. Risk management guides us over a vast range of decision-making, from allocating wealth to safeguarding public health, from waging war to planning a family, from paying insurance premiums to wearing a seatbelt, from planting corn to marketing cornflakes.
>
> *(Bernstein, 1996)*

In a somewhat less utopian vision, sociologist Ulrich Beck describes this condition as the "risk society," in which "dangers are being produced by industry, externalized by economics, individualized by the legal system, legitimized by natural sciences, and made to appear harmless by politics" (Beck, 1992).

In terms of the human processes of society, risk has long been an element affecting differentiation in social class and human health. For example, the phenomenon of environmental racism is driven by the class-based assignment of environmental hazards, which has the additional consequence of permitting such hazards to perpetuate.

Of the perspectives that tell us about people's reaction to risk information, perhaps most prominently there are studies that examine the various psychological mechanisms behind the perception of risk. These are very relevant to understanding the processes of risk communication and how audiences process and understand risk messages.

The Sociological perspective has also been most adept at providing insight into the actions of industries and governments with respect to the technological hazards they create and oversee. A central concept in this area of work is William Freudenburg's "atrophy of vigilance" (Freudenburg, 1992). In this process, the institutional entities responsible for maintaining safety (e.g., the Nuclear Regulatory Commission in the US) enjoy a prolonged period of safety during which atrophy in oversight occurs. This slippage might be associated with technological development or a false sense of security having lead to unofficial relaxation of safety practices (i.e., cutting corners). Then, of course, a major incident occurs. This perspective has been applied to a range of disasters, most notably by Freudenburg to the oil industry disasters in the Gulf of Mexico (Freudenburg & Gramling, 2010).

While the Cultural and Sociological perspectives on risk are rich and offer key insights, the psychologically-based scholarship has been arguably more influential in shaping the practice of risk communication. A sizable volume of research has been compiled focusing on psychological risk perception over the past three decades. But what do we mean by "perception"? Early in the recent growth of this work, the US National Research Council competed a review of risk communication that included a careful consideration of the term "perception". The report argues that risk perception "is more accurately described as the study of human values regarding attributes of hazards (and benefits)." (NRC 1989, p. 53). The critical point made in this understanding is that, in the process of perception, individuals weigh the perceived risks and benefits, whether they are real or imagined. It might be argued that the term perception itself is not optimal, and that judgment would be better (Dunwoody & Neuwirth, 1991), although perception remains dominant in the literature.

For risk communicators, there are two very important perspectives that need to be kept in mind. These perspectives are informed by what has become known as the "psychometric model" of risk perception, and a body of research that examines a phenomenon known as "optimistic bias." These perspectives tell us a great deal about how individuals tend to orient toward risk and risk communications. We can turn to that material here, as informed by previous reviews of the literature (Trumbo, 2002).

To investigate risk perception, researchers surveyed various groups and asked them to respond to a series of nine questions concerning 30 common hazards (Fischhoff, Slovic, Lichtenstein, Read & Combs, 1978). The hazards ran the gamut from the common but fairly dangerous (skiing, swimming pools, auto accidents) to the very uncommon but potentially catastrophic (nerve gas accidents, train derailments). By looking at aspects such as how well a risk is understood, or how much control an individual has over a risk, the researchers used statistical techniques to develop a model showing that reactions to risk can be characterized most compactly in two dimensions—or in a "risk space."

One dimension of risk space is termed dread. This is related to the scale of the risk and the degree to which it harms innocent individuals. Nerve gas accidents and nuclear war have high dread while aspirin and swimming pools do not. The second dimension, termed knowledge, involves how well a risk is understood and how observable its consequences are. On one end of this spectrum are things like electric fields and PCBs, on the other end are automobile accidents. These processes are driven, to some degree, by the availability heuristic, which describes how people ascribe a greater chance of harm to the more vivid and thus memorable hazards they know of, or are informed of, via mechanisms such as the news media. This is why, for example, people often over-estimate deaths in airplane accidents relative to automobile accidents. Taken together, looking at risks in these terms can predict how people will react. Risks that evoke dread, whether they are associated with common or rare hazards, motivate people strongly toward action. Ironically, some of the most common hazards that cause a great number of injuries or deaths, but which do not evoke dread, are commonly tolerated (Morgan, 1993).

This line of investigation has been robust. The two-dimensional model has since been widely replicated and tested in a variety of countries (Slovic, 2000). Others have also expanded risk space into three dimensions using a wide variety of attributes such as the number of people affected or the voluntariness of the risk (Kraus & Slovic, 1988; Morgan, et al., 1985; Mullett, Duquesnoy, Raiff, Fahrasmane & Namur 1993; Slovic, Fischhoff & Lichtenstein 1985).

More recent survey work with the model has addressed the question of how it works with single hazards encountered in "real life," as opposed to the abstract hazards used in the early studies. The early work with the psychometric model wasn't designed to shed light directly on the differences between individuals. The more recent work has done just that, providing a parallel examination of how people differ in their reaction to individual hazards. Many of the same principles apply. However, since human personality and cognitive traits are complex, the amount of explanation provided by these "individual difference" models is somewhat limited. But overall, this work has shown that the psychometric model functions well to describe how individuals react to single, real hazards, and has also provided a reliable set of survey questions for the measurement of this construct (Trumbo, 1999; Trumbo, 1996).

An important perspective in social psychology holds that people process information through two unique mechanisms, cognitive and affective. In an extension of the individual-level approach, Slovic and colleagues have been further exploring the role of emotions, or affect (Slovic, 2010; Slovic, Finucane, Peters & MacGregor, 2004). In this work, individuals are described as possessing an "affect pool," in which images of the world held by the individual, including its hazards, are tagged with emotional markers. As people make judgments, they call on this pool, just as they rely on other mental shortcuts such as how imaginable the risk is, or how similar it is to something else known or experienced. These "dual process" theories have been used to shed some light on individuals' risk perceptions (Slovic, 2010).

Optimistic Bias

While the psychometric model provides insight into the manner in which individuals perceive various hazards as risks—and especially how they react to some hazards versus others—a parallel line of investigation has shown that people are also influenced by a related but unique tendency that has a subtle influence over their orientation toward all hazards and risks. This tendency has become known as "optimistic bias."

In general terms, optimistic bias (or in some circles comparative optimism) is the phenomenon in which individuals see themselves—in comparison to others—as less likely to be harmed by events in the future, or see themselves as being more likely to achieve some goal or status (Burger & Palmer, 1992; Weinstein, 1989). Optimistic bias has been observed in a wide variety of contexts, including risk-taking behaviors such as motorcycle riding, bungee jumping, and smoking (Middleton, Harris & Surman, 1996; Rutter, Quine & Albery, 1998; Weinstein, Marcus & Moser, 2005) and vulnerability to health hazards such as radon (Weinstein, Sandman & Roberts, 1990). Optimistic bias can be strongest in the context of hazards that are infrequently experienced on a personal basis, such as hurricanes (Weinstein, Lyon, Rothman & Cuite, 2000).

A range of circumstances can affect optimistic bias. One does stand out, however. That is the amount of time that has passed since the particular hazard has appeared. An example will help clarify. In work conducted recently by the author, a survey was conducted of US residents living near the coast along the Gulf of Mexico. The survey was conducted in January 2006 immediately in the wake of Hurricane Katrina (Trumbo, Lueck, Marlatt & Peek, 2011; Trumbo, et al., 2010). The purpose of the study was to assess the degree to which physical distance from New Orleans might affect individuals' outlook, or risk perception, for the next hurricane season. While distance turned out to not be a very strong predictor of hurricane risk perception, the study did set the stage for a subsequent survey.

Two years later, there had not been a single hurricane affecting the US Gulf Coast. In January of 2008 the same questionnaire was given to the same individuals in the interest of seeing if their perception of hurricane risk changed over time, following the hypothesis that the two years of quiescence would yield a drop in concern. In addition to a set of questions measuring risk perception, the two surveys also had a pair of questions comparing how individuals viewed their own risk versus the risk of others. This is an indicator of their optimistic bias concerning hurricanes.

Over the two years, these Gulf Coast residents did in fact become markedly less concerned about hurricane risk. Their level of optimistic bias also shifted, with an interesting pattern showing that their assessment of self-risk was lowered along with their assessment of risk to others. The difference between the two remained equal, however. This study therefore illustrates not only the phenomenon of optimistic bias, but also its dynamic, as well as static, nature with respect to changes over time.

Review of Important Factors

Taken together, these perspectives and others have provided researchers and communicators with a broad understanding of the manner in which people tend to evaluate risk information in a biased fashion. Anderson and Spitzberg (2009) provide a summary of these biases and describe how reactions to risk are heightened by a host of factors. Perception of risk is heightened:

- when the hazard is not voluntarily accepted by the individual or is viewed as uncontrollable or irreversible (e.g., groundwater contamination of wells);
- when the form of harm is easy to imagine (such as physical injury), rather than vague or indirect (long-term disease development);
- when the sheer number of people that might potentially be affected is great, such as in a widespread natural disaster;
- when the hazard involved is new or not easily understood (cell phone radiation)—or complex such as the Fukushima Daiichi nuclear disaster;
- when the hazard is nearby (e.g., a hazardous waste incinerator), although this factor can be complicated in the case of ongoing rather than rapid onset hazards that have an economic advantage (for example, individuals who work at a hazardous location may develop tolerance);
- when the hazard poses risk to children (either real, as in car seats, or imagined, as in vaccinations);
- when the risk associated with a hazard is viewed, for any number of reasons, as morally objectionable or fundamentally unfair to those affected, as in the case of terrorism.

Of central concern to risk communicators is that the perception of risk is heightened:

- when the source of risk information is not trusted or, especially, was once trusted but now is not (industry and government face this problem often);
- and when the risk is strongly emphasized in the news media or in other forms of public communication such as social media (the iodine rush on the US west coast following the Fukushima Daiichi nuclear disaster, for example).

It is common to refer to these processes as biasing, which implies that the resulting perceptions are wrong. That was the dominant point of view in the world of risk research and practice for some time. More recently, however, it has come to be recognized that individual and social reactions to risk are based strongly in the values and cultural orientations held by people, which are not easily labelled as right or wrong. For example, parents who resist vaccination for their children over concerns about autism do so in the face of an enormous volume of medical research showing vaccination to be safe, while they also potentially cause harm to public health by not vaccinating their children. But their concern is grounded in a belief that no degree of risk is acceptable in this particular circumstance. In one light, it is easy to label their actions as illogical. In another light, it is difficult to impose a level of acceptable risk on parents concerning their own children—should they be expected to accept one chance in a thousand? One in ten thousand? This particular scenario also highlights the distinction between an individualist versus collectivist cultural orientation. Should one accept individual risk for the greater good of public health?

The complicated manner in which these risk perception factors can intermix is perhaps best illustrated through an example (Trumbo & McComas, 2003; Trumbo, McComas & Besley, 2008). A series of studies of risk perception and information processing was conducted in the context of community cancer concerns. In these cases, concerns are raised by individuals who live in communities where there is a hazard that is thought by some to be associated with a suspected increase in the area's cancer rate. These cases are fairly common in the US, with perhaps as many as 1,200 formal complaints registered annually (Trumbo, 2000). State health agencies respond to these concerns in a variety of ways, ranging from careful multi-year investigations to quick dismissal (Greenberg & Wartenberg, 1991). Even when a case is investigated carefully (about a third are), very few provide any scientific evidence of elevated cancer rates, and virtually none ever provide scientific evidence of a link between cancers and environmental causes (Kase, 1996).

In terms of the definition of risk, these cases involve probability. The epidemiological studies done in these cases often involve small sample sizes (there are few cases of disease or death) that must be analyzed within some geographical boundary determined by the researchers. This leads to a wide margin of error in estimates of cancer rates. The often non-significant findings of epidemiologists are commonly unacceptable to members of the public who are more motivated by emotional factors and disinclined to accept any risk. The situation is further complicated because science does not at present have good tools to link an individuals' cancer to an environmental exposure that may or may not have occurred, at an unknown dose, perhaps a number of years before.

Over the course of this research project, some 35 different cases were investigated using survey methods. The hazards included leaking underground gasoline tanks affecting a very small area, high voltage transmission lines and transformer stations, agricultural chemicals (including those used on golf courses) suspected of entering groundwater, nuclear facilities, a variety of incinerators, military bases, and large-scale industrial chemical complexes. Individuals were asked, for example, whether they had strong emotional reactions to the hazard (fear, anxiety), and they were asked about the manner in which they viewed the hazard cognitively (does the hazard affect many? Is it controllable?).

The common thread among these cases is the harm involved. In each case some form of cancer was suspected to be linked to exposures from the hazards involved. The types of cancers did vary. Leukaemia dominated and presented a number of cases in which there had been childhood deaths. Rare brain cancers were also involved in a number of cases. Breast cancer was also a prominent concern. Regardless of the type of cancer, this type of harm is one that is strongly feared, as it often involves very difficult treatment regimens with the definite prospect of a shortened life or near-term death.

This particular set of studies also serves well to illustrate many of the points made above with respect to the factors that increase risk concerns. Individuals do not voluntarily expose themselves to these hazards, and the effects are seen as uncontrollable and irreversible. Because cancer is involved, the harm is vividly imaginable (Trumbo, et al., 2008). Often the geography of exposure is extensive, affecting a large number of people. The hazards and harms are difficult to understand and are complex. These cases always involve a hazard that is located in near proximity to the people affected. In many cases the harm is most prominent in children, and in all cases the exposure to children is an elevated concern. Often the risk is viewed as unfair or morally objectionable due to the much greater likelihood of toxic emissions to be located in areas of lower wealth and political power.

Of most concern to risk communication, these cases almost always exist in a strongly adversarial framework that pits community activists against an industry or government agency, with the health department often caught in the middle. Trust is a major issue, especially as there usually exists a strong difference between communities and industry in terms of previous experience with such cases and resources to leverage the controversy with regulators and the legal system. Finally, these cases always create significant attention in the news media and through channels of online social networking.

What the studies consistently found—across communities, hazards, and harms—was that the dual-process psychometric model does describe the manner in which individuals viewed risk in these cases. There were strong expressions of both cognitive and affective elements in risk judgments, with those employing cognitive strategies on average indicating greater levels of concern compared to those employing affective strategies. Further, individuals who more strongly used systematic approaches to information processing were more likely to link that with a cognitive basis for risk perception and greater concern, and those using heuristic processing more likely to link to affective perception and lower concern.

One of the unexpected and interesting findings was that careful processing of information and more deliberate cognitive perception increased concern over the hazards. For a good many years the dominant perspective in risk communication was that heightened concern, especially in the absence of hard scientific evidence, could be countered by simply getting people the facts and showing them how to objectively process them, and that emotional factors were responsible for "irrational overreactions" to hazards. At least in these cases, those who were dismissing the hazards as not risky were doing do based on, perhaps, "irrational under reactions."

What this case study description shows in terms of risk communication is the significant complexity that is faced by the professionals doing the communicating. The form of the risk is very complicated, the information environment is complicated and difficult to manage, trust is absent, making effective messages very difficult, and people process messages and make decisions in ways that may seem counter-intuitive to the scientist.

Fortunately, there are ways to successfully communicate risk information.

Approaches to Communicating Risk Information

The now expansive literature on risk communication provides several overarching general principles that have as their foundation the best practices in many domains of communication (Lundgren & McMakin, 2009; Sellnow, Ulmer, Seeger & Littlefield, 2010).

An appropriate place to start such an examination of broad principles is with an established perspective on a definition for risk communication. Because risk communication involves a wide variety of approaches and applications, it must be seen as a multidimensional construct. Krimsky and Plough (1989) provide that there are five dimensions to the concept. First, the nature of the intentions behind the risk message must be considered. Messages can be essentially without goals or can have high expectations for specific outcomes. Second, the content of the risk message is highly variable, running from the more narrow focus of health and environmental messages to wide-ranging social concerns. Third, the audience for which the message is intended must also be considered: sometimes targeted, sometimes not.

A fourth dimension of risk communication involves the source of the information. Frequently this involves scientists and other technical experts, but it can also include a much broader range of sources involving the media and citizen groups, for example. Finally, the manner in which the message travels should be considered. Risk messages can be delivered through very restricted channels, or can flow freely through society.

As researchers have considered the intricacies of risk communication they have identified a number of general tasks, difficulties, and paradoxes that tend to face risk communicators. Fisher (1991) identifies three varieties of challenges. First, the risk communicator must clearly define the objective of the communication, embrace the importance of message evaluation, and realize that no communication effort is ever completely effective. Then, it is critical to make a concerted effort to make the science of the risk assessment accessible to the audience. And finally, the perspective of the audience must be considered and entered into the whole risk equation because public reaction invariably becomes intertwined with the risk condition itself.

One of the most important tasks of the risk communicator is to establish credibility (McComas & Trumbo, 2001; Trumbo & McComas, 2003). Credibility can be hurt when the audience perceives the message to be inconsistent with the facts or inconsistent with previous messages, when the messenger has a reputation for deceit, or when the expert sources appear incompetent or in disagreement. The audience's evaluation of the overall legitimacy of the risk issue also influences credibility.

A consistently difficult task for risk communication involves getting the public's attention. Researchers have found that the individuals who need risk information the most are also often the ones who are least likely to attend to the risk message. Community involvement can be visualized as a pyramid, with the vast majority of individuals being for the most part uninvolved and uninterested (NRC, 1989). Communicators must pay close attention to their choice of communication medium. This almost invariably includes working with the news media. As in any domain, it is important to define the audience carefully in terms of its receptivity to the communication, ability to process the information, and capacity to act on the recommended behaviors. Understanding the audience in terms of demographic characteristics is equally important; especially as risk messages can be differentially understood across age, sex, and racial/ethnic contrasts (Finucane, Slovic, Mertz, Flynn & Satterfield, 2000).

Risk communicators have considered a body of experiences and have developed a "conventional wisdom" approach to risk communication. Johnson and Fisher (1989) identified five items held up to their scientific evaluation. First, details matter, and what may seem like minor differences in the way information is expressed can have a big impact on people's reaction. Comparing risks falls into this domain, for example, comparing a cancer risk against a year of dental x-rays versus eating peanut butter may generate very different reactions, even though the comparisons are technically equivalent. The way that a risk message personalizes the risk has an influence. People react most strongly toward things they can imagine as relevant to themselves. The limitations created by virtue of educational experience affect people's ability to access and process information. This in turn has a strong influence over their reaction to technical information. Information that can't be understood is often less trusted. Dramatic hazards and events are typically reacted to more strongly, especially when they are communicated in a graphic way. And finally, people will take risk information and generalize it, sometime inappropriately, to other hazards that they may perceive as being related.

Best Practices

Risk messages should be crafted to address the following seven elements, which are gleaned from a range of sources and presented without respect to order of importance (Heath & O'Hair, 2009; Lundgren & McMakin, 2009; McComas, 2006; NRC, 1989; Trumbo, 2001).

First, the message must be clear and accessible. The use of technical terms and numbers is often best set aside in favor of more immediate examples and accessible imagery. Clarity can be problematic, however. Oversimplification can lead to misunderstanding, especially when too much emphasis is placed on brevity.

Second, risk communicators need to be aware of, and sensitive to, common misperceptions. Failure here can lead to audience alienation. This becomes especially important when widespread misconceptions need to be dispelled. Misconceptions are best addressed directly and corrected in a non-judgmental way, rather than being written off. When the risk has or could cause death, audience sensitivities become a paramount concern.

Third, risk communicators need to be always aware of the distinction between informing people and influencing them. The public is adept at identifying messages that are designed to influence, and they do not expect (or respect) such messages from official or governmental agencies. There are three conditions under which influence strategies are specifically wrong: when the risk involves public controversy; when the communication strategy approaches deceit; and when the risk only applies to individuals who voluntarily expose themselves to the risk.

Fourth, risk communicators need to show the personal relevance to the individual. Risk analysis needs to be translated into advice for the individual concerning safety measures, political activity, self-evaluation for membership in risk groups, and strategies for further information gathering.

Fifth, risk communicators need to directly address elements of uncertainty. Uncertainty is almost always an important element of risk analysis, management, and communication. Uncertainty over risk arising from imperfect data and expert disagreement should be tackled head on. However, a balanced approach is best. Public distrust develops rapidly when communicators attempt to exaggerate or minimize uncertainty. When the risk involves public policy, differing viewpoints can sometimes seek to take advantage of polar arguments so they may gain sway on an issue. When addressing uncertainty, it is generally recommended that numerical and statistical representations be kept to a minimum and that common and clear language be employed.

Sixth, risk communicators need to be aware that the effective use of risk comparisons is very challenging. Properly used, comparisons can improve understanding of a risk. But comparisons should not be offered as the only way in which to make a decision, but rather as one of many ways. There are also real dangers to effective communication present in the type of risks that are to be compared. A good rule is to maintain similarity in the type of risks compared. For example, comparing the risk of flying 500 miles to the risk of driving 500 miles would be appropriate, while comparing the risk of 20 hours per week exposure to second-hand smoke with eating a peanut butter sandwich per day would not be appropriate. Other techniques for risk comparison that appear to be effective include the comparison of multiple risks rather than just two.

Finally, the risk message should be complete to the greatest degree possible. This is a key ingredient in any risk communication effort, as missing information can often have a greater impact in the long run than information provided early on. There are five critical pieces of information in a complete risk message: the kind of risk present (e.g. health vs. economic); how risks and benefits are intertwined; what options are available; how great the uncertainty is surrounding the risk; and how difficult it is for experts to manage the risk.

All of the valid guidelines that have been developed for risk communicators have as their basis elements of the research literature on risk perception. When these insights are applied to the task of message design a number of factors emerge for consideration by the risk communicator. Six such factors were identified early in the development of professional risk communication in the US National Research Council report on risk communication and have stood the test of time (NRC, 1989):

- People simplify. Information overload is a consequence of the modern world. People cope with this overload by utilizing simplified constructs to evaluate complex problems.
- It is difficult to change people's minds. The desire for cognitive consistency is a fundamental attribute of the human psyche. This can lead people to selectively attend to information that only agrees with previously formed attitudes, to ignore any ambivalence in that information and to become overly attuned to polarized arguments.
- People remember what they see. Thus, people evaluate many risks based on their personal experiences in life and the information they receive from the media. This can lead people to underestimate risks they have not personally experienced.
- People cannot readily detect omissions in the evidence they receive. This especially applies to the anecdotal evidence people gather through their daily lives. It also makes it possible to fool people through the use of omissions, the result of which is typically the eventual loss of all credibility.

- People may disagree more about what risk is than about how large it is. The difficulty of defining risk itself is socially and culturally bound.
- People have difficulty detecting inconsistencies in risk disputes. It requires a considerable level of attention and knowledge to keep abreast of some of the more technical risk arguments. Applying critical thinking skills to those arguments can be even more daunting. Unfortunately, this leaves many members of the public vulnerable to manipulation.

While the manner in which a risk communication is formulated merits considerable attention, the mode through which the message is delivered is also a critical consideration by the risk communicator. Information channel consideration has become an increasingly important task for risk communicators because communication technologies can function uniquely for different audience segments. For example, while the Internet has become nearly ubiquitous, there is still a sizable segment of society with limited access by virtue of socioeconomic and to a small remaining degree geographic factors. The current explosion of social media poses a challenge since relatively little is yet known about its efficacy in the domain of risk communication.

When risk is involved audiences or important audience segments tend to be very active information seekers and will engage with information from a variety of sources and channels. The carefully crafted risk message must function in a very complicated information environment.

Conclusion

The bottom line for the risk communicator is clear: risk communication embodies all of the difficulties and opportunities for failure that plague public communication in general, and risk communication bears a greater burden than many other types of communication efforts. The subject matter is invariably complex, mathematical, controversial, and impacts people's physical safety and health.

In a review of the conventional wisdom of risk communication, Johnson and Fisher conclude that, "In short, the conventional wisdom that risk communication itself is a complicated, hazardous undertaking is quite correct" (Johnson & Fisher 1989, p. 37). But effective risk communication can be achieved. By understanding the audience, the nature of the risk at hand, and the goals of the communication it is entirely possible to craft a message and deliver it in a way that responsibly serves all parties involved to the maximum degree possible.

References

Anderson, P., & Spitzberg, B. (2009). Myths and maxims of risk and crisis communication. In R. Heath and H. O'Hair (Eds.), *Handbook of Risk and Crisis Communication* (pp.205-226), New York: Routledge.

Beck, U. (1992). *Risk society: towards a new modernity*. London, Newbury Park, Calif.: Sage Publications.

Bernstein, P. L. (1996). *Against the Gods: The Remarkable Story of Risk*. New York: John Wiley & Sons.

Burger, J., & Palmer, M. (1992). Changes in and generalization of unrealistic optimism following experiences with stressful events: Reactions to the 1989 California Earthquake. *Personality and Social Psychological Bulletin, 18*, 39-43.

Cline, R., Orom, H., Berry-Bobovski, L., Hernandez, T., Black, C. B., Schwartz, A. G., & Ruckdeschel, J. C. (2010). Community-Level Social Support Responses in a Slow-Motion Technological Disaster: The Case of Libby, Montana. *American Journal of Community Psychology, 46*,(1/2), 1-18.

Douglas, M., & Wildavsky, A. (1980). *Risk and Culture: An Essay on the Selection of Technological and Environmental Dangers*. Berkeley: University of California Press.

Dunwoody, S., & Neuwirth, K. (1991). Coming to terms with the impact of communication on scientific and technological risk judgments. In L. Wilkins & P. Patterson (Eds.). *Risky Business: Communicating Issues of Science, Risk and Public Policy* (pp. 11-30), Westport, CT: Greenwood.

Finucane, M. L., Slovic, P., Mertz, C. K., Flynn, J., & Satterfield, T. A. (2000). Gender, race, and perceived risk: the 'white male' effect. *Health Risk & Society, 2*(2), 159-172.

Fischhoff, B., Slovic, P., Lichtenstein, S., Read, S., & Combs, B. (1978). How safe is safe enough? A psychometric study of attitudes toward technological risks and benefits. *Policy Sciences, 9*, 127-152.

Fisher, A. (1991). Risk communication challenges. *Risk Analysis, 11* (2), 173-179.

Freudenburg, W. (1992). Nothing recedes like success? Risk analysis and the organizational amplification of risks. *Risk, 3*, 1-35.

Freudenburg, W., & Gramling, R. (2010). *Blowout in the Gulf: The BP Oil Spill Disaster and the Future of Energy in America*. Cambridge MA: The MIT Press.

Giddens, A. (1998). *Conversations with Anthony Giddens: Making Sense of Modernity*. Stanford, CA: Stanford University Press.

Greenberg, M., & Wartenberg, D. (1991). Communicating to an alarmed community about cancer clusters: a fifty state survey. *Journal of Community Health, 16*(2), 71-82.

Heath, R., & O'Hair, H. Eds. (2009). *Handbook of Risk and Crisis Communication*. New York: Routledge.

Johnson, F. R., & Fisher, A. (1989). Conventional wisdom of risk communication and evidence from a field experiment. *Risk Analysis, 9*(2), 209-213.

Kahan, D. (2011). *The Cultural Cognition Project* (http://www.culturalcognition.net/). Retrieved Sept. 1, 2011.

Kahan, D., & Braman, D. (2006). Cultural Cognition and Public Policy. *Yale Law & Policy Review, 24*, 155-158.

Kahan, D. M., Jenkins-Smith, H., & Braman, D. (2010). Cultural Cognition of Scientific Consensus. *Journal of Risk Research, 14*, 147-174.

Kase, L. (1996). Why community cancer clusters are often ignored. *Scientific American, 275*(3), 85-6.

Kraus, N., & Slovic, P. (1988). Taxonomic analysis of perceived risk: Modeling individual and group perceptions within homogeneous hazard domains. *Risk Analysis, 8*, 435-455.

Krimsky, S., & Plough, A. (1989). Environmental Hazards: Communicating Risks as a Social Process. *Journal of Communication, 39*(4), 109.

Lundgren, R. E., & McMakin, A. H. (2009). *Risk Communication A Handbook for Communicating Environmental, Safety, and Health Risks.* Hoboken: John Wiley & Sons, Inc.

Martz, H., & Zimmer, W. (1992). The risk of catastrophic failure of the solid rocket boosters on the space shuttle. *The American Statistician, 46,*(1), 42-47.

McComas, K. A., & Trumbo, C. W. (2001). Source credibility in environmental health-risk controversies: application of Meyer's credibility index. *Risk Analysis, 21*(3), 467-80.

McComas, K. A. (2006). Defining Moments in Risk Communication Research: 1996–2005. *Journal of Health Communication, 11*(1), 75-91.

Middleton, W., Harris, P., & Surman, M. (1996). Give 'em enough rope: Perception of health and safety risks in bungee jumpers. *Journal of Social & Clinical Psychology, 15,* 68-79.

Morgan, M. (1993). Risk analysis and management. *Scientific American, 269*(1), 32-41.

Morgan, M., Slovic, P., Nair, I., Geisler, D., MacGregor, D., Fischhoff, B., Lincoln, D., & Florig, K. (1985). Powerline frequency electric and magnetic fields: A pilot study of risk perception. *Risk Analysis, 5,* 139-149.

Mullett, E., Duquesnoy, C., Raiff, P., Fahrasmane, R., & Namur, E. (1993). The evaluative factor of risk perception. *Journal of Applied Social Psychology, 23*(19), 1594-1605.

NRC (1989). National Research Council. *Improving Risk Communication.* Washington DC: National Academy Press.

Picou, J. (2009). Katrina as a Natech Disaster: Toxic Contamination and Long-Term Risks for Residents of New Orleans. *Journal of Applied Social Science, 4*(3), 39-55.

Rutter, D., Quine, L., & Albery, I. (1998). Perceptions of risk in motorcyclists: Unrealistic optimism, relative realism and predictions of behaviour. *British Journal of Psychology, 89*(4), 681-696.

Sellnow, T., Ulmer, R., Seeger, M., & Littlefield, R. (Eds.) (2010). *Effective Risk Communication: A Message-Centered Approach.* New York: Springer.

Slovic, P. (2000). *The Perception of Risk.* London: Earthscan.

Slovic, P., Ed. (2010). *The Feeling of Risk: New Perspectives on Risk Perception.* London: EarthScan.

Slovic, P., Finucane, M. L., Peters, E., & MacGregor, D. G. (2004). Risk as analysis and risk as feelings: Some thoughts about affect, reason, risk, and rationality. *Risk Analysis, 24*(2), 311-322.

Slovic P., Fischhoff B., & Lichtenstein S. (1985). Characterizing perceived risk. In R. Kates, D. Hohenemser, J. Kasperson (eds). *Perilous progress: Technology as hazard.* Boulder, CO: Westview.

Trumbo, C. W. (1996). Examining psychometrics and polarization in a single-risk case study. *Risk Analysis: An Interdisciplinary Journal, 16*(3), 423-432.

——(1999). Heuristic-systematic information processing and risk judgement. *Risk Analysis, 19*(3), 385-393.

——(2000). The nature of public requests for cancer cluster investigations: A survey of state health departments. *American Journal of Public Health, 90*(8), 1300-03.

——(2001). Risk communication. In A. Kent (Ed.). *Encyclopedia of Library and Information Science*. New York, Dekker. *69*(32), 290-325.

——(2002). Information processing and risk perception: An adaptation of the heuristic-systematic model. *Journal of Communication, 52*(2), 367-382.

Trumbo, C. W., Lueck, M., Marlatt, H., & Peek, L. (2011). The effect of proximity to Hurricanes Katrina and Rita on subsequent hurricane outlook and optimistic bias. *Risk Analysis 31*(12), 1907-1916.

Trumbo, C. W., & McComas, K. A. (2003). The function of credibility in information processing for risk perception. *Risk Analysis, 23*(2), 343-53.

Trumbo, C. W., McComas, K. A., & Besley, J. (2008). Individual- and community-level effects on risk perception in cancer cluster investigations. *Risk Analysis, 28*(1), 161-178.

Trumbo, C. W., Peek, L., Marlatt, H., Lueck, M., Gruntfest, E., Demuth, J., McNoldy, B., & Schubert, W. (2010). *Changes in risk perception for hurricane evacuation among Gulf Coast residents, 2006-2008.* Paper presented at the Annual Convention of the Society for Risk Analysis convention. Salt Lake City, UT.

Weinstein, N. D., Lyon, J., Rothman, A., & Cuite, C. (2000). Changes in perceived vulnerability following natural disaster. *Journal of Social and Clinical Psychology, 19*, 372-395.

Weinstein, N. D., Sandman, P., & Roberts, N. (1990). Determinants of self-protective behavior: Home radon testing. *Journal of Applied Social Psychology, 20*, 783-801.

Weinstein, N. D. (1989). Optimistic biases about personal risks. *Science, 246*, 1232-1233.

Weinstein, N. D., Marcus, S. E., & Moser, R. P. (2005). Smokers' unrealistic optimism about their risk. *Tob Control, 14*(1), 55-9.

Zinn, J. (2008). *Social theories of risk and uncertainty: an introduction.* Malden, MA: Blackwell Pub.

7
QUANTITATIVE LITERACY IN SCIENCE COMMUNICATION

Maurice M.W. Cheng, Ka Lok Wong,
Arthur M.S. Lee, Ida Ah Chee Mok

The Idea of Numeracy and Quantitative Literacy

According to the use of the terms introduced in Crowther (1959), a "numerate" person has "a good basic knowledge of arithmetic" and is "able to understand and work with numbers" (Crowther, 1959), while "numeracy" is the "ability to do arithmetic" (Sinclair, 1995). Since Crowther (1959) coined the words "numeracy" or being "numerate," their meanings have gone beyond the mere ability to follow computational rules. There were two aspects of meaning embedded in the original report. These were:

1. An understanding of scientific approaches in studying phenomena;
2. The ability to think quantitatively.

According to the Crowther Report (1959),

Numeracy has come to be an indispensable tool to the understanding and mastery of all phenomena, and not only of those in the relatively close field of the traditional natural sciences. The way in which we think, marshal our evidence and formulate our arguments in every field today is influenced by techniques first applied in science. The educated man, therefore, needs to be numerate as well as literate.

(Crowther Report 1959, Para. 398)

Although numeracy has less connection with science nowadays than it did in this report more than a half century ago, the need for "not only the ability to reason quantitatively, but also some understanding of scientific method and some acquaintance with the achievement of science" (Para. 419), is even more relevant to us now, when science and technology has an increasingly pervasive role to play in the development of our modern society.

The basic stance regarding the need to go beyond arithmetic was reiterated by another important report on British mathematics education (Cockcroft, 1982). In addition to "an ability to make use of mathematical skills which enables an individual to cope with the practical mathematical demands of his everyday life", there is the second attribute of numeracy, namely, "an ability to have some appreciation and understanding of information which is presented in mathematical terms" (Para 39, Cockcroft Report, 1982). The emphasis on the second attribute stems from the belief that "mathematics provides a means of communication which is powerful, concise and unambiguous" (Para 3, ibid).

Bringing together these two educational reports, which have had a significant impact on the British school curriculum, we can appreciate that the need for general numeracy, in close connection with public understanding and communication of science and technology, matters to any fast-developing society.

Similar ideas about the role of numeracy have been emerging in the United States. While British educators have put curricular emphasis on numeracy as described above, their American counterparts have become aware of the increasing variety of social activities in which numerical information and quantitative argument have become prominent (Cohen, 2001). There is a particular concern that citizens should develop some "numerical sophistication," so as to be aware of the issues underlying a quantitative approach to representing and explicating social problems in various political debates. With the development of modern society and the advancement of science and technology, as well as the growing use of calculators and computers in recent decades, the capacity to function comfortably and effectively with quantitative aspects of a wide range of life contexts is essential to one's participation in contemporary society.

Notions of numeracy and quantitative literacy therefore allude partly to education through the school curriculum (mathematics in particular), and partly to the needs of society where science and technology play a significant role. We can appreciate that both terms, while essentially synonymous (with "numeracy" being more British than "quantitative literacy"), cannot easily be reduced to a simple definition.[1] In essence, according to Cohen (2001), "quantitative literacy matters well beyond the sphere of mathematics and science; it is indeed a basic thinking skill parallel to verbal literacy" (p.27). With an emphasis on reasoning and judgment based on quantitative information, quantitative literacy is more than basic arithmetic or even mathematical skills, but is an aspect of "habits of mind" that a student should be developing throughout his or her school education, if he or she is expected to become a citizen fully participating in modern society (Steen, 1997, 2004).

Challenges in Achieving Quantitative Literacy

Despite the need for quantitative literacy and the ability to reason using quantitative information, which have been advocated for several decades, there is a gap between what an educated person should achieve after years of learning school mathematics and statistics and the widespread "innumeracy" amongst the general public (Paulos, 1988, 1995). A couple of examples of "innumeracy" help to illustrate what level of numeracy might be expected of educated citizens.

Representing Incidence Rate vs. Number of Cases

Paulos (1995, p.79) observed that media reporters tend to represent the same phenomena differently in order to serve their own purposes. When they want to downplay a problem, such as the spread of a disease, their common practice is to report the incidence rate (say, 1 in 100,000). When they want to emphasize the severity of the disease, they tend to state absolute numbers of people infected (say, 3,000 cases across the whole of America). Nevertheless, a quantitatively literate reader, with a due recognition of the American population of about 300 million, should be able to see that the two pieces of numerical information are more or less the same. The issue at stake here is not only an understanding of seemingly different information in a numerical sense, but an understanding that people who make use of quantities in their reporting do bring their own value judgments to bear on those uses. Readers should therefore be aware of the values embedded in such representations.

Uncertainty and Risks

With the advancement of medical diagnostic technology, it is likely that citizens will have to deal with a variant of the following medical issue:

Suppose the incidence rate of a disease is 0.2% (or 2 in 1000). A diagnostic test for the disease is 99% accurate (i.e. the test will give a positive result 99% of the times if the subject has the disease, and a negative result 99% of the times if the subject does not have the disease), and you are tested positive. What is the chance that you have the disease?

The answer might seem to be 99%. In fact the chance is less than 20% [3]. The skills required for this analysis do not demand advanced-level mathematics, but simple arithmetic. One could argue that the simple arithmetic and analysis of the situation is too much for a layperson, but a worrying finding was that even medical doctors and other medical professionals did not have a correct understanding of this test result—most of their guesses for the possibility of actually having the disease "were wildly off" (Best, 2004, p.78).

Understanding of Simple Statistical Graphs

As long as statistical graphs constitute an important means to represent statistical data, one's ability to understand and interpret such graphs is undoubtedly a component of quantitative literacy. For example, consider the hypothetical school test item of a number of robberies in two successive years, illustrated by a bar graph Figure 7.1.

As the graph shows the two bars against a truncated scale, the relatively small difference between the two numbers apparently becomes large. Candidates were asked to justify whether a reporter's statement of "a huge increase in the number of robberies from 1998 to 1999" is a "reasonable interpretation of the graph." This was a test item used in the Programme for International Student Assessment (PISA). The performance of 15-year-old students was reported by the OECD (Organisation for Economic Co-operation and Development, 2009): the percentage of correct answers varied from 7% to 41% among the participating countries in 2000; an average of 26% of students answered correctly across all OECD countries. If we acknowledge the commonplace usage of statistical graphs (sometimes manipulated for specific purposes) for general communication, including that in the mass media and official reports, the quantitative literacy reflected by this rather low percentage is not very encouraging.

The last few decades have seen burgeoning discussions concerning numeracy and literacy, and related notions such as quantitative literacy, mathematical literacy, and statistical literacy. Without getting into the confusions arising from different focuses of concerns and different terminologies, we make a simple observation about our present day society: with the growing power of mathematics in quantifying and modeling the world, coupled with the advent

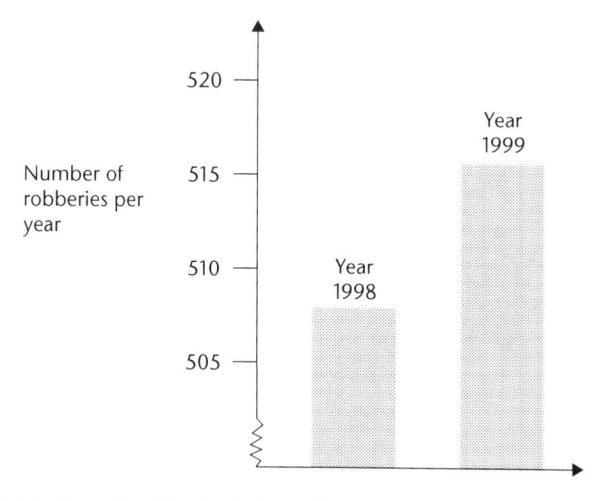

FIGURE 7.1 Number of Robberies in Two Successive Years

of highly effective computer technology, there is an ascendancy of quantitative information (including statistical data, summaries, and graphs) all around us. This is particularly so in a democratic society where citizens are supposed to participate in decision-making for the general good. Reasonable understanding of quantitative information and sensible judgment thereof is therefore crucial.

With a specific concern for scientific literacy, Rutherford (1997, p.64), citing *Science for All Americans*, provided the following characterization of quantitative literacy as a general expectation of scientifically literate persons. People should:

- Understand the nature of mathematics, its role in scientific inquiry and technological progress, and its dependence, in turn, on science and technology;
- Grasp sufficient mathematics to understand important scientific and engineering concepts and certain themes that are not exclusively scientific but are nevertheless widely used in science and technology; and
- Possess quantitative skills sufficient for following scientific stories that appear in the media, participating in public dialogue on issues involving science and technology, and responding critically to claims made in the name of science and technology.

We concur with Rutherford (1997) that the above characterization is not all there is to quantitative literacy, but it certainly highlights the close connections between mathematics, science, and technology that, as the last point in the list indicates, make quantitative skills indispensable. It should be made clear that quantitative skills for the purposes listed above are not in any way sophisticated and advanced. Just as Rutherford considers that it should be "at a level suitable for the educated public," Steen (2004) emphasizes that quantitative literacy is not concerned with sophisticated mathematics, but with a flexible understanding of elementary mathematics that prepares one to apply quantitative skills or ideas to various life contexts.

Achieving Science Communication: The Role of Quantitative Literacy

Although we have highlighted the importance of achieving quantitative literacy among the general public, we believe that governments, the media, the scientific community, and the medical professional are also important stakeholders. While we are aware of the impact of web-based technology that democratizes information and knowledge dissemination, we observe that these stakeholders have also been playing a role in communicating and informing the public on various science-related social issues. By "science communication," we adopt the following definition:

Science Communication may be defined as the use of appropriate skills, media, activities, and dialogue to produce one or more of the following personal responses to science (the vowel analogy)

> **A**wareness, including familiarity with new aspects of science
> **E**njoyment or other affective responses, e.g. appreciating science as entertainment or art
> **I**nterest, as evidenced by voluntary involvement with science or its communication
> **O**pinions, the forming, reforming, or confirming of science-related attitudes
> **U**nderstanding of science, its content, processes, and social factors

Science communication may involve science practitioners, mediators, and other members of the general public, either peer-to-peer or between groups."

(Burns, O'Connor & Stocklmayer, 2003, p.191)

Following contemporary ideas about communication (e.g. Stenning, Lascarides, and Calder, 2006), Burns et al. (2003) do not regard science communication as a unidirectional flow of information from experts directly or through mediators to the laymen or the publics. Rather, meanings are negotiated among all those involved in the communication. For any constructive and beneficial communication, not only did the public enhance their "awareness" of the new aspects of science, but science practitioners and/or the mediators also became more aware of some ways in which the public reacted to the science. Similarly, "understanding" was not supposed to be achieved by the public only. Science practitioners, for example, could in turn gain better insights into the interactions of science and social factors.

We will use three incidents to illustrate how quantitative literacy is essential in achieving effective science communication. We will briefly comment on how actors in these incidents triggered awareness, enjoyment, interest, opinion formation, and understanding of science. We believe that three case studies are vivid and concrete enough to illustrate aspects of science communication (van Dijk, 2011). They are:

1. the communication of the risk of melamine-contaminated milk powder between the Hong Kong government and the public in 2008;
2. the communication of the casualties and injuries between Florence Nightingale and the British government during the Crimean War (1853-1856);
3. the communication of the spread of cholera in London (1850s) between John Snow and the science community.

Although the first incident did not originate in Hong Kong, it affected the general public in the city and caught the attention of media worldwide. Due to its far-reaching impact, it is worthwhile to examine how science communication was conducted. We believe similar food contamination incidents could happen in the future, so as an examination of science communication in this incident would have implications for others.

The relevance of quantitative literacy in science communication is not limited to contemporary society. We use two historic incidents to illustrate how Florence Nightingale and John Snow went about science communication. We selected these cases to illustrate how science communication was conducted among different audiences. They cover communications between government and the public, between science practitioners and the government, and among science practitioners themselves.

Food Contamination—Melamine in Milk Powder

An Introduction to the Issue

Scholarly discussion of public engagement in science has been focused on issues such as energy, environment, medicine, health, reproduction, and sexuality (Blue, 2010). Extensive discussions in the science communication literature of issues such as nuclear energy, radioactivity, pollution and biodiversity, the spread of disease and pandemics, and of *in vitro* fertilization, and of the public understanding of science are manifestations of such focuses. Unlike other issues, however, food has escaped attention. Compared with other socioscientific issues in which there are distinctive and vocal activists, the actors engaging in food production are more numerous but silent. Almost all of them (including us, the authors and readers of this book) are everyday consumers. The actors participate through their consumption practices based on the information they obtain mainly from the media and personal networks (Blue, 2010).

The incident of milk tainted by melamine broke out through the media on 9 September, 2008. It was reported that 14 babies between 6 and 11 months old from Gansu province in China had been admitted to hospital for renal problems over the previous two months (Shen, 2008). We now know that melamine was added to watered-down milk to boost its tested protein content. Consuming melamine can cause kidney stones, renal failure, and death.

The report from mainland China caught the attention of the media in Hong Kong immediately. All major newspapers in Hong Kong subsequently reported the incident. Within a week, Sanlu Group, a national brand of milk powder manufacturer, admitted that their milk powder was contaminated with melamine. It was revealed that two babies had died in relation to the incident (BBC News, 2010). As Hong Kong relied heavily on imported food from mainland China, the incident immediately triggered concerns among the public. This was particularly the case for parents of infants who consumed milk powder on a daily basis. The concerns were reflected in the number of enquiries to the telephone hotline hosted by the Department of Health in the Government of Hong Kong. Between 21 September and 15 October (25 days), the hotlines received around 10,000 enquiries[4]. By mid October, eight children were diagnosed with kidney stones in Hong Kong (Centre for Health Protection, 2008).

At the outset of the incident, parents in Hong Kong were worried about the safety of the milk powder that their infants were taking. An answer to the following question could directly inform their decision-making: "Is the milk powder safe?" The Hong Kong government responded to the incident by checking the melamine contents of milk and dairy products in the market, making the test results public, offering guidelines for the public and medical doctors in relation to the incident, and so on. What was the potential usefulness and effectiveness of these guidelines in addressing this burning question of public concern?

The Consumers' Burning Question

At first glance, the question seems to mandate either a "yes" or "no" answer. However, it is more than a simple closed-end question that has a definite or absolute answer. Whether the milk powder was safe depended on:

1. the melamine level, if any, in the milk powder;
2. the frequency and amount of the contaminated milk powder consumed;
3. the body weight of the baby who might have consumed the contaminated milk powder.

It is apparent that these three factors are inter-related. Whether a baby was likely to be safe depended on how heavy the baby was, and how much melamine was consumed—which in turn was determined by the concentration of the chemical in the milk powder, the total amount of the contaminated milk powder consumed, and the frequency of consumption.

While the answers to the above questions may be potentially clear cut (e.g. there *is* 0.1mg melamine per 1kg of the milk powder, the consumption of milk powder by my baby *is* 20g per day), the question of whether it was safe to consume the milk powder was complicated by the fact that the risk of consuming contaminated milk powder was highly uncertain. That is, even if one consumed the milk powder that contained melamine exceeding the limit, the risk of developing kidney stones or related problems was not certain. Although The Centre for Food Safety of the Hong Kong government advised that "occasional excursion above the safety reference value would have no health consequences provided that the average intake over a long period does not exceed the value" (Centre for Food Safety, 2008, 24 December), such a qualitative response did not entirely meet the needs of the public. In the sections that follow, we will comment on two more specific guidelines offered by the government in addressing public concerns.

Two Safety Standards

Given the complexity of the seemingly simple question, as far as science communication is concerned, there was a need for the scientific community and the government to be selective in the provision of information. At the other side of the communication, it was also important that the parents and consumers could understand some of the most important guidelines in order to make decisions about the consumption of milk powder.

When the incident began, the public was unaware of the safety standard for melamine and the human tolerance of the chemical. It was only as the incident progressed and caught the attention of the public that the media started to report relevant information. The Hong Kong government announced safety information via the web page of the Centre for Food Safety. There were two major safety standards or guidelines:

1. The limits of melamine for infant formula and for other food products were set to be 1mg/kg and 2.5mg/kg respectively.[5]

2. A Tolerable Daily Intake (TDI) was 0.63mg/kg for adult and 0.32mg/kg for children under 36 months.[6]

Although these standards were made known to the public, they had different intended audiences. In general, the limit of melamine for food served as guidelines for food suppliers and food manufacturers in the testing of the safety of the food. Given that all failed samples had been removed from the shelves, it is likely that consumers did not find them directly useful in their decision-making.

TDI seemed to be relevant to consumers, since they could calculate the exact amount of melamine they could tolerate daily based on their body weight. Nevertheless, it is argued that the usefulness of the exactly tolerable amount of melamine per se is limited. It was unlikely that consumers would easily be able to find out the total amount of melamine they ingested daily, along with the food in which they were taking it.

Considering the official communication, it is likely that the information increased the "awareness" and "understanding" of the public in relation to the incident. Nevertheless, it is doubtful whether the information could facilitate the public's formation of informed "opinions" that guided their "personal responses" to the incident. In short, even if the public comprehended these guidelines and the meanings embedded in them, this was unlikely to inform decision-making.

Bringing Safety Standards to Life

The above comments are not intended to suggest that the efforts of the government were futile. We simply point out that the information did not seem to be useful to the general public or to most parents in making decisions. Yet we believe that the information was very important to milk suppliers in Hong Kong, to those who determined the safety of individual brands of milk and related products, and to medical doctors who received suspect cases. Indeed, the government did do more. Based on the limits of melamine and TDI, the government offered more concrete advice for the consumption of milk that was contaminated by melamine. In a press release, information included the following:

> For the High Calcium Low Fat Milk Beverage (1L) sample which recorded a level of 9.9ppm melamine, Dr Chan said it would require a 2-year-old child weighing 9kg to drink around two cups (around 0.5 litres) of the product a day to have exposure reaching the TDI. This product would, however, not usually be given to small children. An adult with average body weight of 60 kg would need to consume about 3.8 litres of the milk per day before reaching the TDI.
>
> *(Centre for Food Safety, 2008, 18 September)*

In the above excerpt, the most heavily contaminated milk found in Hong Kong at that time (10mg/kg of melamine) was used as an example, in order to convey the message that only with a daily consumption of a milk that overshot 10 times the safety limit would anyone reach the TDI. Undoubtedly, such concrete and well-exemplified guidelines gave life to the figures concerning the limit of melamine in food and the TDI. This bears out the suggestion of Lundgren and McMakin (2009) that concrete guidelines are conducive to effective communication and consumers' decision making.

Generalized Guidelines vs. Concrete Examples

There are different kinds of dilemmas in science communication. One of these involves how information is disseminated. Reporting internationally recognized guidelines, such as the limits of melamine and TDI, will facilitate universal standards. Although such reporting may enhance public awareness and understanding of the safety standards, it has very limited value in forming opinions and decision-making. On the other hand, converting the universal standards into concrete scenarios can trigger interest and involvement. However, these concrete scenarios can only speak to these specific cases, and not to others. For example, based on the above quotes from the press release, the information would leave the parents of children younger than two years old with unsolved problems.

   In summary, this case illustrates the complexity of addressing a seemingly simple question. It is suggested that well-exemplified and concrete examples, rather than abstract generalized guidelines that demand substantial further manipulations by the public, are more likely to trigger informed decision-making.

Florence Nightingale and Data Representation

An Introduction to the Issue

Communication of quantitative information can be effectively achieved through representations such as tables and graphics, particularly with the rapid development of friendly and powerful technological tools. Looking back in history, many modern forms of graphical displays may have a much longer and more unusual development than we normally expect (Friendly, n.d). Situated in "the golden age of statistical graphics" in the second half of the 19th century (Friendly, 2008), the story of Florence Nightingale can illustrate a simple yet powerful and innovative use of graphics for political and public affairs (Brasseur, 2005).

Victorian Britain experienced unprecedented social development but the formation of modern cities in this burgeoning industrial era came with unforeseen challenges in public health. During her nursing service in the Crimean War, Nightingale had noticed that it was infectious disease, rather than fatal injuries from the battlefields, that claimed the majority of lives of hospitalized soldiers. More importantly, there was a dramatic drop in hospital mortality when the sanitary conditions were improved. After the war, Nightingale was asked to present a report to Queen Victoria on hospital reform. She made her case through consideration of the causes of war casualties. Table 7.1 records monthly mortalities of soldiers during a two-year period of the Crimean War. The first column records death according to "zymotic" diseases, which is an obsolete term for certain infectious diseases. This is contrasted with the death directly due to wounds and injuries in the battle, displayed in the second column. The third column lists numbers of deaths due to other causes.

However, even provided with such straightforward information in the form of numerical figures, Nightingale still found it difficult to convince the government officials and the general public of the urgent need to improve hospital sanitary practice. What else could she do to draw attention of others to the issues?

Nightingale's Rose Diagram

Accompanying the modern reform of medical and nursing practice, it is perhaps not so well known that The Lady with the Lamp also contributed as a pioneering applied statistician. In an age with no systematic collection of data in hospitals, Nightingale was already aware of the importance of using statistical data for understanding phenomena and conveying messages. In order to highlight the fact that the death of soldiers in the Crimean War was mainly due to poor medical care rather than fatal injuries from battlefields, and that an improvement in hospital sanitation did lower the death toll, Nightingale designed a diagram that is considered to be a classic exemplar in the history of data visualization and applied statistics (see Figure 7.2).

TABLE 7.1 A breakdown of casualties during the Crimean War (*Source:* Nightingale, 1859)

Date	Estimated average monthly number of soldiers	Deaths		
		Zymotic diseases	Wounds & injuries	All other causes
1854				
April	8,571	1	0	5
May	23,333	12	0	9
June	28,333	11	0	6
July	28,722	359	0	23
August	30,246	828	1	30
September	30,290	788	81	70
October	30,643	503	132	128
November	29,736	844	287	106
December	32,779	1,725	114	131
1855				
January	32,393	2,761	83	324
February	30,919	2,120	42	361
March	30,107	1,205	32	172
April	32,252	477	48	57
May	35,473	508	49	37
June	38,863	802	209	31
July	42,647	382	134	33
August	44,614	483	164	25
September	47,751	189	276	20
October	46,852	128	53	18
November	37,853	178	33	32
December	43,217	91	18	28
1856				
January	44,212	42	2	48
February	43,485	24	0	19
March	46,140	15	0	35

Each column of death tolls in Table 7.1 was transformed into monthly mortality rates and represented in the diagrams as a ring of sectors (or wedges), each corresponding to a month, one ring for each year.In each diagram, the area of the red coloured wedges represented the number of deaths that were due to wounds, blue coloured wedges represented deaths from infectious disease, and black coloured wedges represented deaths from any other causes. The diagrams can be considered as a polar-area graph where circular sectors were used to represent values in a data set. Note that it is not the same as a pie chart, since it is the areas of sectors, rather than the angles, that vary and indicate data values. Rather, it is similar to a bar chart but arranged in a circular manner.

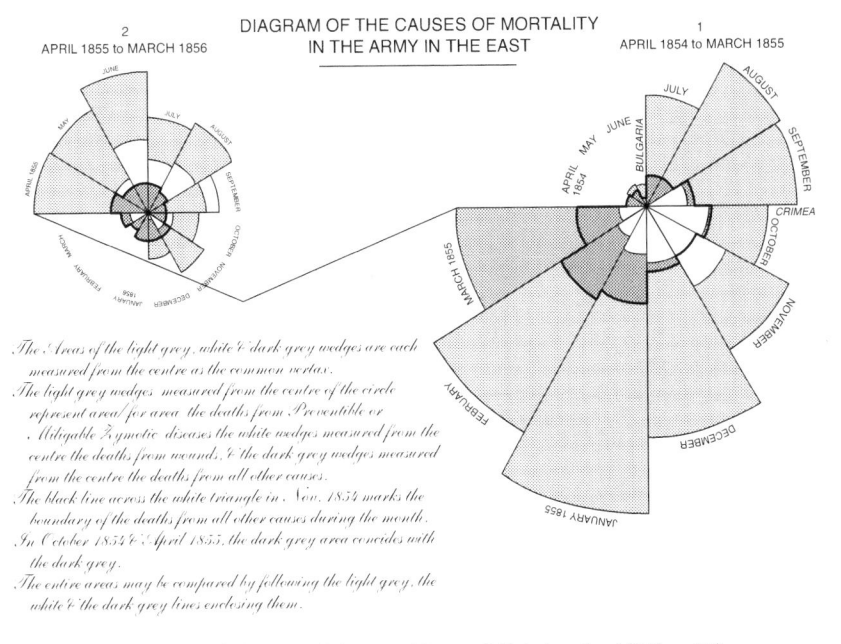

FIGURE 7.2 Nightingale's Rose Diagram (*Source:* Nightingale, 1859, p.19)

The significance of the diagram can be more fully appreciated if we understand how it is used to "present *messages*, not data" (Small, 2010, italics in original). According to Small's analysis, the diagram was carefully designed to highlight the significant drop of deaths due to diseases in the middle of the two years when the military hospital sanitation was greatly improved. This was a major piece of evidence supporting Nightingale's argument for a similar reform in ordinary hospitals, which had not been fully supported by medical officials.

Issues and Concerns with Representing Data Graphically

This case is presented not as a perfect example of graphical display: instead, it helps us to reflect on the use of graphics. There are views that dichotomize graphical presentations as being either "fair" or "biased". According to Rehmeyer (2008), Nightingale's mentor William Farr preferred figures to be shown as dry as possible to avoid "impressions" of facts. However, although it could be qualified as a "fair" representation, the array of figures in the table did not seem to enhance government's "awareness" or change their "opinion" concerning the need to enhance hospital sanitation. It was only Nightingale's Rose that persuaded a change to practice.

In this regard, we would like to highlight that any form of graphical presentation is primarily designed to show selected points of view, by stressing or ignoring certain elements according to the purpose or intent of its creator. Suitable use of graphics can enhance communication of numbers. Inappropriate use, however, may easily confuse or even convey wrong messages. What kinds of graphical representation should be employed and how they should be used can be a matter of science, as well as art, in communication. Today, use of innovative graphics, particularly with the growing popularity of technological tools, can be easily found in the work of professionals including academics, designers, journalists as well as the general public. Presenting numbers in graphical form to ordinary audiences can therefore be mundane technical work or a creative exercise depending on purpose, insight, and skill. The Rose of Florence Nightingale shows us the power of graphics when there is the need to communicate messages based on quantitative information.

John Snow and the Cholera Spread

An Introduction to Cholera and the 1854 London Epidemic

Cholera is an acute infectious disease caused by an intake of food or water contaminated with the bacterium *Vibrio cholerae*. The incubation period ranges from 18 hours to five days. Distinctive symptoms include voluminous watery diarrhoea and vomiting. They can quickly lead to severe dehydration and death if treatment is not promptly given. The cholera bacteria are found in the faeces of infected people. The disease spreads when infected faeces get into the water or food that people consume. Therefore, contaminated water sources are often the cause of pandemic outbreaks. This may be common knowledge today, but it was little known even in the mid-19th century when public health and hygiene as we know it today was not generally practised.

In this section we discuss a classic historical medical detective story, the cholera epidemic outbreak in London in 1854, to see how John Snow convinced the scientific community that cholera was spread by contaminated water. The main tool he used was a spot map (See Figure 7.3). The Golden Square in Soho in London suffered from a cholera outbreak resulting in about 500 deaths in ten days, from 31 August 1854. John Snow carried out an investigation and identified that the source was a contaminated water well. The epidemic stopped after the handle of the water pump was removed. How did John carry out the investigation?

The Contribution of John Snow

Snow began working at the Westminster Hospital, admitted as a member of the Royal College of Surgeons of England in 1838. Being an apprentice surgeon-apothecary, he had observed and studied cholera. He had formulated a hypothesis that the disease was localized at the gut with symptoms of severe fluid loss; and he deduced that the disease was spread by contamination of drinking water based on two earlier outbreaks of cholera in south London in 1849. However, the idea was not well accepted. When the epidemic started in 1853-4 in London districts, Snow realized an opportunity to test his hypothesis (Brody, Rip, Vinten-Johansen, Paneth & Rachman, 2000; Tufte, 1997).

FIGURE 7.3 The Spot Map by John Snow (*Source:* http://www.ph.ucla.edu/epi/snow/highressnowmap.html)

These were the major steps in his investigation:

1. Cholera broke out in Soho in the evening of 31 August, 1854. Snow suspected the water from a community pump-well at Broad and Cambridge streets was contaminated. He tested the water then, but the water looked clear. However, there were organic impurities visible to the naked eye in the next two days.

2. On 5 September, he obtained from the General Register Office a list of 83 deaths which had occurred since 31 August. He plotted the data on a map. The spot map showed that the deaths happened around the water pump. In his report, he wrote:

> On proceeding to the spot, I found that nearly all the deaths had taken place within a short distance of the [Broad Street] pump. There were only ten deaths in houses situated decidedly nearer to another street-pump. In five of these cases the families of the deceased persons informed me that they always sent to the pump in Broad Street, as they preferred the water to that of the pumps

which were nearer. In three other cases, the deceased were children who went to school near the pump in Broad Street ... With regard to the deaths occurring in the locality belonging to the pump, there were 61 instances in which I was informed that the deceased persons used to drink the pump water from Broad Street, either constantly or occasionally ...

(quoted in Brody et al., 2000, p.65)

3.   As a result of Snow's persistent house-by-house, case-by-case detective work, he investigated the cases of the deaths occurring in the locality to the pump and found that 61 used to drink the pump-water, 6 did not drink the water), and there was no information about another 6.Snow described his findings to the authorities on 7 September, 1854. They ordered the removal of the pump-handle, and the epidemic soon ended.

Science Communication: Argument with a Spot Map

The legend of Snow's detective story and the region where his investigation had taken place became famous historical sites in popular media.[7] With simplified descriptions, people might easily believe that the discovery had been arrived by simple observation of the cluster of instances on the map. The visual impact was surely pertinent and effective but it did not reflect the development of the scientific investigation in full.

The spot map was not the only visual representation John Snow used. He also presented the data in a time-series plot (see Figure 7.4). The time series showing the number of cholera cases with respect to the variable of time helps to indicate when the outbreak happened and subsided (Tufte, 1997), but it helps little in the explanation or prediction of a causal theory. Snow's theory was not only argued from the cluster of death cases near the water pump, but also by comparing cases of drinking the water and not drinking the water. Observing that there were no death reports in the brewery of Board Street, Snow interviewed Mr. Huggins the proprietor and wrote:

Mr Huggins believes they do not drink water at all; and he is quite certain that the workmen never obtained water from the pump in the street. There is a deep well in the brewery, in addition to the New River water.

(Snow, 1855a, p.26)

Furthermore, Snow had considered alternative explanations and contrary cases. Snow reported the death of a widow and her niece who were not in the neighbourhood of the Broad street pump, but who had drunk water from a travelling cart which had drawn water from the contaminated pump (Snow, 1855a, p.27).

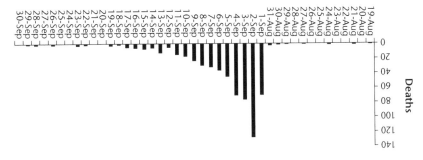

**Deaths from cholera,
each day during the epidemic**

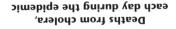

FIGURE 7.4 Time-Series Plot Produced from Data in Snow (1855b)

With respect to how Snow solved the mystery of cholera, Tufte (1997) mentioned the following:

1. Snow's scientific detective work exhibits a shrewd intelligence about evidence, a clear logic of data display and analysis;
2. placing the data in an appropriate context for assessing cause and effect;
3. making quantitative comparison;
4. considering alternative explanations and contrary cases (p.29).

Appropriate choice of representation of data plays a critical role in bringing about the possibility of hypotheses and conclusions. Often, relevant visual displays provide evidence and support for argument that goes beyond texts and words. In the case of Snow, he applied the spot map-making technique skillfully and with meticulous effort. Despite the powerful impact of visual thinking brought about by the spot map, however, the causal theory might not have been established without the factors identified by Tufte (1997). Although we are unable to judge whether the communication initiated by John Snow triggered "enjoyment" among the scientific community, it certainly raised the "awareness" of the scientific community about this new aspect of science, it triggered "interest"—even today, it enhances audience "understanding" of science, and it shaped the "opinion" of the scientific community such that subsequent action was taken to revamp the practices of public health and hygiene.

Conclusion

Quantitative literacy is indispensible for citizens in order for them to fully engage and participate in society. It is imperative to consider which appropriate data representations should be used in the discourse of socioscientific issues.

Taking the incidence of melamine-contaminated milk powder as an example, we have illustrated that an understanding of the meanings, limitations, and assumptions behind various indicators is essential for making everyday judgements. Such understanding is important to enable the public to probe for more relevant information from various sources, and to select the type of safety standard on which they can rely. Given these kinds of demands from the public, it is important that those who play a role in science communication should be aware that the information they offer can trigger a range of responses that lead to informed decision-making. Achieving successful science communication about numbers and quantities is not a straightforward task; we have highlighted the inevitable tension that results from offering generalized guidelines or using concrete examples in informing the public.

Science communication is not limited to communication between governments and the public, it also occurs among various stakeholders such as scientific communities and medical practitioners. We have used two historic incidences to illustrate how graphical representations—a means through which quantities are manifested—can facilitate communication even with informed stakeholders such as other scientists. A key message underlying these two cases is that graphical representations, like all representations, are not impartial reflections or copies of reality. They are embedded with values promoted by those who craft the graphics. In the cases we mentioned, they were crafted to persuade various audiences to make policy decisions, namely, to improve hospital sanitary practices, or to develop measures to curb the spread of disease. School mathematics can draw students' attention to the abuse of statistics or biased representations: in the public discourse of socioscientific issues, however, just as the media can choose to report incidence rates or exact numbers of cases, inevitably those who create graphical representations will be selecting certain messages. In the light of this, the public needs to be aware of the value-laden nature of graphical representations and science communicators should consider carefully how they choose to represent numerical data.

Notes

1   Under the auspices of the National Council on Education and the Disciplines and the leadership of Lynn Arthur Steen, the Quantitative Literacy Design Team, comprised of more than 15 renowned scholars, has made a case statement about quantitative literacy. See Steen (2001) for a more thorough discussion on the idea of quantitative literacy.

2   A similar task was presented by Casscells, Schoenberger, & Graboys (1978). See also Paulos (1988, p.89-90; 1995, p.136-137), Gigerenzer (2002, p.41), and Best (2004, p.77-78).

3   As the incidence rate is 0.2%, on the average, 2 out of 1,000 persons being tested have the disease. Due to a 1% false positive rate, out of those 1,000 persons, on the average, the test yields 10 (falsely) positive results. In other words, out of those 12 persons tested positive, only 2 persons bear the disease. That is, 2/12 = 16.7%.

4   In 2008, the population of Hong Kong was 7,008,900. There were 78,752 live
    births during the year. These figures indicate what 10,000 enquiries would mean in
    the context of the population of the city.
5   The standards came into force in Hong Kong on 23rd September, 2008. It was
    within two weeks of when the incident caught media attention.
6   The TDI here was updated as in September 2008 when the incident broke out.
    World Health Organisation set the TDI to be 0.2mg/kg on 5th December 2008.
7   See for example, "Mike Jay on John Snow and the cholera outbreak of 1854",
    http://www.youtube.com/watch?v=Pq32LB8j2K8

References

BBC News (2010). Timeline: China milk scandal. Retrieved 2nd June, 2011, from
    http://news.bbc.co.uk/2/hi/7720404.stm
Best, J. (2004). *More Damned Lies and Statistics: How Numbers Confuse Public Issues.*
    Berkeley: University of California Press.
Blue, G. (2010). Food, publics, science. *Public Understanding of Science, 19*(2), 147-154.
Brasseur, L. (2005). Florence Nightingale's visual rhetoric in the Rose Diagrams.
    Technical Communication Quarterly, 14(2), 161-182.
Brody, H., Rip, M. R., Vinten-Johansen, P., Paneth, N., & Rachman, S. (2000). Map-
    making and myth-making in Broad Street: the London cholera epidemic, 1854. *The
    Lancet, 356*(9223), 64-68.
Burns, T. W., O'Connor, D. J., & Stocklmayer, S. M. (2003). Science communication:
    a contemporary definition. *Public Understanding of Science, 12*, 183-202.
Casscells, W., Schoenberger, A., & Graboys, T. B. (1978). Interpretation by physicians
    of clinical laboratory results. *New England Journal of Medicine, 299*(18), 999-1001.
Centre for Food Safety (2008, 18th September). Press release: Latest test results of dairy
    product samples. Retrieved 2nd June, 2011, from http://www.cfs.gov.hk/english/
    press/2008_09_18_1_e.html
——(2008, 24th December). Frequently Asked Questions. Retrieved 2nd June, 2011,
    from   http://www.cfs.gov.hk/english/whatsnew/whatsnew_fstr/whatsnew_fstr_
    Test_dairy_product_FAQ.html
Centre for Health Protection (2008). New case of renal stone found in child. Retrieved
    2nd June, 2011, from http://www.chp.gov.hk/en/content/116/14213.html
Cockcroft, W. H. (1982). *Mathematics Counts.* London: Her Majesty's Stationery Office.
Cohen, P. C. (2001). Emergence of numeracy. In L. A. Steen (Ed.), *Mathematics and
    Democracy: The case for Quantitative Literacy* (pp. 23-29). Princeton, NJ: The National
    Council on Education and the Disciplines.
Crowther, G. (1959). *15 to 18: A Report of the Central Advisory Council for Education
    (England).* London: Her Majesty's Stationery Office.
Crowther, J. (Ed.). (1995). *Oxford Advanced Learner's Dictionary of Current English* (5th
    ed.). Oxford: Oxford University Press.
Friendly, M. (2008). A brief history of data visualization. In C. Chen, W. Hardle &
    A. Unwin (Eds.), *Handbook of Computational Statistics: Data Visualization III*
    (pp. 1-34). Heidelberg: Springer.
——(n.d). Data visualization: looking back, going forward. Retrieved 2nd June, 2011,
    from http://www.datavis.ca/

Gigerenzer, G. (2002). *Calculated Risks: How to Know when Numbers Deceive you*. New York: Simon & Schuster.

Lundgren, R. E., & McMakin, A. H. (2009). *Risk Communication: A Handbook for Communicating Environmental, Safety, and Health Risks* (4th ed.). Piscataway, N.J.; Hoboken, N.J.: IEEE Press; Wiley.

Nightingale, F. (1859). *A Contribution to the Sanitary History of the British Army during the Late War with Russia*. London: John W. Parker and Son (Available at http://pds.lib. harvard.edu/pds/view/7420433).

Organisation for Economic Co-operation and Development (OECD) (2009). *Take the Test: Sample from OECD's PISA assessments*. Paris: OECD.

Paulos, J. A. (1988). *Innumeracy: Mathematical Illiteracy and its Consequences*. New York: Vintage Books.

——(1995). *A Mathematician Reads the Newspaper*. New York, NY: Basic Books.

Rehmeyer, J. (2008). Florence Nightingale: the passionate statistician. Retrieved 2nd June, 2011, from http://www.sciencenews.org/view/generic/id/38937/title/ Florence_Nightingale_The_passionate_statistician

Rutherford, F. J. (1997). Thinking Quantitatively about Science. In L. A. Steen (Ed.), *Why numbers count: quantitative literacy for tomorrow's America* (pp. 60-74). New York: College Entrance Examination Board.

Shen, L. L. (2008). Fourteen babies suffered from kidney stones. *Lanzhou Morning Post*, p. A12 (in Chinese). Retrieved 2011-07-27

Sinclair, J. (Ed.). (1995). *Collins COBUILD English Dictionary*. London: HarperCollins.

Small, H. (2010). Florence Nightingale's hockey stick: the real message of her Rose Diagram. Retrieved 2nd June, 2011, from http://www.florence-nightingale-avenging-angel.co.uk/Nightingale_Hockey_Stick.pdf.

Snow, J. (1855a). *On the Mode of Communication of Cholera* (2nd ed.). London (available at http://www.deltaomega.org/snowfin.pdf).

——(1855b). *Report on the Cholera Outbreak in the Parish of St. James, Westminster, during the Autumn of 1854*. London: Churchill (available at http://johnsnow.matrix.msu. edu/work.php?id=15-78-55).

Steen, L. A. (1997). Preface: The new literacy. In L. A. Steen (Ed.), *Why Numbers Count: Quantitative Literacy for Tomorrow's America* (pp. xv-xxviii). New York: College Entrance Examination Board.

——(2004). *Achieving Quantitative Literacy: An Urgent Challenge for Higher Education*. Washington, DC: The Mathematical Association of America.

——(Ed.). (2001). *Mathematics and Democracy: The Case for Quantitative Literacy*. Princeton, NJ: The National Council on Education and the Disciplines.

Stenning, K., Lascarides, A., & Calder, J. (2006). *Introduction to Cognition and Communication*. Cambridge, Mass.: MIT Press.

Tufte, E. R. (1997). *Visual Explanations: Images and Quantities, Evidence and Narrative*. Cheshire, Conn.: Graphics Press.

van Dijk, E. M. (2011). Portraying real science in science communication. *Science Education, 95*(6), 1086-1100.

8

ETHICS AND ACCOUNTABILITY IN SCIENCE AND TECHNOLOGY

Rod Lamberts

Setting the Scene

Ethics is … closely related to politics: it is an attempt to bring the collective desires of group to bear on individuals; or, conversely, it is an attempt by an individual to cause his (sic) desires to become those of his (sic) group.

Russell, 1999, p. 139

Ethical debates—especially those about the *application* of ethics—will be intimately entwined with attempts, both implicit and explicit, to bring people around to a particular point of view. Ethical discussions revolve around assertions of what it means to behave well or badly, correctly or incorrectly. In essence, they are about appropriateness: the appropriateness of ideas and deeds, and the appropriateness of measures used to weigh up ideas and deeds.

Discussions concerning notions of appropriateness are likely to arouse passion and disagreement, as are discussions about accountability. Such discussions are also prone to interesting, but not always helpful, digression. So when presenting a chapter addressing ethics and accountability in science and technology in the context of societal communication and engagement, it is prudent to first set some parameters.

In the sciences, 'ethics' invokes at least the following core aspects:

1. Guidelines for appropriate scientific conduct,

2. The processes and structures that are involved in institutional ethics committees that govern research, and

3. Codes of conduct to which members of professional societies are obliged to adhere should they wish to maintain formal affiliation with the society and preserve the advantages that such affiliation bestows.

These are central to the formal rules and procedures with which scientists must generally comply if they want be seen as scientists. While relevant to the consideration of ethics and accountability in science and technology, it is not these elements that form the focus of this chapter. 'Ethics' is considered more broadly here, in the spaces where science, society, and communication come together in the public sphere. While I will cover some formal—or *formalised*—articulations of ethics, my focus is on the complexity inherent in judgements of accountability made in messy, real-world situations: science as it exists beyond the laboratory and the halls of academe.

Essential also in the context of this book is consideration of the relationship between science and science communication. Should science and science communication be held similarly accountable, and if so, in what contexts? Do they have the same potential to help and hinder? Should they be seen as distinct, yet related, pursuits?

To address this manageably, this section and the next will focus on *science*, ethics, and accountability. Matters concerning the relationship between science and science communication in the context of ethics and accountability will be considered in the last section.

A Brief Note on Terms

'Science' and 'technology' can represent different concepts depending on the nuances under consideration. For the purposes of this chapter however, such distinctions add little. Science and technology will be referred to interchangeably, usually as 'science'. Accountability, responsibility and obligation are, for all intents and purposes here, considered to be synonyms.

Accountability, Ethics and Science

How do we decide for what, and to whom, science and scientists are accountable? And once we have decided this, how might we determine if those obligations or responsibilities have been met?

In making judgements about the appropriateness of the actions of scientists and the processes and outputs of science, we are deciding the extent to which we believe them to be right or wrong. Sometimes it's easy to choose whether something is right or wrong. At other times, much depends on your point of view and the context in which the judgement is being made. Often a simple dichotomy is insufficient, and we need to place things on a continuum from right to wrong, good to bad, necessarily recognizing that much is not clear-cut.

Ethical theories provide systematic and explicit ways to make these judgments. But what are ethics? The study of ethics involves "rational enquiry into, or a theory of, the standards of right and wrong, good and bad, in respect of character and conduct, which ought to be accepted by a class of individuals"

(Maurer, 1997, p.180). In the case of science, 'ethics' tends to refer to notions of appropriate behaviour, frequently made explicit in the codes of conduct of professional societies, or in ethical guidelines that are required by research bodies.

A short note on ethics and laws is also useful here. While laws have ethical substrates, a critical distinction is that laws are something to which all citizens *must* adhere if they wish to avoid punishment meted out under the auspices of the state: we are all obliged to obey the laws of the society in which we live or suffer state-sanctioned consequences. On the other hand, ethics are guidelines to which people can choose to adhere as befits their position or aspirations. The ramifications of being deemed to have behaved unethically are highly dependent on the context in which a behaviour is being judged, and by whom.

Right or Wrong—Common Positions on How to Make Ethical Choices

For centuries philosophers have debated detailed conceptual frameworks that formulate the most meaningful, representative, and practical bases for making ethical judgements. Table 1 summarizes the most commonly proffered and well-debated positions. These provide theoretical and cognitive tools for making decisions about accountability in the sciences and science communication.

TABLE 8.1 Common Ethical Theories and Heuristics for Making Personal Choices

Some common ethical theories

Deontology	Right and wrong is decided against objective moral duties. The right or wrongness of an act is not determined by its consequences, but by the extent to which acts align with duty. Example: 'do the right thing' (regardless of outcome).
Consequentialism	The only way to make judgements of the right or wrong of an action is on its consequences. This can get trickier when we include foreseen and unforeseen consequences. Example: 'the ends justifying the means'.
Intuitionism	As it sounds, using one's immediate reactions or intuition to judge right or wrong. Example: 'going with your gut'.

Some common ethical theories

Normative	Standards of good and bad, right and wrong for a group. This group could be anything: a professional group, a society, a religion. This describes how people within this group ought to think and behave (not necessarily how they *actually* think and behave). The main aim of this branch of ethics is the consideration, study or formulation of norms about conduct and appraisals of character that can be deemed valid. Pursuits into normative ethics often consider ethics in practical situations.
	Examples: it is wrong to plagiarise in an academic environment. Adherence to the Christian 10 commandments is right.
Utilitarianism	Choices between actions are based on creating the greatest number of positive states for the greatest number of people. Places the needs of the many over the needs of the few. What constitutes a 'positive state' is contestable.
	Example: 'taking one (making a sacrifice) for the team'.

Some heuristics for making personal ethical choices

Golden rule	Choosing the action that would result in outcomes that you would be happy to have happen to you. Probably best known as 'do unto others as you would have them do unto you'.
Mentor test	Deciding what to do based on whether you would be happy if your action or decision were judged by an esteemed person in your life such as a parent, teacher, or leader.
News (or publicity, or broadcast) test	Imagining how comfortable you would feel if what you were intending to do were broadcast to your community.
Mirror test	Asking yourself if what you were intending to do would cause you shame: could you look at yourself in the mirror without shame if you acted in the manner you are considering?
Role model test	Considering how you believe a personal role model would act in the situation you find yourself. This is often represented in American popular culture as "what would Jesus do".
"Promise keeping"	Reflecting on your intended action to see if by doing (or not doing) it, are you breaking a promise. A promise doesn't have to be explicitly made. It can be implicit, like certain behaviours associated with jobs.

Reference will be made to these theories and heuristics periodically throughout the rest of this chapter. In doing this, the intention is to provide examples of how ethical constructs might apply to situations, not to exhaustively apply them all, nor to imply that the interpretations presented here are the only ones possible. I strongly encourage discussion, debate, and critique of these examples.

Ethical Relativism, Ethical Subjectivism and Self-Interest

Ethical relativism adopts the position that judgements about that which is right/wrong, appropriate/inappropriate, or ethical/unethical are only meaningful relative to the culture in which such judgements are made. This can be appealing in our post-modern world, especially when doing studies in comparative culture. However appealing this view might seem in principle though, it is not particular useful—nor realistic—when trying to formulate practical guidelines for how to behave ethically.

According to Singer (2011) the idea that ethics should only be viewed from the relative perspective of a particular society "has most implausible consequences". Here is one of his examples—animal experimentation—to illustrate this:

- Society A (my society) disapproves of animal experiments
- Society B (another society) approves of animal experiments
- My saying animal experimentation is wrong is really just saying my society (society A) disapproves of animal experimentation
- Someone from society B saying animal experimentation is right is doing the same thing
- Both of us are expressing the truth, from an ethically relativist position, so there is no point in contesting, because there is no contest.

This kind of reasoning gets us nowhere in trying to apply ethics to practical situations.

When you move from ethical relativism to ethical subjectivism (that all ethical positions are subjective, not just those based on societal or cultural differences), a significant difficulty with the application of this broadest articulation of ethical relativism, as Singer points out, is its "... inability to account for ethical disagreement" (p.7).

With ethical subjectivism, if I say "animal experiments are wrong", that is no more or less true than you saying "animal experiments are right". From a subjectivist position, both statements are true. Not only is there no factual disagreement to resolve, there would be no ethical way to resolve it if there were.

We often speak about ethics as if there exist some external, objective standards of right and wrong against which to judge behaviours. In reality though, in doing this we are making assertions about our personal beliefs, desires and preferences, and masquerading them as ethical debate. Singer (2011), however, asserts that any argument justifying behaviour in terms of ethical standards cannot be made purely in terms of self-interest. Behaviour that advantages one individual and no other is not amenable to support based on ethics because:

> Self-interested acts must be shown to be compatible with more broadly based ethical principles if they are to be ethically defensible, for the notion of ethics carries with it the idea of something bigger than the individual. If I am to defend my conduct on ethical grounds, I cannot point only to the benefits it brings me. I must address myself to a larger audience.
>
> *(Singer 2011, p.10)*

So ethical positions require some consideration of the effects an action may have on others. Singer notes that a common aspect of ethics as articulated over the centuries is that

> … ethics is in some sense universal … the justification of an ethical principle cannot be in terms of any partial or sectional group. Ethics takes a universal point of view. This does not mean that a particular ethical judgment must be seen as universally applicable. Circumstances alter cases. … What it does mean is that in making ethical judgements, we go beyond our own likes and dislikes.
>
> *(Singer 2011, p.11)*

This position—that for an argument to be 'ethical' it must at least include considerations beyond the self—forms one of the two ethical generalisations that apply to the rest of this chapter.

The other generalisation is this. When we apply ethics to considerations about the appropriateness of science and science communication, what is most practically important is being aware of the criteria we personally employ to make our judgements. Agreeing or disagreeing is less important than being able to articulate, and defend, one's position.

To What, or Whom, are Scientists Accountable?

Arguably, a scientist is first and foremost accountable to science itself, which in practice is represented by scientific peers: their community of practice. Scientists are at a minimum obliged to not fabricate, falsify or plagiarise, the three big evils of scientific dishonesty (see for example, Martinson, Anderson & de Vries, 2005), and must subject their work for validation via peer scrutiny.

But the way society might hold science accountable for its actions, or inactions, is not always congruent with the scientific community of practice. This section presents some instances where the world beyond science may hold science or scientists accountable for their decisions and actions. The examples are predominantly drawn from climate science. Throughout this section, consider the extent to which science and scientists should be held accountable, and how you might form an ethically robust position for judging the appropriateness of the suggestions being made.

Informing, Warning, or Helping Society

At face value, it sounds reasonable to assert that science has an obligation to inform choices in society, and can/should be held accountable based on the extent to which it does this. Ethically, this is perhaps a deontological position: science has a duty to society. The mechanisms and processes via which society might hold science accountable, however, are far from straightforward, as are the circumstances under which they should do so. Opinion also varies about what might constitute an acceptable penalty, should scientists fail a test of accountability.

Turning to climate change, Pidgeon and Fischhoff (2011) are unambiguous in discussing scientists' obligations to society: "Climate scientists bear a heavy burden: potentially, the fate of the world lies partly in their hands." (p.35). They continue: "Whatever the barriers [to action on climate change] might be, scientists' obligation is to provide the information that is necessary, if not sufficient, for informed choices." (p.38). In declaring this obligation, arguably these authors are implying there is an implicit promise made between science and society.

For Grothman and colleagues (2010) "Although several factors contribute to societal inaction in resolving environmental problems, scientists must recognise that they share part of the blame." (p.284). According to these authors, scientists have an obligation to inform us and must accept some of the blame when we fail to act on their information.

Former Caltech president and Nobel Laureate David Baltimore is emphatic about the obligations (and also the rights) of scientists in general, and in the climate discourse in particular (Baltimore, 2006). Speaking in the context of increasing discord among scientists in the US in 2006, with specific reference to cases where NASA climate science was ignored and muzzled by Bush administration policy agendas, Baltimore argues that

Instead of resignedly shrugging their shoulders whenever such a case of scientific manipulation arises ... scientists need to recognize the potency of the threat this governmental philosophy represents to the long-cherished independence of US science.

(Baltimore 2006, p.891)

By 'recognize', Baltimore means act. He is throwing down the gauntlet, in essence saying that scientists have an obligation to do something about the situation. He is also suggesting (but *only* suggesting) they should be held accountable somehow if they do not. There is an implication of punishment and accountability here, perhaps considered defendable under a utilitarian ethic?

These examples argue that scientists working in the climate change arena have obligations to society, and indeed to science itself. If we accept that such responsibility exists, what then? Should particular scientists be held accountable if they do not meet such obligations? And if they are, to make such accountability meaningful, there should be consequences if obligations are not met. Should we be satisfied with somehow reprimanding 'science' for not meeting such obligations, or do we go even further?

Should we, for example, take it as far as the Italian legal system did in 2011 when scientists from The Great Risks Commission were put on trial for manslaughter because a judge determined they

> gave inexact, incomplete and contradictory information about whether smaller tremors in L'Aquila six months before the 6.3 magnitude quake on 6 April [2009], which killed more than 300 people, should have been viewed as warning signs of the subsequent disaster.
>
> *(Batty, 2011)*

Did the Italian legal system take it too far? Regardless of one's position, this raises questions about how scientific advice might be incorporated into society as a whole. How might we judge what science and scientists have contributed, the extent to which they are obliged to do so, and what should happen if they don't?

If we take a normative view, then we need know how scientists as a group believe, and articulate, the way they *ought* to behave, and judge them accordingly. Alternatively, might it be better to trust our 'gut', and decide using our intuition, or first reactions to the appropriateness of an act? This might allow for a more flexible approach to judgements of right and wrong, based on individual cases. It is never straightforward.

Politics and Policy

In these early decades of the 21st century, science and politics are possibly nowhere more intimately entangled than in the world of climate change. In this context, Maibach and Priest (2009) are very direct: "Gone are the days when science could pretend to operate independent of social policy or in a realm miraculously devoid of politics" (p.300). With the stakes so high, and the arguments so fierce, examples in this realm exemplify crucial issues in the conversation about ethics and accountability in science (and by association, science communication).

For Pidgeon and Fischhoff (2011), "Realizing the practical value of climate-related research means ensuring that diverse policymakers and the public understand the risks and uncertainties relevant to the decisions that each faces." (p.55). Much climate-related research is of practical value to policy-makers and through them, society at large. Policy-makers necessarily have to rely on scientists not only to inform them, but to do so in ways that are relevant and intelligible. Compelling though such assertions may be, it is complex and difficult to judge the extent to which scientists meet their obligations to inform policy-makers.

As Will Grant describes in Chapter 3, Roger Pielke (2007) outlines four idealised roles science might play in policy and politics. In theory, once we decide which role we think is most appropriate, we have a starting position from which to decide for what scientists should be held accountable.

Is it really so simple though? For example, should the role occupied by scientists change depending on the issue at hand? Is it ever all right to be more than a Pure Scientist? Perhaps you agree with Pielke that in the climate change space, scientists have an obligation to become Honest Brokers? Is it in fact possible to be such a creature, especially if the scientists have strong emotional connections to the issue they are representing and communicating to political actors?

Clearly, debating the extent and types of responsibility science has in the political and policy world raises many ethical questions. Pielke offers what is essentially a deontological approach, defining ideal roles by considering how scientists might most suitably behave. Discussion about the relative merits of these roles highlights in particular the question of the appropriateness of advocacy and persuasion, not just with policy-makers, but with society at large.

Persuasion, Advocacy and Framing

Persuasion involves presenting information with the intent of influencing an outcome, pushing an agenda, or forcing an opinion. Advocacy is related—though it tends to involve promoting a *single* particular interest or position—and encompasses elements of persuasion. The distinction in practice may not always be clear, and in the literature referred to in this chapter, the terms are often used interchangeably.

Framing is more nuanced than persuasion and advocacy. It can be used persuasively and might be applied when advocating, but is predominately a communication technique which serves to enhance the understanding of facts and ideas by varying the emphasis of a message. It differs from 'spin'—something PR, advertising, and marketers may use—in that spin actively aims to misrepresent.

When advising policy-makers and the public at large, is there room for persuasion and advocacy by, or on behalf of, science and scientists? Is it ever ethical to step beyond the Pure Scientist role, even for policy areas of such great concern as climate change? Fischhoff (2007) thinks not, arguing that scientists should aim to be non-persuasive in public climate communications. According to him,

Nonpersuasive communication lets the science speak for itself. It recognizes that reasonable individuals may reach different conclusions—even if it is undertaken in the hope that most individuals will make similar, desired choices (e.g., commitment to energy efficiency). If it fails, then persuasive communication may be needed. However, such advocacy comes at a price, turning scientists into peddlers rather than arbiters of truth. Advocacy must be very effective to compensate for eroding scientists' status as trusted observers and reporters.

(Fischhoff 2007, p.7208)

His main concern with scientists getting involved in public advocacy is that it means taking, and defending, a position following the norms of politics rather than those of science. While public advocacy might include basing arguments on facts, it does not necessarily mean using *all* the facts, or giving appropriate weight to all facts that are relevant to the position being defended. For Fischhoff, the danger when getting deeply involved in public advocacy is "... winning battles over what science *says*, while losing the war over what science *is*" (p.7205). [emphasis added]

On critical matters such as climate change, should scientists be so coy? We could argue that they are failing in their responsibility to society if they do not try to push for courses of action that address problems identified by their research. Lynas (2005), for example, acknowledges that public advocacy can diminish the position of trust people afford scientists. However, in the case of big issues like climate change, he suggests that

scientific objectivity should not condemn scientists to the political equivalent of a Trappist vow of silence. If our future is under threat, scientists have a duty to say so. Where, for example, is the volcanologist who refuses on the grounds of 'neutrality' to warn of impending eruptions? Or the tropical meteorologist who sits tight-lipped with his models and satellite data while a Category 5 storm spins towards a vulnerable Caribbean coastline?

(Lynas 2005, p.25)

So do we adopt a more consequentialist position, arguing that it is the severity or magnitude of the issue and its effects, rather than a person's behaviour as a scientist, that should dictate how we judge appropriate ways to act? In severe cases, do the ends justify the means? Perhaps so, but then, what constitutes "severe enough"?

Lynas' position becomes more compelling in examples like those proffered by Oreskes and Conway (2010) who cite decision theory in explaining why the US has generally failed to act on climate change. If knowledge is—or is perceived to be—uncertain, the best option (and most likely choice) is to maintain the status quo. They go on to argue that although science doesn't provide certainty, it has formed a solid basis for decision-making in many areas of human endeavour (like medicine and space-exploration). Therefore they believe that advocating for action based on the accumulated weight of climate science is necessary to push past the public perception of the evidence as uncertain, in order to overcome societal inertia against changing the status quo. Clearly they believe scientists have a responsibility to speak up and take a stance.

This position is supported in the vast literature on ethical risk communication. Lundgren and McMakin (1998) present three circumstances under which the use of persuasion is not only justified, but required. For them, one should use persuasion when:

1. some members of an audience are in immediate danger of injury or death;
2. people who may be at risk are not the ones who have control of the action or activity that exposes them to it (e.g., pregnancy smoking and the fetus); or
3. people in the audience have asked to be persuaded.

All of these apply to the climate change scenarios painted by the overwhelming weight of scientific evidence, so does this mean the normative ethics of risk communication should be applied to assessments of accountability among scientists? If overt persuasion or advocacy seem too much, we still have 'framing': communication that *can* include all the facts, but applies nuance according to the emphasis placed on the presentation of these facts.

Framing means that the same evidence can be presented through different filters. For example, again in the climate change context, Hulme (2009) suggests there are six major frames through which climate change is presented in public discourse. These are climate change as:

1. market failure;
2. technological risk;
3. global injustice;
4. over consumption;
5. mostly natural, and;
6. planetary "tipping points".

Each of these represent ways into the climate discourse, emphasising different elements, aspects or angles depending on the agenda, or the interests, of the audience (Hulme, 2011). They all can legitimately lay claim to calling on the weight of relevant scientific evidence to support their position, with message framing being the critical difference.

Applying appropriate frames can put climate change science information in terms that are more meaningful to those with whom one seeks to communicate. Groffman, et al. (2010) go further, arguing that scientists and communicators have an *obligation* to research, understand, and then apply useful frames. For example, this could mean reframing climate challenges as opportunities, such as new technology for clean energy, or the potential for economic growth.

Perhaps this foray into persuasion, advocacy and framing leaves you thinking that the position of the pure scientist is the most honest and least harmful way for science to support decision-makers. It is seductive to argue that the most ethical path to take is to be open about science processes, upfront about their failings, and hand-over the information to decision-makers to do with it as they will.

Openness and Sharing

In 2005, a *Nature* editorial presented a strong statement concerning openness, interacting with the media, and the role of scientists on matters of scientific uncertainty (Baltimore, 2005). The editorial pushed back against assumptions by political leaders that people cannot cope with uncertainty and that presenting scientific evidence as uncertain might strike panic among the public or induce a state of decision paralysis. *Nature* argued that leaders should "assume a degree of maturity on the part of the media and public, and represent the state of the science, risks and all" (p.1).

The editorial went on to assert not only that scientists had an obligation to help journalists with analyses that were "for public consumption", but further, that

learned societies [such as the Royal Society] have a particular responsibility in such circumstances. They should not wait to be called on by government to provide analysis. Rather, where the stakes for public interests are high, they should promptly convene a working group charged with delivering a statement of the state of relevant evidence, within a few days if necessary. Above all, they should provide the media with succinct statements and readily reproducible graphics that clearly illustrate not only the conclusions drawn by the working groups but also the state of the uncertainties.

(Baltimore 2005, p.1)

While laudable—and perhaps appealing—many complexities are not addressed here. The statement is at best idealistic, at worst, politically naïve. It says nothing about how such a situation could be realised. Should the community, for example, hold learned societies accountable based on their track record of identifying critical areas of scientific uncertainty and then preparing scientifically robust materials that are also easily used by media outlets and digested by non-scientific publics? If so, what happens to those societies that are found wanting? Should they lose funding, or perhaps be publicly shamed?

Ethical quandaries abound even in apparently more straightforward situations that only consider small groups of people. For example, critical questions arise when obligations in medical research require that results are supplied to participants after they participate in experimental trials. MacNeil and Fernandez (2006) suggest that the potential benefits of doing this include:

- positive impacts on the future health of participants;
- increased public awareness of the research and its outcomes; and
- demonstrating to participants the direct results of their participation.

There are negative impacts as well, as noted by Dixon-Woods, Jackson, Windridge and Kenyon (2006). For example, participants may:

- be reminded of a difficult time in their lives;
- find out they were in the control group, or that the element of the trial they were in didn't work; or
- discover they are at increased risk of health problems.

So even something apparently straightforward and overtly reasonable as offering people the results of study in which they participated can be ethically fraught when trying to determine the right course of action, and how best to judge its success.

The Public Face of Science

There are those who suggest that one way to inspire and maintain trust in science is to be upfront about science's uncertainties and problems, as well as its more powerful conclusions (see for example Pidgeon & Fischhoff, 2011 or Schneider, 2009). In considering how science might best do good for society, and how it might be held accountable for its actions, even this is problematic. There are myriad misbehaviours perpetrated by scientists beyond the standard "fabrication, falsification and plagiarism" categories of scientific misconduct, as Martinson, Anderson and Vries (2005) reveal. These authors surveyed more than 3200 early and mid-career scientists in the USA, and their respondents admitted to many unethical activities including:

- excluding data from analyses based on 'gut feeling'
- caving in to pressure from funding sources to make methodological changes; and
- withholding methods and/or result details in publications.

Worryingly, the results of this study made it into the mainstream media, and not just in the US. In Australia, the Australian Broadcasting Corporation under the heading "One-third of scientists admit fudging research" (ABC, 2005) gave a quick précis of Martinson and colleagues' findings, including a comment by a senior Australian academic playing down the findings: "Some of the behaviours that are described here are serious but I think quite a large number of them are probably not as bad as they sound". So what are people to make of this?

Martinson et al. (2005) provide cases of scientists behaving unethically in the context of their responsibility to science both as a process *and* as a community of practice, but what about scientists' overarching responsibility to society at large? For example, in responding to the survey, did participants do the right thing by science, or indeed, society? Arguably, exposing these 'white lies' tainted the reputation of science in general, and may render robust, credible, science-based positions presented in the public eye less trustworthy. Of course, it may also have made science—and in particular *scientists*—seem more human, less remote, and therefore possibly more sympathetic and *more* trustworthy. Either is possible, but how do we judge what was the right thing to do in this situation? A deontological position would nudge towards openness regardless of costs to science. A consequentialist view might encourage silence. Then of course, the perspectives of individual scientists could vary depending on many factors, some of which are outlined in the table in the opening section of this chapter.

Returning to climate change, Nisbett (2010) believes that "Climate change initiatives can and should contribute meaningfully to feelings of trust and efficacy". Pidgeon and Fischhoff (2011) note (as do Groffman et al., 2010) that few institutions are as trusted as science, and with this trust comes an obligation not to betray it. In keeping with these sentiments, Schnieder (2009) writes of the obligation of scientists to state, should the platform allow it, where their scientific assessment of risk ends and their personal values come in.

So once more there are decisions to be made, and the consequences are not always clear. Worse, the same actions may have both positive *and* negative effects, possibly in equal measure, depending on how they are judged.

Ethics, Accountability, Science, and Communication

Arguably, the process of science communication is a facilitator of science's obligations to society, identifying where the needs are, to whom particular sciences will be more relevant and how best to connect relevant science with those who need it, or whom science needs. In providing such facilitation, effective, evidence-based science communication enterprises put science through an ethical and reality filter. That is, identifying and defining what is desirable and realistic in the public space, what is appropriate, and how appropriateness might be judged.

There is no magical formula for this, nor will every situation lend itself to using the same criteria for assessing accountability, or the extent to which obligations have been met. But if there is one thing science communication research and experience provides, it is information about publics, audiences, and people.

Science communication research in particular addresses a critical obligation inherent in Pidgeon and Fischhoff's (2011) reminder: "There is no way to know what information people need without doing research that begins by listening to them. Such listening protects against simplistic solutions to complex problems." (p.39). Such research could equally be about publics, policy-makers or scientists themselves.

Groffman and colleagues (2010) also appear to be calling for science communication research and practice here:

The public and decision makers need more than information and technical knowledge—they need mental frameworks, or models, for "connecting the dots" between otherwise apparently isolated events, trends, and policy solutions. These linkages make it easier for them to recognize the connection between their everyday lives, specific values, and various environmental problems.

(Groffman, et al., 2010, p.287)

Once more, the domain of the science communication researcher and practitioner is indicated—though not by name—in this instance as 'connecting the dots'. Is this then an implicit obligation incumbent upon the community of science communicators? Are we in fact the ones who should be held accountable when science fails to meet public expectations or requirements? And is it via the application of ethical theories that we might best be judged?

Miah (2005) goes so far as to say that communicating science with non-science audiences *is* in fact communicating ethics. By this he means that it is far more the ethical, rather than the technical, dimensions of science with which people engage in the non-expert domain. According to Miah

... one has only to pick up any piece of scientific journalism to appreciate that the salient aspects of science, technology and medicine have less to do with the technical details, than with the ethical, social, moral, and political implication of that technology. The public(s) are concerned primarily for what science 'means', rather than for how it 'works'.

(Miah 2005, p.416)

This position contrasts profoundly with many of those presented earlier in this chapter from the perspective of the sciences. Miah argues for nuanced, interpreted versions of science to be the primary goal of science communication in public spaces rather than merely the non-persuasive provision of science facts. But he asserts strongly that scientists are not the ones to do this. He believes that ethicists are the best people to communicate ethics, just as scientists are the best people to communicate technical details of their science. Extending Miah's argument that experts should be employed to communicate in their areas of expertise, perhaps once more we should see science communicators as the ones to be held accountable for the effects of public-science interactions?

This view, though interesting, is extreme. Nevertheless, what is being reinforced directly and implicitly by many of the authors referred to in this chapter is that for science, appropriately fulfilling its contestable obligations to society is not something that can be done by scientists alone.

In the climate arena, Pidgeon and Fischhoff (2011) advocate for teams made up of relevant coordinated experts (for example climate scientists, decision scientists, program designers, and communication and social science specialists) as the best way to be effective, and that their

> … coordination must maintain a rhetorical stance of non-persuasive communication, trusting the evidence to speak for itself, without spin or colouring. Although there is an important place for persuasive communication, encouraging individual behaviours and public policies, it must be distinct, lest scientists come to be seen as inept politicians. If climate scientists passionately offer dispassionate accounts of the evidence, it will preserve their uniquely trusted social position and avoid the advocacy that most are ill-suited to pursue by disposition, experience and training.
>
> *(Pidgeon & Fischhoff 2011, p. 39-40)*

From this we could suggest that the science communicator has a responsibility to engage in persuasive, science-based debates in ways that ill-befit scientists, but are still sustained by science at their core. In this case it would be incumbent upon scientists to engage with science communicators when their evidence suggests matters of societal concern. Considering science and scientists' responsibilities to society can be a theoretically interesting and engaging exercise but pragmatism is required when communicating science in situations where protagonists' interests, priorities and ways of describing and interacting with the world differ from those of the sciences. Specifically, what is required is pragmatism in the way science is incorporated into the greater societal landscape, and with this, pragmatism in the way science communication enterprises are conceived, conducted, and evaluated.

Using ethical benchmarks to decide what science is accountable for, and the extent to which it has met obligations according to that with which it is being held accountable, will only be helpful to some degree. No single ethical position will be appropriate for defining or assessing the appropriateness of the acts of science, scientists and science communicators *sui generis*. Context is critical, and the goals of the science or communication enterprise must be taken into account, as must the perspectives from which obligations and responsibilities are being considered.

Formal categories and theories from the field of ethics offer us systematic tools for making decisions about the extent to which science and science communication might be meeting their obligations to the world. However, in *applying* ethics to real world situations, practicalities should be borne in mind. If the application of ethics becomes mired too deeply in theoretical debate, there will be no action.

Science communication has a potent role to play in all manner of societal-scientific discourses but science and science communication will do this best only by recognizing where their strengths and weaknesses lie and partnering with other experts to make up for their own shortcomings.

References

ABC. (2005). One-third of scientists admit fudging research. *ABC News online.* Retrieved June 28, 2011, from http://www.abc.net.au/news/newsitems/200506/s1388075.htm.

Baltimore, D. (2005). Editorial: Responding to uncertainty. *Nature, 437*(7055), 1.

——(2006). Editorial: Science under attack. *Nature, 439*(7079), 891.

Batty, D. (2011). Italy earthquake experts charged with manslaughter. *The Guardian.* Retrieved from http://www.guardian.co.uk/world/2011/may/26/italy-quake-experts-manslaughter-charge.

Dixon-Woods, M., Jackson, C., Windridge, K., & Kenyon, S. (2006). Receiving a summary of the results of a trial: a qualitative study of participants' views. *British Medical Journal, 332*(January), 206-209.

Fischhoff, B. (2007). Nonpersuasive communication about matters of greatest urgency: climate change. *Environmental Science & Technology, 1*, 7204-7208.

Groffman, P., Stylinski, C., Nisbet, M., Duarte, C., Jordan, R., Burgin, A., et al. (2010). Restarting the conversation: challenges at the interface between ecology and society. *Front. Ecol Environ, 8*(6), 284-291.

Hulme, M. (2009). *Why we disagree about climate change: understanding controversy, inaction and opportunity.* Cambridge: Cambridge University Press.

——(2011). You've been framed: six new ways to understand climate change. *The Conversation.* Retrieved July 5, 2011, from http://theconversation.edu.au/youve-been-framed-six-new-ways-to-understand-climate-change-2119.

Lundgren, R. E., & McMakin, A. H. (1998). *Risk communication: A handbook for communicating environmental, safety and health risks* (2nd ed.). Columbus, Ohio: Battelle Press.

Lynas, M. (2005). Get off the fence. *New Scientist,* 25.

MacNeil, S. D., & Fernandez, C. V. (2006). Offering results to research participants is ethicaly right but not fully explored. *British Medical Journal, 332*(January), 188-189.

Maibach, E., & Priest, S. H. (2009). No more "business as usual:" addressing climate change through constructive engagement. *Science Communication, 30*(3), 299-304.

Martinson, B. C., Anderson, M. S., & Vries, R. de. (2005). Scientists behaving badly. *Nature, 435*(9 June), 737-738.

Mautner, T. (1997). *Dictionary of philosophy* (2nd ed.). London: Penguin.

Miah, A. (2005). Genetics, cyberspace and bioethics: why not a public engagement with ethics? *Public Understanding of Science, 14*, 409-421.

Nisbett, M. C. (2010). Investing in civic education about climate change: what should be the goals? *Big Think blog.* Retrieved May 5, 2011, from http://bigthink.com/ideas/24578.

Oreskes, N., & Conway, E. M. (2010). *Merchants of Doubt.* New York: Bloomsbury Press.

Pidgeon, N., & Fischhoff, B. (2011). The role of social and decision sciences in communicating uncertain climate risks. *Nature Climate Change, 1,* 35-41. Retrieved from www.nature.com/natureclimatechange.

Pielke, R. A. (2007). *The honest broker: Making sense of science in policy and politics.* New York: Cambridge University Press.

Russell, B. (1999). Science and ethics. In C. Pigden (Ed.), *Russell on ethics.* New York: Routledge.

Schneider, S. (2009). *Science as a contact sport: Inside the battle to save the earth's climate.* Washington D.C: National Geographic Society.

Singer, P. (2011). *Practical Ethics* (3rd ed.). New York: Cambridge University Press.

9
BELIEFS AND THE VALUE OF EVIDENCE

Michael J. Reiss

Background

Science, above all, is about the production and testing of empirically grounded theories that have objective validity and generate knowledge that is of value both for enhancing our understanding of the material world and for enabling us to control and harness aspects of it. Such a straightforward notion, though, led almost inevitably to the deficit model of misunderstandings of science which, while rejected in the science communication community (Holliman, Thomas, Smidt, Scanlon & Whitelegg, 2009), continues to be the first port of call for many scientists and for the general public alike.

At the same time, the continuing influence of religion on public life has led many in the science community to seek to distance science from religion. A common refrain among such scientists is to state that belief has no part in science, being instead a peculiar and unimpressive element of the religious life (e.g. Dawkins, 2006).

The theory of evolution provides a distinctive stage on which to investigate the importance of belief in science. By examining why some individuals find the scientific consensus concerning evolution attractive, while others reject it so decisively, one can study the extent to which objective evidence is sufficient to persuade people of the truth of the standard scientific account. My intention is to produce an analysis that benefits all those concerned with science communication about evolution in today's diverse societies, whether in schools, in science museums, or in religious settings.

The Importance of Belief for Religion and for Science

The word 'belief', as with almost all words that are worth exploring in any depth, has a number of related meanings and uses (cf. Austin, 1979; Kenny, 1992; Wittgenstein, 1953). Fundamentally, though, it is about acceptance (both intellectually and in terms of implications for practice). 'I don't believe it is going to rain today' means that I do not accept that such an event is at all likely to occur, and that I am unlikely to take precautions against wet weather. Of course, a belief may be held with varying degrees of conviction. Although, as I write this (14 January 2012 in Cambridgeshire, UK), I do not, in fact, believe that it is going to rain today (it is a sunny afternoon and the BBC weather forecast predicts no rain), I wouldn't wager my academic reputation on the assertion, whereas I would with regards to the propositions 'Fairies do not exist', 'There is no validity in horoscopes', and 'All humans are the descendants of much simpler and smaller organisms that lived over a thousand million years ago'.

Consider now the statement 'God exists'. This is not (really) a scientific statement. Although some people uttering such a statement might be prepared to accept that certain empirical circumstances (a life of continued misery, an absence of any personal divine revelation or answers to petitionary prayers, a growing conviction that there is nothing before, behind, or beyond that that is material and self-evident) would eventually lead to its refutation, others uttering such a statement would continue to hold it whatever evidence failed to support it or apparently refuted it (cf. the dark night of the soul). Now consider the statement 'The Quran was revealed to Muhammad by the angel Gabriel'. In principle this statement, as with any historical statement, admits of empirical support or refutation. The reality, though, is that the events in question lie so far back in time, from 610 to 632 CE, that there is no real chance of their veracity being objectively tested, so acceptance of the truth of the statement tells us more about people's Muslim beliefs than whether or not the events actually occurred.

We can conclude from this that the words 'belief' and 'believe', as in 'I believe in God' and 'I believe in the theory of evolution', are used in non-identical ways in science and in religion, even if there may be some overlap (the overlap being the reason why '(really)' appears in my assertion above that 'God exists ... is not (really) a scientific statement'). This is hardly surprising. Indeed, statements of belief in other areas of knowledge often cannot be reduced to scientific statements. I believe that the internal angles of any flat triangle add up to $180°$ but the truth of this statement is arrived at differently in mathematics (i.e. through logical proof—cf. Euclid's *Elements*, Book 1, Proposition 32) than it would be if it were a scientific statement along the lines 'All vertebrates have four limbs', to test which one would look at large numbers of vertebrates. In mathematics, it doesn't help (except when teaching pupils about the truth of the proposition) to corral large numbers of triangles and then carefully measure and sum their internal angles. Moral philosophy is another example of a discipline where conclusions are arrived at non-scientifically. While science plays the central role in determining the age at which a developing human fetus can feel pain, science alone doesn't enable one to determine the best abortion policy.

The Scientific Consensus Concerning Evolution

As with any large area of science, there are parts of what we might term 'front-line' evolution that are unclear, where scientists still actively work attempting to discern what is going on or has gone on in nature. But much of evolution is not like that. Evolution is a well-established body of knowledge that has built up over 150 years as a result of the activities of many thousands of scientists. The following are examples of statements about evolution that lack scientific controversy:

- All of today's life on Earth is the result of modification by descent from the simplest ancestors over several thousand million years;
- Natural selection is a major driving force behind evolution;
- Evolution relies on the inheritance of genetic information that helps its possessor to be more likely to survive and reproduce;
- Most inheritance is vertical (from parents), though some is horizontal (e.g. as a result of viral infection);
- The evolutionary forces that gave rise to humans do not differ in kind from those that gave rise to any other species.

For those, such as I, who accept such statements and the theory of evolution, there is much about the theory of evolution that is intellectually attractive. For a start, a single theory provides a way of explaining a tremendous range of observations: for example, why it is that there are no rabbits in the Precambrian, why there are many superficial parallels between marsupial and placental mammals, why monogamy is more common in birds than in fish, and why sterility (for example, in termites, bees, ants, wasps, and naked mole rats) is more likely to arise in certain circumstances than in others. Indeed, I have argued elsewhere that evolutionary biology can help with some theological questions, including the problem of suffering (Reiss, 2000).

Rejecting Evolution

The theory of evolution is not a single proposition that a person either wholly accepts or wholly rejects. Consider, for example, the issue as to whether all of today's life arose from inorganic precursors. I have no doubt (my language is intentional) that it did, but I suspect (we lack decent social science survey data) that worldwide this is a minority opinion. Comparably, creationists accept microevolution (the genetic changes that happen over time within populations), but may not accept macroevolution (genetic changes that happen at and above the level of species).

There is a whole set of non-religious reasons why someone may actively reject aspects of the theory of evolution. After all, it seems to defy common sense to suppose that life in all its complexity has evolved from non-life. And then there is the tremendous diversity of life we see around us: it hardly seems reasonable to presume that koala bears, great white sharks, earthworms, grasses, flesh-eating bacteria and the editors of this book all share a common ancestor—yet that is what mainstream evolutionary theory holds.

It is, though, for religious reasons that many people reject evolution. Creationism exists in a number of different forms, but something like 50% of adults in Turkey, 40% in the USA, and 15% in the UK reject the theory of evolution and believe that the Earth came into existence as described by a literal (i.e. fundamentalist) reading of the early parts of the Bible or the Quran, and that the most that evolution has done is to change species into closely related species (Lawes, 2009; Miller, Scott & Okamoto, 2006). Christian fundamentalists generally hold that the Earth is nothing like as old as evolutionary biologists and geologists conclude—as young as 10,000 years or so for Young Earth Creationists. For Muslims, the age of the Earth is much less of an issue.

Allied to creationism is the theory of intelligent design. While many of those who advocate intelligent design have been involved in the creationism movement, to the extent that the US courts have argued that the country's First Amendment separation of religion and the State precludes its teaching in public schools (Moore, 2007), intelligent design can claim to be a theory that simply critiques aspects of evolutionary biology rather than advocating or requiring religious faith. Those who promote intelligent design typically come from a conservative faith-based position (though there are atheists who accept intelligent design). However, in their arguments against evolution, they typically make no reference to the scriptures or a deity but argue that the intricacy of what we see in the natural world, including at a sub-cellular level, provides strong evidence for the existence of an intelligence behind this (e.g. Meyer, 2009). An undirected process, such as natural selection, is held to be incapable of explaining all such intricacy.

How Should Science Communicators Respond? Lessons from BSE and GMOs

The response by science communicators to the range of positions held about evolution needs, I assert, to take account of the following:

1. Among scientists, the theory of evolution is held to be a robust, well-established and, at its core, a scientifically uncontroversial theory.

2. Within biology, evolution occupies a central place. There is much in biology in the absence of evolution that is worth studying and has been discovered, but an evolutionary framework is what enables biologists to

give meaning to the diversity of life that we see around us and to situate today's life in an historical context:

3.   In common with many scientific theories, evolution is not easy to understand. It has counter-intuitive elements and, in addition, is actively rejected by many people for religious reasons.

In making recommendations as to how evolution might be communicated (whether on television, in museums, in schools or elsewhere), we can learn from communication efforts about bovine spongiform encephalopathy (BSE) and genetically modified organisms (GMOs)

BSE is a disease of cattle. Causing progressive degeneration to the brain and spinal cord, so that infected animals eventually have trouble standing, it is commonly known as 'mad cow disease'. The disease became particularly prevalent in the UK in the early 1990s. Eventually, this was shown to be because cattle infected with the causative agent, a type of prion (a misfolded protein), were included in food fed to other cattle. Initially this merely provoked widespread feelings of disgust—cattle are not only not cannibalistic, but herbivorous—and that this was all a waste of money; farmers of destroyed cattle were reimbursed. However, concerns about human health were soon raised.

Initially, such concerns were dismissed by the UK government. In what is now seen as a classic instance of how not to handle public communication about science, civil servants, advisors, and government ministers determined that the best way forward was to reassure the public that it was still perfectly safe to eat meat from cattle. Attempts by scientists and others to argue that there might be a risk were rebuffed (Irwin, 2009). A 1990 photograph of the Minister for Agriculture, Fisheries and Food, John Selwyn Gummer, attempting to feed his four year-old daughter, Cordelia, a beefburger has become an enduring visual icon.

This tidy way of dealing with the problem unravelled when, as sometimes happens with science, the science changed. The government was forced to admit that there was a possibility that the rise in the number of cases of a horrible disease, new variant Creutzfeldt-Jakob disease (nvCJD) might be due to BSE. Indeed, this proved to be the case, and for a number of years there were worst-case scenario predictions of the deaths of hundreds of thousands of people from nvCJD in the UK.

The main lesson for communication about evolution that can be drawn from the BSE affair is that public trust is easily lost when over-confident assertions are made about the conclusions of science. The lesson to be drawn from GMOs might be thought to be the same, but it is subtly different in that in addition to near-inevitable uncertainties about safety given the rise of a new technology, there were other objections to genetic engineering.

The typical form of genetic engineering entails moving the DNA from one species to a second species so that the cell machinery of the second species makes the gene product of the first. A widespread application entails moving genes for herbicide resistance into crop plants. When the growing crop is then sprayed with the herbicide in question, the crop plants survive whereas competing weeds are killed. As agronomists have been altering the genes of crop plants for many decades via a whole range of technologies, many crop scientists were bemused as widespread objections to GMOs, particularly in Europe, became evident in the 1990s.

Aside from safety concerns—and, unlike BSE, time has not so far shown such fears to have been realised—other reasons why many objected to GMOs centred on the unnaturalness of the technology. Even those of no religious persuasion sometimes referred to the use of the technology as 'playing God'. Interestingly, while there were those who objected to genetic engineering on theological grounds, there were many who argued the opposite (Reiss & Straughan, 1996). It was also perhaps unfortunate from the perspective of those in favour of the technology that the first major battle in Europe was fought over bovine somatotropin (bST).

bST is a naturally occurring hormone, produced in the pituitary glands of cows. Back in the 1930s it was known that injecting cows with bST could boost milk yields. In the early 1990s Monsanto brought a genetically engineered form of bST to market. This was shown to boost milk yields, typically by about 10%. However, its use outside of the USA was fiercely resisted and, to this day, its use is not permitted in the European Union. While there were some concerns about possible risks to human health, more important were perceived risks to the health of the cows (they are more likely to develop mastitis) and, perhaps most importantly, the view, first, that milk is 'natural' and such naturalness is contaminated by genetic engineering and, secondly, that consumers have a right to choose whether or not they wish to consume genetically modified foods, and that once bST became widely used in Europe, it would be virtually impossible to keep out of the food chain.

My experience of giving large numbers of talks in the 1990s on genetic engineering (to farmers' groups, women's institutes, church groups, and others) and participating in public debates was that a whole concatenation of objections to genetic engineering came together: a presumption that the countryside, and so farming, was and should be natural; concerns about human health; distrust of government, big business and commercial science; concerns about animal welfare and the environment; resistance to the USA way of life; and a fear that things were simply moving too fast with today's realities increasingly resembling science fiction. My experience of participating in the UK National Consensus Conference on Plant Biotechnology (1994), the Advisory Committee on Novel Foods and Processes (1998-2001) and the GM Science Review Panel (2002-04) was that, as time went on, many of the participating scientists became increasingly exasperated. I have some sympathy with this point of view, but the part of me that is a school teacher knows that there it is generally counter-productive to blame learners for not learning what one wants them to learn. Effective science communication is more about providing the appropriate environs within which learning can take place than it is blaming people when they don't learn what one wishes they would.

Communicating Evolution in School Science

Few countries have produced explicit guidance as to how schools might deal with the issues of creationism or intelligent design in the science classroom. One country that has is England (Reiss, 2011). In the summer of 2007, after months of behind-the-scenes meetings and discussions, the then DCSF (Department of Children, Schools and Families) Guidance on Creationism and Intelligent Design received Ministerial approval and was published (DCSF, 2007). The Guidance points out that the use of the word 'theory' in science (as in 'the theory of evolution') can mislead those not familiar with science as a subject discipline, because it is different from the everyday meaning, when it is used to mean little more than an idea. In science the word indicates that there is a substantial amount of supporting evidence, underpinned by principles and explanations accepted by the international scientific community.

The DCSF Guidance goes on to state: ''Creationism and intelligent design are sometimes claimed to be scientific theories. This is not the case as they have no underpinning scientific principles, or explanations, and are not accepted by the science community as a whole'' (DCSF, 2007) and then goes on to say:

Creationism and intelligent design are not part of the Science National Curriculum programmes of study and should not be taught as science. However, there is a real difference between teaching 'x' and teaching about 'x'. Any questions about creationism and intelligent design which arise in science lessons, for example as a result of media coverage, could provide the opportunity to explain or explore why they are not considered to be scientific theories and, in the right context, why evolution is considered to be a scientific theory.

(DCSF, 2007)

This seems to me a key point and one that is independent of country, whether or not a country permits the teaching of religion (as in the UK) or does not (as in France, Turkey, and the USA). Many scientists, and some science educators, fear that consideration of creationism or intelligent design in a science classroom legitimises them. For example, the excellent book Science, Evolution, and Creationism published by the US National Academy of Sciences and Institute of Medicine asserts ''The ideas offered by intelligent design creationists are not the products of scientific reasoning. Discussing these ideas in science classes would not be appropriate given their lack of scientific support'' (National Academy of Sciences and Institute of Medicine, 2008, p. 52).

As I have argued (Reiss, 2008), I agree with the first sentence of this last quotation but disagree with the second. Just because something lacks scientific support doesn't seem to me a sufficient reason to omit it from a science lesson. Nancy Brickhouse and Will Letts (1998) have argued that one of the central problems in science education is that science is often taught 'dogmatically'. With particular reference to creationism they write:

> Should student beliefs about creationism be addressed in the science curriculum? Is the dictum stated in the California's *Science Frameworks* (California Department of Education, 1990) that any student who brings up the matter of creationism is to be referred to a family member or member of the clergy a reasonable policy? We think not. Although we do not believe that what people call 'creationist science' is good science (nor do scientists), to place a gag order on teachers about the subject entirely seems counterproductive. Particularly in parts of the country where there are significant numbers of conservative religious people, ignoring students' views about creationism because they do not qualify as good science is insensitive at best.
>
> *(Brickhouse & Letts, 1998, p. 227)*

Subsequently, Thomas Nagel (2008) argued that so-called scientific reasons for excluding intelligent design (ID) from science lessons do not stand up to critical scrutiny. With reference to the USA he concludes:

> I understand the attitude that ID is just the latest manifestation of the fundamentalist threat, and that you have to stand and fight them here or you will end up having to fight for the right to teach evolution at all. However, I believe that both intellectually and constitutionally the line does not have to be drawn at this point, and that a noncommittal discussion of some of the issues would be preferable.
>
> *(Nagel, 2008, p. 205)*

It seems to me that school biology lessons should present students with the scientific consensus about evolution, and that parents should not have the right to withdraw their children from such lessons. Part of the purpose of school science lessons is to introduce students to the main conclusions of science—and the theory of evolution is one of science's main conclusions. At the same time, science teachers should be respectful of any students who do not accept the theory of evolution for religious (or any other) reasons. Indeed, nothing pedagogically is to be gained by denigrating or ridiculing students who do not accept the theory of evolution.

My advice for science teachers is not to get into theological discussions, for example about the interpretation of scripture. Stick to the science and if you are fortunate enough to have one or more students who are articulate and able to present any of the various creationists arguments against the scientific evidence for evolution (e.g. that the theory of evolution contradicts the second law of thermodynamics, that radioactive dating techniques make unwarranted assumptions about the constancy of decay rates, that evolution from inorganic precursors is impossible in the same way that modern science disproved theories of spontaneous generation), avoid getting into a ping-pong debate with such students. Instead, use their contributions to get the rest of the group to think rigorously and critically about such arguments and the standard accounts of the evidence for evolution.

My own experience of teaching the theory of evolution for some thirty years to school students, undergraduate biologists, trainee science teachers, members of the general public and others, is that people who do not accept the theory of evolution for religious reasons are most unlikely to change their views as a result of one or two lessons on the topic, and others have concluded similarly (e.g. Long, 2011). However, that is no reason not to teach the theory of evolution to such people. One can gain a better understanding of something without necessarily accepting it. Furthermore, some studies suggest that careful and respectful teaching about evolution can indeed make students considerably more likely to accept at least some aspects of the theory of evolution (Winslow, Staver & Scharmann, 2011).

Communicating Evolution in Science Museums

Science museums have long had exhibits about evolution. Tony Bennett (2004) examines the history of museum displays about evolution. He looks at nineteenth century studies in geology, palaeontology, natural history, archaeology, and anthropology and "trace[s] the development, across each of these disciplines, of an 'archaeological gaze' in which the relations between past and present are envisaged as so many sequential accumulations, carried over from one period to another so that each layer of development can be read to identify the pasts that have been deposited within it" (Bennett, 2004, pp. 6–7). Bennett concludes that evolutionary museums "are just as much institutions of culture as art museums" (p. 187).

In one sense this is obvious—museums and galleries have to make selections about what to display and how to narrate such displays, and these are clearly cultural decisions whether one is referring to art, evolution, mathematics or any technology. However, whereas a visitor to an art gallery is unlikely to presume that what is being viewed is the only reading possible, a visitor to a science museum might presume that they are being presented with objective fact.

Monique Scott too has produced a book about evolution in museums (Scott, 2007). Scott's work, unlike Bennett (2004), is more to do with the now than with history. Using questionnaires and interviews, she gathered the views of nearly 500 visitors at the Natural History Museum in London, the Horniman Museum in London, the National Museum of Kenya in Nairobi, and the American Museum of Natural History in New York. Perhaps her key finding is that many of the visitors interpreted the human evolution exhibitions as providing a linear narrative of progress from African prehistory to a European present. As she puts it:

> Progress narratives persist as an interpretive strategy because they still function as a conceptual crutch. They are nearly ubiquitous in popular culture (can you imagine human evolution without imagining the cartoonish images of humans evolving single-file toward their destiny?) and they stand largely unchallenged in museum exhibitions which conventionally move case-by-linear-case from Africa to Europe. Many museum visitors, particularly Western museum visitors, rely upon cultural progress narratives—particularly the Victorian anthropological notion that human evolution has proceeded linearly from a primitive African prehistory to a civilized Europe—to facilitate their own comprehension and acceptance of African origins. Overwhelmingly, museum visitors relate to origins stories intimately, and in ways that satisfy or redeem the images they already have of themselves.
>
> *(Scott, 2007, p. 2)*

So what might one hope that science museums would consider when putting together exhibitions about evolution? Even if we presume that a museum decides to concentrate on the mainstream scientific account of evolution, eschewing any debates with creationism or intelligent design, there are still a myriad of decisions (conscious or otherwise) that those putting together an exhibition need to make:

- To what extent does one favour mammals and birds (beloved by many visitors) over the less spectacular but sometimes more informative invertebrates and other species?
- How much does one oversimplify (e.g. over the story of the evolution of the horse)? Too little oversimplification and the typical visitor is going to learn almost nothing, overwhelmed by difficult detail. Too much oversimplification and what our visitor learns may be no better than a reinforcement of error (cf. Scott's point above about a linear narrative of progress);
- To what extent should the curator(s) concentrate on scientific consensus and to what extent should they address scientific controversy, for example over the importance of punctuated equilibria and Lamarckism in evolution, and the relationship between micro- and macroevolution?
- To what extent should the social and cultural contexts of evolution be addressed (e.g. the reception by Victorian society and in France of Darwin's *On the Origin of Species* in 1859)?
- Can sensitive topics like intelligence, race, and sexual behavior be examined?

- How can today's digital technologies and dialogic possibilities (e.g. Science Museum, 2012) enhance learning for visitors both in the museum and away from it?
- How can museums use concepts such as 'upstream engagement' (Sulgeo & Wilsdon, 2009) to inform their provision in this area?

Of course, comparable decisions are made by school teachers all the time but science museum exhibitions and other presentations of science cannot, unlike classroom teachers of science, rapidly alter their presentations to take account of the particular learners in front of them (Reiss, 2012).

Communicating Evolution in Religious Settings and in Religious Education Lessons

There is much to be said for scientists, clergy, and others in religious settings talking positively about the theory of evolution. Such conversations can go some way to counteracting the perception in some people's minds that religions are inevitably against evolution—a perception that has some persistence, despite frequent denials from many religious leaders, and is buttressed by creationist museums (e.g. http://creationmuseum.org/) and zoos (e.g. www.noahsarkzoo farm.co.uk/). Indeed, perhaps somewhat optimistically, I would ask those running creationist places of learning to make one concession to evolution. I do not expect them to promote evolution but it is reasonable to ask them to make it clear that the scientific consensus is that the theory of evolution and not creationism is the best available explanation for the history and diversity of life. Of course, it is perfectly acceptable for those running creationist institutions to critique evolution and to try to persuade those visiting such institutions that the standard evolutionary account is wrong. But just as science teachers with no religious faith should respect students who have creationist views, so creationists should not misrepresent creationism as being in the scientific mainstream. It is not.

Finally, we can consider how evolution and creationism might be addressed in religious education (RE) lessons, for countries that have such lessons (or address at least some of the issues in history, social studies, humanities, citizenship or whatever). In England, the DCSF and QCA (Qualifications and Curriculum Authority) published a non-statutory national framework for RE and associated teaching units that include a unit asking 'How can we answer questions about creation and origins?' (QCA, 2006). The unit focuses on creation and the origins of the universe and human life, as well as the relationships between religion and science. This is a carefully written 23-page guide. Along with its non-evaluative stance towards the various positions, what strikes me as a science educator is the high expectations of students it has. For example, in answer to the question 'Is the universe designed? Who could have designed it?' it is suggested that teachers of 13–14 year-olds should:

Give the pupils opportunities to explore, through a website, DVD or written text (see 'Resources'), a range of different answers to these questions, including answers given by members of different faiths. These answers should include the views of creationists, evolutionists, advocates of intelligent design and philosophers of religion, such as Anselm, Thomas Aquinas, Blaise Pascal and Francis Bacon.

(QCA, 2006, p. 16)

We can note that this non-evaluative stance towards the various positions takes place in a context where, since the late 1950s in England and Wales, teachers of religious education in schools have abandoned a form of religious education where the inculcation of Christianity was a central aim, and in which Christianity was often presumed to be the sole framework within which life found meaning and moral direction. Nowadays a more pluralist vision is adopted (e.g. Jackson, 2004) in which students are enabled to develop, clarify and refine their own views about matters religious, while being introduced to a range of positions. This is very different to the position in science where the presumption, whether implicit or stated, is nearly always that the scientific understanding of the world is either a valid one or the valid one (e.g. Atkins, 2006).

Conclusion

The theory of evolution lies at the heart of modern science. By considering how evolution might be communicated in a range of settings, including school science lessons, science museums and religious settings, we can explore how science communication can tackle issues of belief and the value of evidence. Sensitive teaching about evolution can serve as a fine introduction to the practices of science and the extent of scientific knowledge.

In communicating the science of evolution, it should be made clear that there is a strong scientific consensus, and has been for many years, that the Earth is several billion years old, that life arose though natural processes from inorganic matter, and that all organisms are related by virtue of common descent. In some settings this is all that need be said. In others, it will be appropriate to deal with objections to the theory of evolution. Some such objections are nothing to do with religion but the most frequent ones are. As the history of shortcomings in dealing with public reaction to BSE and GMOs shows, it is not effective science communication to blame people when they don't learn what one wants them to. It is possible to communicate evolution in a way that is both true to science and respectful of those who do not accept it.

References

Atkins, P. (2006). Atheism and science. In P. Clayton, & Z. Simpson (Eds.), *The Oxford handbook of religion and science* (pp. 124-136). Oxford: Oxford University Press.

Austin, J. L. (1979). *Philosophical papers, 3rd ed.* Oxford: Oxford University Press.

Bennett, T. (2004). *Pasts beyond memory: Evolution, museums, colonialism.* London: Routledge.

Brickhouse, N. W., & Letts, IV, W. J. (1998). The problem of dogmatism in science education. In J. T. Sears, & J. C. Carper (Eds.), *Curriculum, religion, and public education: Conversations for an enlarging public square* (pp. 221-230). New York: Teachers College, Columbia University.

Dawkins, R. (2006). *The God delusion.* London: Bantam Press.

DCSF (2007). *Guidance on creationism and intelligent design.* Retrieved from http://webarchive.nationalarchives.gov.uk/20071204131026/http://www.teachernet.gov.uk/docbank/index.cfm?id=11890.

Holliman, R., Thomas, J., Smidt, S., Scanlon, E., & Whitelegg, E. (Eds.) (2009). *Practising science communication in the information age: theorizing professional practices.* Oxford: Oxford University Press.

Irwin, A. (2009). Moving forwards or in circles? Science communication and scientific governance in an age of innovation. In R. Holliman, E. Whitelegg, E. Scanlon, S. Smidt, & J. Thomas (Eds.), *Investigating science communication in the information age: Implications for public engagement and popular media* (pp. 3-17). Oxford: Oxford University Press.

Jackson, R. (2004). *Rethinking religious education and plurality: Issues in diversity and pedagogy.* Abingdon: Routledge.

Kenny, A. (1992). *What is faith? Essays in the philosophy of religion.* Oxford: Oxford University Press.

Lawes, C. (2009). *Faith and Darwin: Harmony, conflict, or confusion?* London: Theos.

Long, D. E. (2011). *Evolution and religion in American education: An ethnography.* Dordrecht: Springer.

Meyer, S. C. (2009). *Signature in the cell: DNA and the evidence for Intelligent Design.* New York: HarperCollins.

Miller, J. D., Scott, E. C., & Okamoto, S. (2006). Public acceptance of evolution. *Science, 313,* 765-766.

Moore, R. (2007). The history of the creationism/evolution controversy and likely future developments. In L. Jones, & M. J. Reiss (Eds.), *Teaching about scientific origins: Taking account of creationism* (pp. 11-29). New York: Peter Lang.

Nagel. T. (2008). Public education and intelligent design. *Philosophy & Public Affairs, 36,* 187-205.

National Academy of Sciences and Institute of Medicine (2008). *Science, evolution, and creationism.* Washington, DC: National Academies Press.

QCA (2006). How can we answer questions about creation and origins? Learning from religion and science: Christianity, Hinduism, Islam and Humanism—Year 9. Retrieved from http://webarchive.nationalarchives.gov.uk/20110813032310/http://www.qcda.gov.uk/libraryAssets/media/qca-06-2728_y9_science_religion_master.pdf.

Reiss, M. J. (2000). On suffering and meaning: an evolutionary perspective. *Modern Believing, 41*(2), 39-46.

——(2008). Teaching evolution in a creationist environment: an approach based on worldviews, not misconceptions. *School Science Review, 90*(331), 49-56.

——(2011). How should creationism and intelligent design be dealt with in the classroom? *Journal of Philosophy of Education, 45,* 399-415.

——(2012). Learning out of the classroom. In J. Oversby, (Ed.), *ASE Guide to Research in Science Education*, Association for Science Education, Hatfield, pp. 91-97.

Reiss, M. J., & Straughan, R. (1996). *Improving nature? The science and ethics of genetic engineering*. Cambridge: Cambridge University Press.

Science Museum (2012). *Talk science … contemporary science discussion for the classroom*. Retrieved from www.talkscience.org.uk/.

Scott, M. (2007). *Rethinking evolution in the museum: Envisioning African origins*. London: Routledge.

Stilgoe, J., & Wilsdon, J. (2009). The new politics of public engagement with science. In R. Holliman, E. Whitelegg, E. Scanlon, S. Smidt and J. Thomas (Eds.), *Investigating science communication in the information age: Implications for public engagement and popular media* (pp. 18-34). Oxford: Oxford University Press.

Winslow, M. W., Staver, J. R., & Scharmann, L. C. (2011). Evolution and personal religious belief: Christian university biology-related majors' search for reconciliation. *Journal of Research in Science Teaching, 48*, 1026-1049.

Wittgenstein, L. (1953). *Philosophical investigations*. Oxford: Basil Blackwell.

PART IV

Informal Learning

10

HELPING LEARNING IN SCIENCE COMMUNICATION

John K. Gilbert

The Importance of Science and Technology

Science has direct impacts, every day, on all aspects of our lives. On our personal lives: how we strive to remain fed, warm, and healthy for as long as possible. On *our social lives:* how we physically move from place to place, relate to, and communicate with our fellow humans. On *our economic lives:* how we are able to acquire the resources to remain fed warm and healthy. In short, science and technology have become *major cultural institutions for humanity.*

Despite this importance, and although the subjects are included in some way in the compulsory education of young people in many, if not most, countries, the extent of voluntary continuation of such studies is less than most governments want (High Level Group on Science Education, 2007). Almost all universities offer a broad range of courses in science and technology leading to degrees, with varying uptakes. Training focused on the needs of specific industries is widely available. However, very little effort is made, beyond that of the splendid 'Open Universities', to engage the adult population in systematic non-specialist learning about these major cultural institutions. Why is this? Certainly, as Lindy Orthia argues in Chapter 5, people's attitudes to learning may be shaped by the negative attitudes to science, as such, that they hold. However, another reason is that such learning is found to be very difficult by many young people. The memory of this difficulty carries over into adult life, placing science outside the range of possible personal interests.

Why is Learning about Science Found to be 'Hard' by Many People?

Formal science education, that which is bounded by a prescribed curriculum, became established in the schools and universities of Western Europe and North America in the mid-1800s. The intention was to identify and educate a relatively small number of individuals to lead the rapidly evolving research and development activities underpinning the many new engineering-based industries. Those highly-motivated individuals had little trouble in remembering the myriad facts and abstract concepts that were shaping science as we know it today. As time passed, the range of people to whom science was taught gradually expanded. Today it is thought that all young people should learn some science: 'scientific literacy for all' is the prevailing motto. However, what is taught—the curriculum content—is still largely governed by the assumed needs of 'pre-specialist students', rather than those of the great majority. Many students cannot see why they are being taught this content: adults remember their sense of alienation when young.

This sense of alienation is compounded when the relevance of what is taught is considered. As science developed rapidly in the 19th and 20th centuries, the range of detailed facts and explanatory concepts increased exponentially. This ever-expanding range was progressively included in formal science curricula, which rapidly became overloaded with fragmentary information which had to be recalled for assessment purposes. Of course, many of the basic concepts used in science were established in the 19th century, and their importance is undiminished. A clear focus on them is, however, obscured by a fog of surrounding facts. Moreover, the newer, exciting themes in modern science such as electronics, genetics, ecology, and astronomy all press for inclusion, and places have not been found for these quickly enough. Although some curriculum content is removed from time to time, the abiding memory of school for many is of having to learn too much material, much of it not evidently relevant to today's world and the challenges people face within it.

Perhaps the biggest source of difficulty for many people lies in the assumptions made by their teachers about how learning—especially in science—takes place. Three models of learning have come to be successively used in formal education: the transmission model, the personal constructivist model, and the social constructivist model.

The transmission model, which was universally used before printing was widely available, became consolidated as the dominant model underpinning general education after World War II, it having been used successfully for basic military training during that conflict. It assumed that a learner did not know a particular topic, having either never learnt it or having forgotten it. The teacher's task was to present material in a highly logical way, the students' task was to memorise a 'copy' of that material, their success in doing so being assessed by regular testing for recall. Failure to learn, in these terms, was evidence of a lack of 'intelligence' or idleness: the former leading to a withdrawal of future learning opportunities, and the latter to punishment of some sort. Success received praise and progression to further topics, the whole enterprise being justified by 'behavioural psychology'. This approach gradually lost its pre-eminent position as it was found that what was learnt was very often a distorted version of what was intended: these were the so-called 'misconceptions' or 'alternative conceptions'. Although behavioural psychology is unable to explain the existence of misconceptions, the transmission model still underpins much teaching today, including that in science education.

There are several *constructivist* models of learning. What they have in common is the assumption that the student is mentally active, the process of learning being the *construction,* the development of new, or the adaption of existing, networks linking the ideas under consideration with what is already known. A seminal figure in this approach was Jean Piaget, who theorised that the capability to construct these mental networks developed with age in a qualitatively step-wise manner (see, for example, Piaget, 1969). Although Piaget's approach to experimental work has proved to be of enduring value, many examples were identified of young people showing an advanced understanding of ideas when much younger than Piaget predicted. Although his theories are much less widely used today as the basis of research into learning, they still have an implicit following in classroom teaching, where some students are still labelled as 'not ready' to learn abstract ideas. This approach never seemed to have much significance for adult learning, where the 'highest' level in Piaget's developmental scheme was just assumed to have taken place.

A later development was that of *personal constructivism,* associated with the name of George Kelly, which suggested that there was no natural limit to what could be learnt by an individual (See, for example, Kelly, 1955). Whilst Piaget's theory interpreted misconceptions/alternative conceptions as arising from inadequate mental development, Kelly's theory afforded them the status of having arisen from different forms of adaption to the presence of prior knowledge during learning. This approach opened the way to developing ways of 'correcting' these conceptions, which have been found in people of all ages and levels of formal and informal education. However, both Piaget's and Kelly's theories viewed mental interaction between individuals, mediated by speech, as of only second-order importance in learning.

The *social constructivist model,* however, closely associated with the name of Vygotsky (Vygotksy, 1962) places great emphasis on the social context in which learning takes place. The learning of an individual is heavily influenced by social interaction with the teacher and other students, as well as with the representations of the knowledge being learnt that are available, such as text, diagrams, video and audio, and practical work. The individual learns not only what to think but how to think by these interactions, which are appropriately called an 'apprenticeship in learning' (Rogoff, 1990). Inevitably, it is the nature of these interactions that controls the quality of what it learnt, particularly the model of thinking presented by the teacher, whether directly or indirectly. A key issue, and one to which I will return shortly, is the understanding acquired by the student of the language used in what is being studied. The social constructivist model subsumes the personal constructivist model: it is the individual who learns, but what is learnt and how this takes place is substantially influenced by the social circumstances in which it takes place.

Making Learning Easier

The universal use of teaching methods based on personal or social constructivist models is the key to successful learning. They will make learning easier for individuals of all ages in all educational circumstances, both formal (see Chapter 11) and informal (see Chapter 12). This should lead to improved attitudes towards, and performance in, science education at school and university level. In Chapter 11, Sean Perera and Sue Stocklmayer discuss how this can be done through the adoption of inquiry-based learning. This ought to lead, in the longer term, to an increased interest in science communication by adults in general. These techniques are already well-established in adult education.

However, there are several other measures that can be taken to ensure that these constructivist approaches are successful. The first of these is to teach through the medium of situations that are of direct and readily recognisable personal, social, and economic, importance to the learners. This will ensure that the ideas learnt are actually of direct use in the world of today and tomorrow. The second is to make sure that the learners fully understand the often complex and specific language of science. The third is to ensure that both the situations enquired into and the language used to do so are fully understood, this being helped if material is encountered in different *modes of representation* drawn from the range of the *concrete/material*, the *verbal*, the *visual*, the *symbolic,* and the *gestural*. Each of these issues is taken up in turn below. Finally, the question is addressed: how we can be sure that high-quality learning has taken place?

Making Learning Easier: Dealing with Familiar and Important Situations

A key approach to making learning easier is that, rather than teaching individual concepts in isolation, divorced from their significance for engineering and their societal implications as parts of technologies, they are addressed within 'situations'. Examples of this broad class are: complex objects (for example, a mobile 'phone); a place (e.g. a manufacturing plant); a problem (e.g. the provision of pure water); a contemporary event (e.g. an accident at a nuclear power plant); or an historical event (e.g. the 'discovery' of the Americas). To be of educational value, the particular example chosen for study must be—or must readily come to be—familiar to the learner. This ensures that their relevance, whether personal, social, or economic, can be readily grasped. A situation becomes a 'context for study' when an understanding of it is sought as a result of systematic inquiry. Building on the ideas of Duranti and Goodwin (1992, p.3):

- A context can be defined as a spatial, temporal, or social situation of personal, social, economic, or cultural significance. This significance must be recognised by students as being of importance in some specific way;
- Particular events in, or aspects of, a context can readily be the subject of study by students. This would necessarily involve the acquisition and use of language;
- The language used in that study must be, in part at least, specific to that situation;
- The study must evoke the use of a student's relevant prior knowledge.

The use of contexts as a basis for study is an effective way of identifying and thoroughly understanding those concepts that are key tools in the scientific and allied enterprises. The issue of 'excessive factual content' is circumnavigated by doing so. Using the approach is not easy, for there is no prescription for the identification of potential contexts. Doing so will arise from a consideration of everyday life, from knowledge of the history of, and recent developments in, science, from news reports of contemporary social and political events. Whilst adults are already widely acknowledged as only being willing to learn about what they are specifically interested in, the expansion of this approach to schools, which is gradually taking place, should lead to educational continuity between childhood and adulthood.

To be effective, context-based learning poses challenges to both teacher and students. The sustained engagement with one context—implying the use of valuable 'curriculum time'—requires that the teacher is very familiar with all the events and explanatory concepts involved in it. The students must be willing to engage with this in-depth experience, which will contrast with how they may usually have been taught. There must be a reciprocal focus on the context itself—what is happening—and on the emergent understanding of the explanatory concepts—why that is happening. Where these conditions are met, there is evidence (Bennett, 2003) that:

- students' interest in science is sustained or increased;
- students can more readily see the relationship between science concepts and the world-as-experienced;
- the learning of key concepts is at least as effective as in conventional courses;
- teachers feel more 'in control' of classroom events than when just giving accounts of the uses of concepts after they have been taught by assumption of the transmission model.

However, it does not seem that the use of context-based courses alone leads to improved willingness to continue with the study of science and technology (Bennett, 2003, p.114). The other major dimensions to making learning easier must also be in operation for this to take place.

Making Learning Easier: The Language Used Must be Meaningful

Language is, self-evidently, of central importance in the learning and teaching of any subject. Contrary to popular belief, words do not just encode concepts for transmission from the teacher to the learner. Rather they are used by the teacher to create concepts in the mind of that learner. Of course, both the teacher and the student hope that something very like the teacher's interpretation will be acquired by the student. However, there are several potential general linguistic impediments that must be overcome or avoided:

- Learning science tends to involve the acquisition of many new words. For example, Merzyn (1987) found this to be 6-8 per lesson for 14- to 16-year-olds. These have to be defined when introduced to avoid being misunderstood;
- Many of these words were created specifically for use in science such as 'volt', 'rhizome', 'enthalpy'. Although not as problematic for students as was once thought, they do all need defining upon first introduction;
- The use of words from everyday life can have a different meaning in science, for example, 'force' or 'weight'. The physical circumstances ('everyday' compared with 'science') in which a word is used may be not be appreciated by students, leading to misconceptions. Those circumstances must thus be carefully explained to students, preferably by practical demonstration;
- The meaning of logical connectives such as 'thus', 'because', or 'however', vital in the conduct of explanation in science, may not be understood. Gardner (1975) found that they make the comprehension of written text more difficult for 15-year-olds. Although surely a theme for general language education, this potential problem should not be overlooked by science teachers;
- Words in common usage, although used in the same way in science, may just not be familiar to students. For example, a large number of words were found 'difficult' in this way by many UK secondary school students, such as 'abundant', 'adjacent', 'contrast', 'incident', 'composition' and 'contrast' (Cassels & Johnstone, 1985, p.14). Again, this is primarily a matter for general language education.

There is one language matter that transcends issues associated with particular words. This is concerned with the use of metaphors and analogies which, having a central role in both science (e.g. Galaburda,1997) and engineering (e.g. Ferguson, 1993), are widely used in science communication at all levels and with all audiences (Duit, 1991). A metaphor is a generalised comparison between two entities that can be represented as 'An X is a Y'. Put another way, something that has to be understood is considered the same as something to which it seems to have a great similarity. For example, 'An atom is a planetary system'.

These idealisations are useful in that they identify things that can then be compared in detail in the form of an simile i.e. 'An X is *like* a Y', e.g. 'An atom is like a planetary system' (the Bohr model). The issue then is: what is the scope of 'like'? Hesse (1966) split a potential analogy into three parts: 'positive' relationships, which are those parts which do seem to be comparable; 'negative' relationships, those parts which do not seem comparable; and the 'neutral' analogue, which are those parts about which a decision cannot be reached. An unfortunate choice of metaphor and/or an inaccurate valuation of its components can lead to misconceptions. There are thus two issues here that make science communication more successful. The first is the careful choice of metaphors that are used: the source of them must be very familiar to the learners. The second is an understanding of how analogies work and experience by learners in analysing and using them.

Words, whether spoken, read, or written, are capable of representing, of communicating, a high proportion of the meaning of the ideas of science. There are other *modes of representation*, each with a definite capacity for conveying meaning, which are often associated with the use of words. The greatest success in communicating science, in the learning of it, will arise when a suitable combination of these modes is used.

Making Learning Easier: Improving the Visualisation of Ideas

Thinking is often said to involve 'visualising': what is, might, or could be, meant by an idea. Alas, this commonly used word has acquired two distinct, but related, meanings. As 'external representation', a visualisation refers to a version of something that is being learnt that can be shared with other people through common access to words, graphs, diagrams, pictures. As 'internal representation', a visualisation refers to the mental model that is created in the mind by an individual. External representations are invaluable tools in education: for teachers, in order to convey meaning; for students, in order to explore and consolidate their understanding. Internal representations are the tools for all learning, not least in science, where the making and testing of predictions about future events is the mainstay of the enterprise. The two activities of perceiving an external phenomenon, and the creation of an internal representation of it, share the same brain mechanisms. A model for the acquisition, mental manipulation, and use of internal representations has been produced by Paivio (1986). It divides external representations into two groups: the verbal, i.e. words received either orally or in written form; and the non-verbal, meaning gestures, images (of objects, diagrams, graphs etc), sounds, or touch. These are processed by, and stored in, distinct types of structures in the brain.

Within the verbal system, the meanings of individual words are stored separately. However, structures can be formed between these separate stores which enable complex phenomena or relationships to be understood. For example, the meaning of 'circulatory system' is composed of the separate meanings of the words 'heart', 'arteries', 'veins', whilst a whole can be created from these components.

Within the non-verbal structure, the meanings of other types of external representations are also stored separately. Again, structures can be formed between them such that complex phenomena are understood. For example, different sub-types of diagrams of the circulatory system can produce a full explanation of it by their integration into a whole. Perhaps most importantly, links can be also formed between the two types of store (verbal and non-verbal), so that the different representational capabilities of the two systems can be brought to bear on one phenomenon. The output of the two systems, whether verbal or non-verbal, is informed by the structures that have been created within and between them.

The learning of science is easier if both the verbal and non-verbal systems are exploited to the full in the provision of educational opportunities. This can only be done if the capabilities of the various modes of representation are known. There are five distinct general modes of external representation, each with particular form representational capacity: the verbal, the concrete/material, the visual, the gestural, and the symbolic. Establishing good practice in each of them is made complicated, for two reasons. First, there are several sub-modes of each general mode. For example, the visual mode includes photos, sketches, diagrams, flow charts. Second, on many occasions, the uses of several modes are integrated together, e.g. discussing a concrete model or interpreting a graph. However, good practice in the use of each of them must be established before their most effective combination can be evaluated. Each of these is now taken in turn.

Making Learning Easier: The Use of the Verbal Mode

Those issues that arise whenever words are used—the meanings of individual words or word types and the operation of analogies—have already been discussed above. However, good practice must be adhered to in each of the ways that words are externally represented, in speech, in reading, in writing, if learning is to be made easier.

A visit to very many classrooms will immediately demonstrate that the great majority of the talking is done by the teacher. Mortimer and Scott (2003) have identified four types of classroom talk:

- *Non-interactive authoritative.* The teacher talks: the students listen. This type is the mainstay of the transmission model of education, where a copy of the teacher's knowledge is to be passed to each student. It is the dominant model in university lectures—an appropriate title—particularly where there are many students present;
- *Interactive authoritative.* The pattern in this type, found in many schools is as follows: the teacher initiates an interaction by asking a question; a student responds, often with a word, phrase, or incomplete sentence; the teacher evaluates that response. This cycle is repeated, with variation made by the teacher, until a student produces the anticipated answer;
- *Non-interactive dialogic.* Here individual students can raise questions and provide answers to them. The teacher evaluates these questions and answers. This is found in many classrooms;
- *Interactive dialogic.* Here both the teacher and the students enter into something approaching a normal conversation, where any of the participants can raise a question, provide an answer, or evaluate an answer. This is only found in some classrooms.

The use of the non-interactive dialogic and the interactive dialogic approaches will make it more likely that individual students will form internal representations of the ideas being addressed: learning is made easier. However, these approaches do require that the teacher has both expert subject knowledge in the topic being studied and the skill to manage interactions with and between students. Particular skills include ensuring the use of a suitable mix of 'open' questions—those where the precise answer is unknown—and 'closed' questions—those to which there is a unique answer; and ensuring that a suitable number of 'higher order' questions are asked—those that require the 'analysis', 'synthesis' or 'evaluation' of information, not merely its recital.

Although students are often required to have textbooks, little systematic use seems to be made of reading for the communication of science. This is a serious omission, for reading can give the student access at a controlled pace to *argumentation*, the process by which the status of individual facts in respect of theories is established. A structure for arguments has been proposed by Toulmin (1958) and its place in science communication evaluated by Erduran (2008). In summary, a valid form of argument has five components: a statement of the claim that is being made; data that supports that claim; warrants, which are the links being made between the data and the claim; backing, which is evidence from elsewhere that supports the claim being made; and a statement and refutation of possible rebuttals, that is of situations where the claim might not be true. Easier learning of science is facilitated by books placing an emphasis on argumentation, rather than on simple assertion. This will ensure that the student is required to mentally engage with the text. Adequate high-quality time should be made available for that reading.

The learning of science does not often require the student to write, the formulaic 'writing-up' of laboratory work being the exception. This omission neglects an opportunity to support learning, for the act of composing text requires the student to express their internal representations of ideas in external form and hence to clarify them. Writing can do this by being in one of three genres: writing to learn, so that key concepts come to be understood by expressing them; writing to reason, that is for the construction of arguments; and writing to communicate, that is having a sense of purpose and of audience. The skills of writing for science communication are only gradually developed, a truism to which the authors of chapters in this book will no doubt attest!

Making Learning Easier: Using the Concrete/Material Mode

The concrete/material mode is most commonly used to represent a model of a physical structure. The spatial or temporal relationships between the entities of which it is composed can then be emphasised. A wide range of analogies can be used. Examples of this mode are: the 'ball-and-stick' model of DNA; the 'coloured dye' model of the human vein/artery system; the 'exploded' representation (a 'model') of a planned new building. The attractions of this mode are that:

- the three-dimensional nature of the original is retained;
- sight—perhaps the most valuable of the senses—is employed;
- aspects of the original can be emphasised or suppressed, dependent on the focus of attention;
- the original is brought within the scope of human vision if the original is very large (e.g. a factory) or very small (a virus).

The key to the successful use of the concrete/material mode is that the 'conventions of interpretation' linking a particular representation to the original phenomenon must be known in detail by students. This can be done by a combination of personal experience in handling and interpreting particular examples and by direct instruction about the representational system.

This dual approach is needed because, as the versatility of this mode are so great, a large number of sub-modes have come into use, each with different representational emphasis and hence a different use of analogy. A good example is the range used to represent molecular structures (see Figure 10.1). The three major sub-modes ('Open', 'Space Filling', and 'Orbital') each enable aspects of a model of the molecule to be represented ('+') or not ('-')

System	Open			Space Filling			Orbital		
Subsystem	Ball-and-Stick		Skeletal		Molecular		Ionic			
Similarity	+	–	+	–	+	–	+	–	+	–
Nature	★		★		★		★		★	
Size	★			★	★		★		★	
Shape		★		★	★		★		★	
Angles	★		★			★		★	★	
Surface		★		★	★		★		★	
Texture		★		★	★		★			★

FIGURE 10.1. Scope and Limitations of Analogue Status for Concrete/Material Representational Subsystems Used in Chemistry (*Source:* Gilbert, Reiner & Nakhleh, 2008, p.10, *Visualization: Theory and Practice in Science Education.* Dordrecht: Springer.)

Making Learning Easier: Using the Visual Mode

The visual mode includes a wide range of sub-modes that provide a graduated spectrum of representational capabilities. At one end are 'photographs', at the other 'graphs', with many versions of the 'diagram' in between them.

Photographs are a convenient way of showing what a phenomenon actually looks like when met and, for that reason, they are widely included in textbooks. The communicational value of any given photograph will be heavily influenced by the caption that accompanies it and by its relation to the text in which it is embedded. Pozzer and Roth (2003) found four forms of relationship in biology textbooks:

- *The decorative.* A photograph is included without a caption or direct reference to it in the text;
- *The illustrative.* The identity of the object or objects in the photograph is stated in the caption but there is no reference to it in the text;
- *The explanatory.* The identity of the object or objects in the photograph is stated in the caption and some explanation of them is given in the caption, but the surrounding text does not relate to that explanation;
- *The complementary.* Some explanation of the photograph is provided in the caption, which is not present in the text, yet which adds something to the text.

Although the equivalent of 'captions' is provided verbally in video, these four alternatives are found there. Improving learning from photographs requires not that the photographer had great skill in composing the 'shot', but that the result is included in text in the 'complementary' form.

The opposite end of the spectrum of visual representation is the use of 'the graph'. A wide variety of sub-modes is available, including the Venn form, the block graph, as well as the conventional line graph. They will support learning if the student knows the 'code of interpretation' for each form and if the component parts of each particular example are clearly labelled. Again, the use of a graph will be enhanced if it picks up and complements the text in some way.

In between the photograph and the graph lie the infinite variations of the notion of 'diagram'. A diagram can be thought of as consisting of images of objects linked together. The images of objects can range from photographs, through simple sketches of objects, to symbols representing them. The links can be either implicit or explicit and refer to the dimensions of direction, time, or causality. Diagrams can thus contain a great deal of information in a condensed form. However, they will best support learning if the conventions governing their use are clearly known to students.

Making Learning Easier: Using the Gestural Mode

Gestures are movements of the hands and arms used when some information is being presented. They do seem to be important in learning, but little research has been undertaken into them (see Radford, Edwards & Arzarello, 2009). Any example seems to serve one of four purposes, namely to: (1) point to a real or imaginary object that is under discussion; (2) draw a physical analogy of something that familiar in order to explain the object under discussion; (3) indicate trends, for example the repetition of events; or, (4) add emphasis to what is said. Most learners—and indeed their teachers—will not be consciously aware of gestures used during explanations. Being aware should enhance the learning taking place.

Making Learning Easier: Using the Symbolic Mode

Two symbolic modes of representation are often met in science education but less often in science communication. The first of these is the mathematical equation. The learning of science can be helped in three ways: (1) making sure that learners have a full understanding of the particular form being used; (2) using the simplest form when providing an explanation; and (3) explaining mathematical equations in words and not leaving them to 'speak for themselves'. The second type of symbolic mode is the chemical equation. The main problem encountered in using this mode is that the genre has evolved over the last century or so, and textbooks still use a miscellany of these forms. There is an international convention now in place about the writing of equations (the IUPAC system) and learning is undoubtedly supported if this system is always adhered to and students are fully conversant with it.

Making Learning Easier: Clarity in Explanations

All the suggestions made above should make the learning of science easier. However, the one issue that will have the greatest impact is if the learner has a clear vision of the *purpose* of the communication, of the type of explanation that is being sought. Obscurity of purpose is sure to lead to poor quality learning.

In any communication there are one or more of six distinct types of explanation. They are:

- *Contextualised explanations.* These answer the questions: 'What phenomena are being explained?' Some parts of the world-as-experienced are separated from the continuum of experience and given names;
- *Intentional explanations.* These answer the questions: 'Why should these phenomena be investigated?' The explanations will probably concern their importance as the basis for technologies that seem needed;
- *Descriptive explanations.* These answer the questions: 'What are the properties of these phenomena?' These will identify properties of interest, the ways that they may be measured, and the range of values typically obtained for them;
- *Interpretative explanations.* The questions here are: 'Of what are these phenomena composed?' The answers will invoke entities not usually directly visible to the human eye;
- *Causal explanations.* The questions 'Why do these phenomena behave as they do?' will produce answers of the type that 'these causes, involving previously identified descriptive explanations, leads to those properties, either deterministically or statistically'. These explanations are the type most valued by philosophers;
- *Predictive explanations.* The production and testing of predictions is the basis of science. If answers to the questions 'what properties will these phenomena have under other, specified, circumstances?' produces predictions that are substantiated by experimental work, then the associated descriptive and causal explanations will have proved to be of value.

(Gilbert, Boulter & Rutherford, 2000)

Thus, all attempted science communications should make it clear what type(s) of explanations they will provide.

How Can You Tell if a Science Communication has been Effective?

Attempts to learn always require time, concentration, and mental effort. How can a learner, having made such an attempt, evaluate whether it was worthwhile? I suggest the posing of three questions as criteria:

- Can I now explain what I set out to explain?

- Can I transfer my explanations, with modifications, to situations that seem like the one that I have studied?
- Am I sufficiently interested in what I have learned to try to learn more about the subject in the future?

Positive responses to all three questions will suggest that the science communication provided was successful.

References

Bennett, J. (2003). *Teaching and Learning Science*. London: Continuum.

Cassels, J., & Johnstone, A. (1985). *Words that matter in science*. London: Royal Society of Chemistry.

Duit, R. (1991). On the role of analogies and metaphors in learning science. *Science Education* , 75(6), 649-672.

Duranti, A., & Goodwin, C. (1992). *Rethinking context: Language as an interactive phenomenon*. Cambridge: Cambridge University Press.

Erduran, S. (2008). Methodological foundations in the study of argumentation in science classrooms. In S.-A. Erduran, *Argumentation in Science Education* (pp. 47-70). Dordrecht: Springer.

Ferguson, E. (1993). *Engineering and the mind's eye*. Cambridge, MA: MIT Press.

Galaburda, A. (1997). Profiles, Part 1: Faraday, Maxwell, Einstein. In T. West, *In the mind's eye* (pp. 97-129). Amherst, NY: Prometheus Books.

Gardner, P. (1975). Logical connectives in science: A summary of the findings. *Research in Science Education* , 25(3), 9-24.

Gilbert, J. K., Boulter, C. J., & Rutherford, M. (2000). Explanations with models in science education. In J. Gilbert, *Developing Models in Science Education* (pp. 193-208). Dordrecht: Kluwer.

Hesse, M. (1966). *Models and analogies in science*. Notre Dame University, Indiana: Notre Dame University Press.

High Level Group on Science Education. (2007). *Science now: A renewed pedagogy for the future of Europe*. Brussels: European Union.

Kelly. G. A. (1955). *The psychology of personal constructs*. New York: Norton.

Merzyn, G. (1987). The language of school science. *International Journal of Science Education* , 9(4),483-9.

Mortimer, E., & Scott, P. (2003). *Meaning making in secondary science classrooms*. Maidenhead, Berks.: Open University Press.

Paivio, A. (1986). *Mental representations: A dual coding approach*. Oxford: Oxford University Press.

Piaget, J. (1969). The child's conception of the world. Totowa, NJ: Littlefield, Adams and Co.

Pozzer, L., & Roth, W.-M. (2003). Prevalence, function, and structure of photographs in high school biology textbooks. *Journal of Research in Science Teaching*, 40(10), 1089-1114.

Radford, L., Edwards, L., & Arzarello, F. (2009). Introduction. *International Journal of Mathematics Education*, (70) 91-95.

Rogoff, B. (1990). *Apprenticeship in thinking: Cognitive development in social context.* Oxford: Oxford University Press.

Toulmin, S. (1958). *The uses of argument.* Cambridge: Cambridge University Press.

Vygotksy, L. (1962). *Thought and Language.* Cambridge, MA: MIT Press.

11

SCIENCE COMMUNICATION AND SCIENCE EDUCATION

Sean Perera and Susan Stocklmayer

Introduction

The scientist and science communicator Michael Faraday once remarked that an education system which fails to inspire students about science "must have been greatly deficient in some very important principle." This chapter argues that the element lacking in many science classrooms is an understanding of effective principles of science and technology communication. Focusing on inquiry-based pedagogy, we will highlight the challenges for science teachers to implement reform directives. We suggest that a constructivist framework for professional development, based on established science communication principles, may assist in implementing inquiry in the science classroom.

The important role of inquiry and hands-on experience in learning about science has been implicit for decades, even centuries. Indeed, it might be said to have originated with the alchemists, who believed that their apprentices should have laboratories in which to find out more about the world of chemistry:

That youth may not be idely trained up in notions, speculations, and verbal disputes, but may learn to inure their hands to labour, and put their fingers to the furnaces, that the mysteries discovered by Pyrotechny and the wonders brought to light by Chymistry may be rendered familiar unto them ... Unless they have Laboratories as well as Libraries, and work in the fire, better than build Castles in the air.

(John Webster, surgeon and alchemist, 1654, p. 106)

Real attention to the role of open-ended practical work in the modern era, however, may be said to have begun in the US with the introduction of Physical Science Study Committee (PSSC) Physics in the late 1950s. Spurred on by the

Cold War, the drive to recruit more students into physics and engineering led to the assembly of a very senior group of physicists who wished to counteract not only the Soviet scientific threat but:

> ... a rising tide of irrationalism and suspicion among the general public that, they believed, directly threatened the continued health and advancement of science in the United States. While military conflict opened the door to reform, it was this cultural conflict that fundamentally shaped its substance.
>
> *(Rudolph, 2006, p.2)*

To counter this "tide of irrationalism" scientists felt that they should work to provide intellectual training to achieve a "more rational citizenry" (Ibid, p.2). The program was expressly targeted at elite students: not only at future scientists but future politicians, lawyers, and business leaders. It incorporated elements of history, experimentation, and media: a "massive Teachers Guide ... brilliantly simple and original experiments" and "wonderful films" (French, 2006). It was, to sum up, well ahead of its time. Many prominent US physicists give credit to the program for their subsequent careers. It ran to seven editions.

PSSC Physics came into being at a time when there was growing concern that the public's view of science needed to be changed. After the euphoria of the first space explorations, scientists and educators began to worry that support for science was weakening. Hence the introduction of PSSC Physics and, on the other side of the Atlantic, Nuffield Science.

The introduction of the Nuffield Foundation Science Teaching Project in Great Britain in 1962 was a response to numerous socio-economic, intellectual and national influences which saw the need to reform school science and, in particular, to establish an official agency responsible for the science curriculum (Waring, 1979). The Ministry of Education proposed a Curriculum Study Group. A major campaigner was the Science Masters' Association, and to a lesser degree the Association of Women Science Teachers[1]. It was serendipitous that the Chairman of the Advisory Council of Scientific Policy which deliberated on these issues at the time was, in fact, a Trustee of the Nuffield Foundation. This resulted in the Nuffield Foundation donating £250,000 for a long-term programme to improve science teaching, which subsequently became known as the *Nuffield Project*.

Nuffield Science aimed to be intellectually exciting with a focus on the nature of science. It too had extensive teacher guides, an emphasis on practical exploration, questioning techniques, and inquiry learning.

> There was to be a stress on encouraging attitudes of critical inquiry and on developing ability to weigh evidence, assess probabilities and become familiar with the main principles and methods of science. The teaching strategy ... was to avoid telling pupils all the answers and instead to give them time and opportunity to learn by working out scientific problems for themselves.
>
> *(Meyer, 1970, p.1)*

Neither PSSC nor Nuffield Science has survived in their original format and, although both have had considerable influence on subsequent curricula, they proved to be too challenging both for teachers and students. Teachers and students were used to something different and found it very hard to adjust. Reflective comments from those involved with the design of PSSC included: "I was forced to recognize that only a few of the participants were qualified to pick up the ball and run with it … some of them [were] simply out of their depth … Some of my participants were probably better off teaching a traditional course …" (French, 2006, p.2).

Linderfield (2006) asked: "So why did the PSSC course disappear? First of all, it was too tough … It became a course for the elite" (pp. 1–2). Rigden (2006) has commented that, despite its very happy and positive beginnings, "PSSC failed to capture a sizeable share of the [textbook] market because PSSC required both teachers and students to think" (p.2). It became known as a "hard" course, in contrast to "those introductory physics textboks that have enjoyed wide adoption, [in which] equations drive the content" (Ibid, p.2).

In the UK, an evaluation of students who had studied Nuffield Science for three years, compared with a control group taught by "traditional" methods, has an echo of these comments. The expectation that students would 'discover' scientific ideas by practical experience was not realized. "The course is too hard for the average GCE pupil" and "unrelated to the pupil's personal social environment" (Meyer, 1970, p.295).

Meyer made many other comments, both negative and positive, including his finding that as students progressed, those who had been successful at Nuffield Science had enhanced thinking skills. However, considering the program in the light of contemporary views of the role of science at that time, we highlight an interesting speculation that pupils had perhaps been "over-trained to show interest in problem-solving and to be hypercritical of even generally accepted scientific facts":

Perhaps this is a healthy counter to the admitted over-emphasis on facts for facts sake prevalent in former teaching. There is, however, the danger of increasing cynicism and doubt about the value of our cultural heritage and hence even of the values of our whole social order if this objective is over-emphasized at the expense of other objectives. Of course a certain amount of cynicism is healthy, especially in science, but this must be developed alongside a respect for past achievement and an understanding and appreciation at the vast resource of knowledge that man has accumulated during his past 30,000 years. (p.293)

The inquiry method was, therefore, in some sense threatening.
We have included this extensive discussion of these two atypical programs because they represented a challenge to the traditional methods of the day.

PSSC and Nuffield Science were not only influential in the USA and the UK. The British Commonwealth countries also tried Nuffield Science, and the ideas of PSSC physics proved far-reaching. Based on a laudable push for inquiry-based learning, they nevertheless eventually failed because neither teachers nor students really understood how to implement them. They never really "caught on" and were seen as somewhat anarchic. Other textbooks from that period reveal that expectations for science education at that time were generally grounded in a mathematical, rote-learning approach which did not encourage critical thinking. These texts retained their position as the preferred style.

Science Communication and Education in the 1980s and 1990s

By the 1980s, traditional science texts were still very much the dominant sources for science education. "Hands-on" was not a widely used expression, and practical work was generally not open-ended or exploratory. School examinations were about solving routine problems and remembering routine facts. The ideas of inquiry-based teaching and learning thus lay dormant for many years. Even in the second decade of the 21st Century, many websites on this topic still include sections entitled, "What is inquiry-based teaching" and "Why do it?"

Considering the area of science and technology *communication*, we note that in parallel with this emphasis on factual knowledge in formal education, the 1980s saw the birth of the movement known as the "Public Understanding of Science" (PUS). In the UK and in many other countries this was:

> … a time when the scientific community feared public indifference more than animosity. There were no votes in basic research, and funding was shrinking – or at least static, which researchers tend to perceive as the same thing. A better-informed public would be more inclined to support science, and the technologies it helps generate. And increasing public understanding of science would be good for recruiting future researchers, and good for the economy.
>
> *(Turney, 2002, p.1)*

The PUS movement gave rise to the 'deficit model' of the public, discussed at length elsewhere in this book. Suffice to say here that the essence of this model is that the onus rests with the public to learn and understand more science. The publication of a seminal paper by Durant, Evans and Thomas (1989) in *Nature* emphasised the lack of understanding on the part of the public when questioned on a variety of scientific facts drawn from several disciplines and pitched at the level of middle school science teaching. It has been shown elsewhere (Stocklmayer & Bryant, 2012) that these questions are themselves highly questionable. Nevertheless, the result of this quiz (in which the public did not do too well) was that considerable efforts were made to preach the scientific message more widely, and this has continued in almost every country of the world ever since.

Part of these efforts in the 1980s included an attempt to reinforce the nature of science in school curricula (Gregory & Miller, 1998). One such example was *Project 2061*. Designed by the American Association for the Advancement of Science in 1985, this Project was intended to develop a scientifically literate American citizenship by the return date of Halley's Comet in 2061. The project stemmed from the report entitled *'Science for all Americans'* which defined a 'science-literate' person as

one who is aware that science, mathematics, and technology are interdependent human enterprises with strengths and limitations; understands key concepts and principles of science; is familiar with the natural world and recognizes both its diversity and unity; and uses scientific knowledge and scientific ways of thinking for individual and social purposes'.

(AAAS, 1985)

This definition—and its emphasis on understanding and knowledge—reflected the thinking underlying the PUS movement of the day.

The 1980s, however, also gave rise to the Science Centre movement as part of the outreach efforts of PUS. At the start of that decade, there were very few such centres: notable examples were the San Francisco Exploratorium, the Science Museum in London, the Ontario Science Centre, Questacon in Canberra, and the Singapore Science Centre. There has been massive growth: The Toronto Declaration issued at the 5th World Congress of Science Centres in 2008 states: "Each year, 290,000,000 citizens actively participate in the exhibitions, programs, events and outreach initiatives organized by 2,400 science centres worldwide" (Ontario Science Centre, 2008). Since the birth of science centres, teachers have consulted them for workshops, assistance in demonstrations and activities, and so on. On their part, science centres have always offered resources and enrichment programmes to both teachers and students. Their effect has been evident and far-reaching and continues today. It is our contention that this move to communicate science in an interactive, approachable environment, together with recent substantial changes to views of science communication with the public, have paved the way for a renewed emphasis on inquiry learning in schools.

In the 1990s, there were several authors who sought to change the deficit view of the public, especially where scientific controversies caused an erosion of public trust. Gradually, people came to understand that the public was not an amorphous mass which was deficient in knowledge and learning (Gilbert, Stockmayer & Garnett, 1999). A new view of the relationship between science and the many faces of the public began to emerge. At the same time, constructivist learning became commonly recognized as the appropriate path for science teaching (although at the start of the 1990s the term was hotly debated and not well articulated for effective employment in the classroom). The term "hands-on, minds-on" began to be used in science centres, in recognition of the importance of building on prior knowledge and experience.

Science Communication and Education Today: A Time for Change

In 2000, an important report was formulated by the UK Government which influenced thinking about science and the public in most countries practising and teaching Western science (House of Lords, 2000). It effectively accused the PUS movement of being top-down and arrogant, and advocated a different approach to the relationship between science and society:

> Although scientists are a minority of the population, democratic citizenship in a modern society depends, among other things, on the ability of citizens to comprehend, criticise and use scientific ideas and claims .... the applications of science raise, or feed into, complex ethical and social questions ....
>
> *(Section 1.11)*

The emphasis had thus shifted from factual knowledge to an understanding of the impact and implications of science in an increasingly technological world. This Report stimulated a flurry of funded programmes, aimed at engaging the public, rather than educating them. At the same time, new questions were raised about the purposes of science education. These questions derived directly from the realization that the majority of the public were not engaged with science, and that previous attempts to "educate" the uninterested had largely failed:

> ... the contents of the science curriculum have not evolved accordingly. In today's world, a 'healthy and vibrant democracy' needs a public 'with a broad understanding of major scientific ideas who, whilst appreciating the value of science and its contribution to our culture, can engage critically with issues and arguments which involve scientific knowledge'.
>
> *(Section 6.15)*

The report recommends that the curriculum should be adjusted "to incorporate more technology and more 'ideas about science'". This would facilitate "more time for looking behind and beyond the facts into their sources and implications, and the processes of which they are a part" (Section 6.16). Inquiry learning is, therefore, firmly back on the agenda. It is evident, however, that the early difficulties attached to open-ended inquiry learning in the classroom are still an impediment to its implementation. We need, therefore, to answer the question: What, exactly, is being asked of teachers?

Inquiry-based Pedagogy

Inquiry-based teaching and learning are characterised by the responsibility placed on the learner, which promotes learning experiences that have personal significance (Simon & Johnson, 2008). To achieve this outcome, teachers should focus less on transmitting information and more on facilitating active learning experiences for their students. National science education directives in many countries and many cultures now emphasise inquiry-based pedagogy. The Australian *National Framework for Professional Standards for Teaching* (MCEETYA, 2003), for example, implicitly links this approach to future engagement with science: teachers should develop in their students critical thinking, creativity, and the ability to solve complex problems (among other skills). Similar directives to science teaching appear in: the USA *National Science Education Standards* (NRC, 1996); science curricula of the various Canadian provinces such as the Alberta *Focus on Inquiry* in 2004 and *The Ontario Science Curriculum* in 2008; *Kurikulum 2004* in Indonesia; the Singapore *Lower Secondary Science Syllabus* in 2008; and the Sri Lankan *Science and Technology Curriculum* in 2004, to mention just a few. In general, these curricula have been developed in parallel with a changing view of the role of the public in science and technology. This learner-centred approach is closely allied with constructivist learning, which stipulates that new knowledge should be built on existing understandings, so that actual learning translates into a more active, personally meaningful exercise. Learning science through inquiry requires students to scaffold (build a connected network of) prior knowledge, construct understandings, evaluate alternative conceptions, apply ideas in a socio-cultural context, and engage in open-ended questions, co-operative learning, and reflection (Shymansky et al., 1997), all of which are elements of a model constructivist classroom.

The Problem with Implementing Change

While science education reform directives agree that the teaching and learning of school science needs to move towards inquiry-based approaches, there is also unanimity that science teachers are the key in effecting this change (see, for example, Lumpe, 2004). There is a large body of evidence, however, to suggest that the pedagogy many teachers currently use to teach science defeats the aims of this reform. Osborne (2006), for instance, states quite pointedly that, in spite of numerous efforts to reform science education, teachers have failed to incorporate inquiry-based pedagogy in the classroom:

Four decades after Schwab's (1962) argument that science should be taught as an *enquiry into enquiry*, and almost a century since John Dewey (in 1916) advocated that classroom learning be a student-centred process of enquiry, we still find ourselves struggling to achieve such practices in the science classroom.

(p. 2)

Instead, more traditional models of pedagogy are practiced in science classrooms internationally. As well as the transmission mode of delivery, traditional teaching approaches predominantly feature abstract, difficult, and decontextualized content. In Australia, for example, traditional models of pedagogy are still practiced as widely as they were almost three decades ago (see Rennie, Goodrum & Hackling, 2001; Staer, Goodrum & Hackling, 1998). Perera (2011) also identified the language of science and the apparent Western ownership of science as problematic; these factors alienate non-Western learners by depriving them of belonging to, and ownership of, scientific culture.

Many authors see "ownership" of science as an important factor in a successful relationship between science and society (for example, Gilbert, Stocklmayer & Garnett, 1999; Huber & Moore, 2001; Rennie & Stocklmayer, 2003). As long as students see science as irrelevant and divorced from everyday experiences, they are unlikely to become scientifically engaged citizens. In the classroom, this implies that students' beliefs and opinions are afforded respect and consideration, which is an important factor in inquiry learning but one that is often overlooked.

Teachers' reluctance to adopt science education reform does not stem from their lack of information about inquiry-based pedagogy. For example, studies in the US (Cox-Petersen, 2001) and Israel (Eylon, Berger & Bagno, 2008) have revealed that teachers rate inquiry-based instruction as a better approach to teaching science. The teachers were also aware that student-centred approaches are more effective than traditional models of teaching. An Australian study, for instance, found that almost all (98.6%) of the teachers they surveyed believed that inquiry-based student-centred teaching had the potential to improve students' scientific aptitude (Harris, Jensz & Baldwin, 2005). The inconsistency arises because pedagogical change involves more than simply adopting teaching methods that have been proven to be effective. Teachers' reluctance to adopt pedagogy based on inquiry, and their persistence with conventional forms of teaching, is a manifestation of certain beliefs which they strongly hold (Bandura, 1986; Bybee, 1993).

Changing to accommodate inquiry learning, therefore, makes far-reaching demands on teachers, as Black and Aitkin (1996) explain:

> … teachers will have to change almost every aspect of their professional equipment. They will have to reconsider themselves entirely: not only the structures of their material and their classroom techniques, but even their fundamental beliefs and attitudes concerning learning.
>
> *(p.63)*

The need to change beliefs which govern teachers' classroom practices plays a fundamental role in science education reform (Van Driel, Beijaard & Verloop, 2001). Such beliefs may include: thinking that students come into the classroom with blank minds (*tabula rasa*); belief in the power of the textbook; the effect of early training (the "apprenticeship of observation" (Lortie, 1975)); the influence of the culture and existing teaching philosophy in the school; the need to teach to the examination; and, last and perhaps the most pervasive and difficult of all, ambivalence and fear about lack of control associated with student-centred inquiry being actually implemented in the classroom.

Teachers need support to accommodate change, and changes to teachers' pedagogy cannot be addressed superficially by mere fine-tuning of teachers' competencies or their preferred pedagogy (Rennie et al., 2001). Rather, teachers require professional development that influences the fundamental premises on which they base their practical reasoning about teaching in specific situations (Shulman, 1990). There are presently two main forms of science teacher professional development: formal preservice training offered through university-based education programs, and in-service programs. In this chapter we address the latter.

In-service Professional Development Programs

The UK Council for Science and Technology (2000) has reported that there have been on-going efforts to support science teachers professionally for decades. While earlier efforts, especially those in the post-Depression era, emphasised "topping-up" teachers' content knowledge, assuming that such an approach would improve their pedagogy, more contemporary models recognise that in-service professional development needs to do more than this. As Tytler (2007) points out, based on a review of science education in Australia, the support that teachers receive needs to probe deeper to address their personal convictions:

> The scale of the challenge in moving to a system which is focused on a very specific view of science content, and with many teachers long used to a transmissive pedagogy, should not be underestimated. What is required in order for many teachers to make the change is a new set of beliefs about the nature and purposes of science education. Also required is a new set of teaching and learning skills that give more agency to students, and open up the possibility of new knowledges being produced, rather than simply rehearsals of well-known knowledge elements. These are significant changes, beyond the reach of simple content delivery models of professional development.
>
> *(p.60)*

While it is widely agreed that teachers need to develop a greater awareness about science and that professional development programs offered to science teachers need to change focus, there is ambiguity about the means by which these changes need to be facilitated. The "top-up" mode had echoes of the deficit model—just tell them what is lacking and things will change. A newer model of effective communication emphasises the importance of a dialogic approach, respecting existing beliefs, with an emphasis on relevance and role modelling.

Helping teachers to become comfortable with the challenges endemic to inquiry-based classrooms can only be achieved when teachers themselves can make sense of this approach and view differently what is important in science learning, in ways that are personally meaningful. To communicate in this way resembles closely the Personal Awareness of Science and Technology (PAST) model that has been described by Stocklmayer and Gilbert (2002). This model emphasises the need to actively engage with previous experiences in order to advance (or change) an individual's personal awareness about a particular scientific concept. For in-service professional development, this implies modelling a method that informs scientific understandings by linking experiences that evoke teachers' previous awareness about relevant scientific concepts. It is possible here to draw a parallel to Yager's (1991) description of the characteristics of professional development within a model of constructivist learning:

- Conceptual change in teachers is most helpfully considered in terms of whether or not new ideas are intelligible, plausible, fruitful, and feasible;
- Whenever possible, in-service training must model, but not mimic, the strategies and ideas being advanced;
- Different groups will enter in-service training with different levels of relevant knowledge and experience; and
- Those conducting the in-service training must be sensitive to their own needs to undergo conceptual change.

(p.57)

It is important to note that these recommendations for in-service professional development closely resemble recommendations made with regard to inquiry-based learning for students. Researchers believe that in-service based on constructivist principles would better equip teachers to teach science by inquiry, through having faced similar learning experiences, skills and thinking as their own students (Loucks-Horsley, Love, Stiles, Mundry & Hewson, 1998):

> First, by becoming a learner of the content, teachers broaden their own understanding and knowledge of the content they are addressing with their students. Second, by learning through inquiry ... and experiencing the process for themselves, teachers are better prepared to implement the practices in their classrooms.

(p.49)

In-service training which is based on constructivist approaches has the advantage of empowering teachers to take ownership of their teaching because they take charge of their own learning, thus elevating teachers from a much despised technician-like approach to their professional practice (Davis, 2001). Such forms of in-service training can also model features of inquiry-based pedagogy, such as opportunities for collaborative learning, hands-on investigations and other practical aspects of science.

A Framework for In-service Training Based on Principles of Effective Science Communication

A study of hands-on professional development workshops offered to high school science teachers was conducted at the Australian National University in Canberra (Perera, 2010). These workshops had a mix of activities modelling various aspects of inquiry learning such as concept mapping, cooperative learning, demonstrations, games, problem solving, predict-observe-explain, model building, and so on. There were also theoretical discussions about constructivism, misconceptions, and the use of analogies. The workshops were positively evaluated by participants, and were found to have transferable benefits in the classroom. Analysis of the method of facilitation and presentation of the workshops, and reasons for their success, indicated that there were strong elements of constructivism that, to a large degree, exemplified current ideas about good science communication practice. These fell into three distinct stages. (Figure 11.1).

The three stages—*To examine, To inform,* and *To scaffold*—recognised, respectively: communications within the workshops which investigated the scientific awareness the teachers brought to the workshops; increased that awareness by active construction; and last, contextualised teachers' awareness with regard to their daily experiences. Since scaffolding enabled the teachers to construct further scientific awareness and understandings, this last stage left the teachers open to new awareness, (hence that margin of the diagram is denoted with a dotted-line).

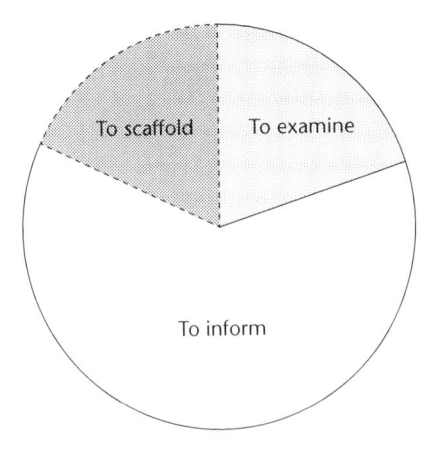

FIGURE 11.1 A Framework for Professional Development Modelled on Constructivist Principles.

Science and technology communication practices that underpinned delivery of workshop content, by facilitating inquiry-based constructivist learning principles, are identified as important elements of this framework (Perera, 2010). The teachers' experiences of the workshops specifically mentioned engaging communication practices: interactive dialogues, narratives, simple "hands-on and minds-on" experiments, animated and purposeful oral delivery, and easy to understand language. These practices are distinct from science education, although they are sometimes encountered in that context, and more accurately find their origin in science communication. Aspects of each stage that relate to exemplary science communication practice are given in Table 11.1.

TABLE 11.1 Science Communication Practices as Exemplified in Professional Development Modelled on Constructivist Principles.

Stages of constructivist learning	Science communication practices
Stage 1: *to examine*	Non-threatening activities (such as discussions of likely alternative conceptions held by students) to become familiar with the teachers' scientific awareness.
Stage 2: *to inform*	• Tactile and identifiable demonstrations that use simple and inexpensive equipment. • Historic narratives that describe actual human involvement in scientific achievements. • Rhetorical devices to punctuate oral delivery, such as visual imagery, metaphors, *gedanken* experiments. • Theatrical devices augment oral delivery, such as varied intonation and empathetic body language.
Stage 3: *to scaffold*	References and linkages to commonplace and current occurrences, such as events and objects from teachers' daily lives, or current issues in science.

Conclusion

It is clear that the introduction of inquiry learning requires a fundamental change in the way that science is taught. In Chapter 10, John Gilbert has outlined a variety of modes of learning that are important for relevant and interested engagement. If students are to continue to be engaged beyond school science, to become life-long learners in science, classroom teaching will need to be different. This can only happen if practising teachers, as well as those currently undergoing pre-service, are assisted in implementing these challenging ideas. Science *communication* has embraced the need for change, and is re-examining the goals for public engagement. There is now a universal recognition that a one-way process founded in the deficit model is unlikely to have successful outcomes. Science *education* also needs to acknowledge this, and to recognise that the only way to achieve change is to allow teachers the time to explore their own beliefs and communication practices.

The workshops have demonstrated that it is possible to advance scientific awareness and understandings amongst teachers in ways that are personally meaningful. Effective science and technology communication practices are critical in this regard. In turn, teachers modelling these practices with their students will facilitate the kind of learning that fosters a continuing engagement with science.

To summarise these practices, we can do no better than to quote (from Porter & Friday, 1974), two outstanding early scientists, who were also great communicators, born a century apart (Figure 11.2). Michael Faraday's (MF) skills as a scientist led him to share his discoveries, through lectures and demonstrations, with an adoring public. Sir Lawrence Bragg (LB), Nobel Laureate, who was an excellent lecturer, also wrote extensively for his fellow scientists on what it means to be a good communicator. (We apologise for the historic gender bias in these quotes.)

On finding out what your audience knows:

A lecture is made or marred in the first ten minutes. This is the time to … remind the audience of things they half know already … I have listened to so much splendid material lost to the audience because the lecturer failed to realise that it did not know what he was talking about, whereas, if the precious first ten minutes had been spent on preparation, he would have carried his listeners with him for the rest of the talk. (LB)

On keeping it simple

How many 'main points' can we get over in an hour? I think the answer should be 'one'. (MF)

On being informing and engaging

A flame should be lighted at the commencement and kept alive with unremitting splendour to the end. (MF)

On the importance of relevance

A guiding principle of the popular lecture is that of starting with something with which the audience is thoroughly familiar in everyday life, and leading them further with that basis. (LB)

On the techniques of theatre

The utterance should not be rapid and hurried, and consequently unintelligible, but slow and deliberate, conveying ideas with ease … (MF)

We come back, it seems to me, to the essential feature… which is the emotional contact between lecturer and audience. (LB)

On humour

Above all, jokes have a marked and enduring effect. (LB)

On provoking curiosity

One wishes to give the audience the aesthetic pleasure of seeing how puzzling phenomena became crystal clear when one has the clue and thinks about them in the right way. So make sure the audience is first puzzled ... (LB)

On demonstrations

... the difference between the description of an experiment and the actual witnessing of it are as great as the difference between looking at a foreign country on the map and visiting it. (LB)

On the importance of language

He may be honestly trying to avoid technical language but it goes further than that. He has to put himself in the place of the intelligent layman and realise that ideas and experiences so familiar to him are unexplored country to his listener.

On respect for one's audience

His whole behaviour should evince a respect for his audience... he should give them full reason to believe that all his powers have been exerted for their pleasure and satisfaction. (MF)

On making it entertaining

... the generality of mankind cannot accompany us one short hour unless the path is strewed with flowers. (MF)

FIGURE 11.2 Some Communication Principles from the Writings of Michael Faraday (MF) (1791–1867) and Lawrence Bragg (LB) (1891–1971) (modified from Porter & Friday, 1974).

Faraday's and Bragg's thoughts on science communication techniques are as fresh today as when they were written. This does not, however, mean that they have achieved universal currency. Indeed, their ideas about respect and consideration for the audience and for successful engagement are being newly reframed in the context of science communication today. Just like inquiry learning, however, their message is that good ideas never really go away.

Notes

1   Both Associations merged in 1963 to establish the Association for Science Education.

References

American Association for the Advancement of Science, The. (1985). *Project 2061—Science for All Americans*. Oxford University Press, http://www.literacynet.org/science/all.html, Retrieved 5.7.11

Bandura, A. (1986). *Social foundations of thought and action: A social cognitive theory*. Englewood Cliffs, NJ: Prentice-Hall.

Black, P., & Aitkin, J. (1996). *Changing the subject: Innovations in science, mathematics and technology education*. London: Routledge.

Bybee, R. W. (1993). *Reforming science education*. New York: Teachers College Press.

Cox-Petersen, A. M. (2001). Empowering science teachers as researchers and inquirers. *Journal of Science Teacher Education, 12*(2): 107-122.

Davis, S. K. (2002). "Change is hard": What science teachers are telling us about reform and teacher learning of innovative practices. *Science Education, 87*(1): 3-30.

Durant, J. R., Evans, G. A., & Thomas, G. P. (1989). The public understanding of science. *Nature, 340*, 11-14.

Eylon, B. S., Berger, H., & Bagno, E. (2008). An evidence-based continuous professional development programme on knowledge integration in Physics: A study of teachers' collective discourse. *International Journal of Science Education, 30*(5): 619-641.

French, A. P. (2006). *PSSC 50 years later. Discovering the PSSC: A personal memoir*. American Association of Physics Teachers, http://ajp.aapt.org/, p.1-3. Retrieved 23.6.11.

Gilbert, J. K., Stocklmayer, S. M., & Garnett, R. (1999). *Mental modelling in science and technology centres: What are visitors really doing?*. In S. Stocklmayer and T. Hardy (Eds.), Proceedings of Learning Science in Informal Contexts. Canberra: Questacon. pp. 16-32.

Gregory, J., & Miller, S. (1998). *Science in Public: Communication, Culture and Credibility*. New York: Plenum Trade.

Harris, K.L., Jensz, F., & Baldwin, G. (2005). *Who's teaching science? Meeting the demand for qualified science teachers in Australian secondary schools*. Report prepared for the Australian Council of Deans of Science. http://www.usq.edu.au/resources/who's_teaching_science.pdf, Retrieved 4.4.2007.

House of Lords (2000). *Third report of the select committee on science and society*. London: House of Lords.

Huber, R.A., & Moore, C.J. (2001). A model for extending hands-on science to be inquiry based. *School Science and Mathematics, 101*(1): 32-41.

Indonesian Ministry of Education. (2004). Guidelines for Physics Teaching. In *Kurikulum 2004*. Jakarta: Curriculum Centre. pp. 11-13.

Lindenfeld, P. (2006). *PSSC 50 years later. From New Brunswick to Tirupati with PSSC*. American Association of Physics Teachers, http://ajp.aapt.org/, p.1-2. Retrieved 23.6.11.

Lortie, D. (1975). *Schoolteacher: A sociological study*. Chicago: The University of Chicago Press.

Loucks-Horsley, S., Love, N., Stiles, K. E., Mundry, S., & Hewson, P. W. (1998). *Designing professional development for teachers of science and mathematics*. 2nd Edition. Thousand Oaks, CA: Corwin Press.

Lumpe, A. T. (2004). Editorial. *Journal of Science Teacher Education*, 15(3): 173.

Meyer, G. R. (1970). Reactions of pupils to Nuffield Science teaching project trial materials in England at the Ordinary Level of the General Certificate of Education. *Journal of Research in Science Teaching*, 7, 283-302.

Ministerial Council on Education, Employment Training and Youth Affairs. (MCEETYA). (2003). *A National Framework for the Professional Standards for Teaching*. Melbourne: Author.

Ontario Science Centre. (2008). *Toronto Declaration*. 5th World Congress of Science Centres. Ontario: June, 2008, http://www.ontariosciencecentre.ca/aboutus/torontodeclaration.asp Retrieved 5.7.11.

Osborne, J. (2006). *Towards a science education for all: The role of ideas, evidence and argument.* Paper presented at the 2006 ACER Research Conference: Boosting Science Learning - What will it take? Canberra, Australia.

Perera, S. (2010). *Exploring the implication of science communication practices on a model for teacher professional development: Serving up the Pierian Waters.* (PhD thesis). Canberra: The Australian National University Digital Repository: http://research.anu.edu.au/access/

——(2011). Science teachers from non-Western backgrounds challenged by Western science: A whole other ball game. *The International Journal of Science in Society. Volume 2,* (2): pp.11-22.

Porter, G., & Friday, J. (1974). *Advice to lecturers: An anthology taken from the writings of Michael Faraday and Lawrence Bragg.* Sussex: The Royal Institution.

Rennie, L., Goodrum, D., & Hackling, M. (2001). Science teaching and learning in Australian schools: Results of a national study. *Research in Science Education, 31*: 455-498.

Rennie, L. J., & Stocklmayer, S. M. (2003). The Communication of Science and Technology: Past, Present and Future Agendas. *International Journal of Science Education, 25,* 759-773.

Rigden, J. S. (2006). *PSSC 50 years later. With PSSC, teachers and students had to think.* American Association of Physics Teachers, http://ajp.aapt.org/, p.1-3. Retrieved 23.6.11.

Rudolph, J. L. (2006). *PSSC 50 years later. PSSC in historical context: Science, national security and American culture during the Cold War.* American Association of Physics Teachers, http://ajp.aapt.org/, p.1-3. Retrieved 23.6.11.

Shulman, L. S. (1990). *Paradigms and programs: Research in teaching and learning.* Volume 1. New York: Macmillan.

Shymansky, J. A., Henriques, L., Chidsey, J. L., Dunkhase, J., Jorgensen, M., & Yore, L. D. (1997). A professional development system as a catalyst for changing science teachers. *Journal of Science Teacher Education, 8*(1): 29-42.

Simon, S., & Johnson, S. (2008). Professional learning portfolios for argumentation in school science. *International Journal of Science Education, 30*(5): 669-688.

Staer, H., Goodrum, D., & Hackling, M. (1998). High school laboratory work in Western Australia: Openness to inquiry. *Research in Science Education, 28*(2), 219–228.

Stocklmayer, S. M., & Bryant, C. (2012). Science and the public: What should people know? *International Journal of Science Education Part B: Communication and Public Engagement, 2,* 81-101.

Stocklmayer, S., & Gilbert, J. K. (2002). New experiences and old knowledge: Towards a model for the personal awareness of science and technology. *International Journal of Science Education, 24*(8): 835-858.

Turney, J. (2002). *Understanding and engagement: The changing face of science and society.* Wellcome Trust. http://www.wellcome.ac.uk/News/2002/Features/WTD004756.htm. Retrieved 24.6.11.

Tytler, R. (2007). *Australian Education Review: Re-imaging Science Education.* Australian Council for Educational Research. Melbourne: Australian Council for Educational Research. http://www.acer.edu.au/documents/AER51_ReimaginingSciEdu.pdf Retrieved 14.12.2007.

UK Council for Science and Technology. (2000). *Science teachers: A report on supporting and developing the profession of science teaching in primary and secondary schools.* London: Author.

van den Berg, E. (2001). Impact of inservice education in elementary science: Participants revisited a year later. *Journal of Science Teacher Education, 12*: 29-45.

Van Driel, J. H., Beijaard, D., & Verloop, N. (2001). Professional development and reform in science education: The role of teachers' practical knowledge. *Journal of Research in Science Teaching, 38(2)*: 137-158.

Waring, M. (1979). *Social pressures and curriculum innovation: A study of the Nuffield Foundation Science Teaching Project.* London: Methuen & co Ltd.

Webster, J. (1654). *Academiarum examen, or the Examination of Academies.* http://books.google.com/books?id=wIBBAAAAcAAJ&printsec=frontcover#v=onepage&q&f=false. Retrieved 24.6.11.

Yager, R.E. (1991). The Constructivist learning model: Towards real reform in science education. *The Science Teacher, 67(1)*: 44-45. (Republished in 2000).

12

THE PRACTICE OF SCIENCE AND TECHNOLOGY COMMUNICATION IN SCIENCE MUSEUMS

Léonie J. Rennie

Introduction

Learning about science (including technology and other cognate disciplines) in informal environments is a continuous, life-long process. Unlike schooling, where some blocks of the timetable are labelled "science" and thus inform students what they are learning about, science permeates everyday experiences in ways that are unlabelled and implicit. Family members who apply sunburn lotion outdoors, search for fossils in beachside cliffs, don 3-D glasses in movie theatres, spread fertiliser on garden beds, pump up a basketball to just the right firmness, or recharge the batteries of their smart phone, are all using science even though they may be unaware of its significance in the activity at hand. In such cases, people are participating in science voluntarily, in ways they have learned to do, because the experiences or their outcomes are inherently interesting or useful to them. However, there are many opportunities to participate in everyday activities where the science is not implicit and the experience is deliberately designed not only to make science visible, but to communicate it explicitly to those members of the public who engage with it. These more formal sources of science learning can be clustered into three groups (Rennie, 2007a). There are museums and similar institutions with an educational purpose, sometimes called designed spaces (Bell et al., 2009), in which people can pursue their interests and engage in science-related experiences; community and government organisations that provide science-related educational programs to all levels of the public, including after-school activities and programs for adults and senior citizens; and print and electronic media, including prepared programs, such as science documentaries, depositories of knowledge available for online searching, and software for web-based information exchange.

Thinking about the wide variety of ways in which people can engage in science in these diverse contexts requires pushing the notion of learning science well beyond the limits of cognitive concepts, and reaching into the realms of interest, enthusiasm, motivation, and the social context of learning. Broadening the boundaries of learning means embracing a broad view of the nature of the science communicated in everyday environments. When the communication of science is intentional, how is that science communicated? How, for example, do informal sources of learning about science, such as museums, zoos, botanical gardens, environmental and other interpretative centres, and print, visual, and electronic media, actually present their science? Are there common elements among the range of science-related activities with which people can engage successfully? Do people experience these activities in the same way or in different ways? Answers to these questions come from looking at both the nature of the science to be communicated and the ways that people engage with that science.

The communication of science and technology in everyday, informal environments is characterised by choice. People may choose to notice and engage in various opportunities to learn about science, or they may not. If they do choose to engage, then they are generally in control of how they interpret the science information offered. However, the science that is on offer is not usually science as the scientists see it. Instead, these informal sources of learning present their information in narrative or story form. Developing that science story requires selecting, packaging, and presenting science information in such a way that the intended audience is motivated to engage with it, can understand and make use of it, according to their own needs, interests, and experience.

This chapter explores how such science stories are developed and provides examples to illustrate how the stories are interpreted. To keep the chapter manageable, science museums and similar institutions are used as examples of informal environments where science communication may take place.[1] Following an exploration of the typical agendas of science museums and their audiences, the chapter moves to a discussion of how science stories are developed as a means of communication through exhibits and the factors that affect how the audience responds to them. Two research examples are provided to illustrate both intended and unintended audience response. The chapter concludes with a discussion of the issues exhibit designers need to consider about the effective communication of science.

An Exploration of Agendas

Because choice is so central to the process of science communication in informal environments, successful communication requires that an informal organisation, such as a science museum, understands its audience well enough to present science stories that are interesting and relevant, and with which the audience is willing to engage. This does not happen automatically. Informal organisations have a purpose for their existence, and visitors have reasons for attending them.

In other words, each has an agenda and there needs to be a "coming together" of these agendas to enable a two-way interaction between the source of science information and its audience, so that dialogue can develop and effective communication can occur. What kinds of agendas are typical?

The Statutes of the International Council of Museums (ICOM, 2007) define museums as follows: "A museum is a non-profit, permanent institution in the service of society and its development, open to the public, which acquires, conserves, researches, communicates and exhibits the tangible and intangible heritage of humanity and its environment for the purposes of education, study and enjoyment" (Article 03-3). In terms of communication about science, traditional natural history museums, art galleries, and specialist museums devoted to particular events, persons, or cultures, can all have a role to play if they choose to bring the scientific aspects of their collections or displays to the fore.

Most commonly, we think of science museums and centres, environmentally-based organisations, such as aquaria, botanical and zoological gardens, and other interpretive centres as places where science is most likely to be communicated. These places usually have an advertised purpose, or mission statement, that involves education particularly relating to science. For example, the mission statement of my local science centre is "to increase interest and participation in science and technology." My local state museum has a mission "to inspire people to explore and share their identity, culture, environment and sense of place, and to experience the diversity and creativity of our world"; being a natural history museum, a good deal of this exploration will be related to science and technology. So, one part of the agenda of science and other museums is to raise awareness of science and technology and their place in our world, and to provide opportunities for people to explore, engage with, and learn about them.

Another part of the agenda of informal institutions concerns funding. Science museums and similar institutions rely on visitors for their existence. Some have entrance fees, some are sponsored or have particular exhibits sponsored, many have government support, most have a combination of such sources of support, and some are entirely government funded. Unless they can offer visitors a value-for-money experience, these institutions will fail to attract sufficient visitors and risk losing financial support from sponsors or government. Pressures for accountability to their public are moving even the larger, collections-based museums from being curatorial and researcher-driven institutions, to become more educational and audience-driven. As educators have more involvement with exhibit development, museums increasingly become sources of educational, social, and cultural change (Scott, 2003). Thus the agenda becomes increasingly concerned with attracting, as well as serving, their audience. These institutions may advertise their offerings through the web, via brochures and the press, and in order to encourage visitors, they often focus on family fun and entertainment rather than learning. In fact, the "edutainment" element, particularly of science centres, has often been criticised as focusing on the entertainment side to the detriment of education. There is a fine line to be trod for these institutions!

Visitors also have an agenda in mind when they choose to visit a museum. When asked about the purpose of their visit, most say they want to have fun and be entertained. Visitors in family groups may also justify their visit in terms of an educational opportunity for their children. Adult visitors usually do not state their own education as a primary purpose of their visit, but it is often their secondary purpose (Falk, Moussouri & Coulson, 1998; Shields, 1993). There is a long history of research which supports the idea that learning happens, if not during the visit, then often afterwards, when other experiences reinforce the visit experience to result in new understanding (for a review, see Rennie, 2007a). Students' field trips to science museums are a particular form of visitation, and there may be a mix of agendas, with the teacher having a more educational agenda in mind than the students do. Nevertheless, numerous reviews of research on the outcomes of field trips (for example: DeWitt & Storksdieck, 2008; Rennie & McClafferty, 1995; Rickinson, et al., 2004), demonstrate that valuable learning, and thus communication, can result when the field trip is well integrated with the school curriculum (Anderson, Lucas, Ginns & Dierking, 2000) and students have some input into the purpose and nature of the field trip (Griffin, 2004).

Communicating Science in Museums and Science Centres

What is it about science that might be communicated in informal institutions such as museums and science centres? In the context of science education, Hodson (1998) wrote about

learning science—acquiring and developing conceptual and theoretical knowledge; *learning about science*—developing an understanding of the nature and methods of science, an appreciation of its history and development, and an awareness of the complex interactions among science, technology, society and environment; and *doing science*—engaging in and developing expertise in scientific enquiry and problem solving'. (*p. 191, italics in original*)

Realistically, given the constraints of institutional agendas, visitors are most likely to have opportunities to *learn science* via engagement with exhibits and exhibitions that are designed mainly to communicate knowledge. Wellington (1990, p. 250) suggested that in science centres, most of this will be "knowledge that", rather than "knowledge how" or "knowledge why", but importantly, "knowing that" is a precursor to learning how and why. Many fewer exhibitions are designed to communicate something about nature of science, its development. These enable visitors to *learn about science* and to engage with some of the affective aspects, attitudes, and values relating to science. Exhibits can also provide opportunities to *do science*, albeit vicariously or through "thought experiments." How might this communication take place?

Building on the work of others (such as Layton, Jenkins, Macgill & Davey, 1993), I have suggested a model that describes the communication of science as an interactive process in which science information is selected, packaged, and presented by a variety of media, and then interpreted in multiple ways by a heterogeneous public (Rennie, 2007b). In the context of a science museum or similar institution, the starting point is the science that is familiar to scientists; the product and process of scientific research. This is the science that represents the scientists' interpretation of real-world phenomena within the context of the scientific community in which they work, expressed in the terms, culture, and values of that community. Enabling the lay public to access that science requires it to be reinterpreted, or reconstructed, into a form more easily understood. In museums, this reconstruction results in an exhibit, or a series of exhibits, designed for the intended audience. I use the term "science-related story" to describe the result of this process of deconstructing and then reconstructing the target science information, following Milne (1998) who pointed out that, "once ideas are presented selectively in science we are no longer telling the facts. We are instead telling a story" (p. 176). When a visitor engages with the science-related story underpinning a museum exhibit, what is understood and learned is likely to be unique to the individual, because the visitor-exhibit interaction depends on the personal and social context of that individual, including his or her background knowledge, experiences, interests, and motivation for engagement. Falk and Dierking (2000) describe this experience more fully in their Contextual Model of Learning.

Visitor-Exhibit Interaction and the Communication of Science

Visits to science museums provide opportunities for dialogue about science. The focus of this dialogue is primarily the science-related stories told by the exhibits. In collections-based displays, the museum-to-visitor communication is based around interpreting a generally static exhibit, such as an object or diorama. Assistance for the visitor is provided through visual means, such as text labels, that are sometimes supplemented by film or video loops *in situ* or on hand-held devices, or by aural means, either through portable headsets (audio-tours) or short, button-activated explanations through mounted speakers. In interactive science museums, where almost all exhibits are purpose-built to demonstrate some concept or phenomenon, the visitor is expected to participate actively in the exhibit experience through touch or other sensory responses to promote action from, or interaction with, the exhibit. Usually text signage assists with directions and interpretation of the concept or phenomenon.

The effectiveness of dialogue between the visitor and exhibit is determined by the willingness of the visitor to engage with the exhibit and to persist in interpreting the outcome. The resultant communication may be aided (or not!) by the quality of the exhibit, the attendance of explainers, and by social interaction among the visitors themselves. If we consider the quality of an exhibit in terms of its ability to promote dialogue, then it must first engage the visitor. There is a large background of research that endeavours to define a successful exhibit, and there is consensus on a number of features. Perry (1989), for example, concluded that the exhibit should promote curiosity; induce feelings of competence, confidence and control; offer a challenge; provide opportunities for play and enjoyment; and enable communication through social interaction. Semper (1990) suggested that exhibits offer multiple modes of learning and thus cater to a wide range of visitors. Allen (2004) introduced the notion of "immediate apprehendability" (that is, it is immediately obvious what one has to do to engage with an exhibit) as an important attribute of a good exhibit.

Social interaction is another factor that can promote dialogue between the visitor and the exhibit and thus enhance communication. Interaction with an explainer, docent, or other guides, such as teachers on field trips, can be differentially helpful. Effective guides are those who encourage visitors themselves to engage and explore, rather than tell or lecture to them, thus dampening their opportunities to learn. Family groups are important learning institutions (Ellenbogen, Luke & Dierking, 2004). McManus' (1994) description of families as teams of "hunter-gatherers" is an apt metaphor as members "forage" for interesting exhibits and share their interesting finds with others. Parents who question children and scaffold their thinking can enhance the child's learning experience (Crowley, et al., 2001).

Research into the Learning Outcomes of Museum Visits

Research into the impact of museum visits began well over a century ago and has accelerated dramatically over the last several decades. Not surprisingly, methods and their theoretical base follow contemporary approaches. Current theoretical frameworks tend to combine a constructivist and socio-cultural approach to learning. Falk and Dierking's (2000) Contextual Model of Learning in the museum environment has received significant attention because it offers a sensible combination of physical, social, and personal factors as determinants of the outcome, and importantly, emphasises the aspect of time: learning does not necessarily have to happen during the visit, it can occur a considerable time later when visitors are reminded of their visit by a new experience (Stocklmayer & Gilbert, 2002).

In Rennie and Johnston (2004) we analysed the difficulties faced by researchers in the informal environment. The elements of choice, transiency of visit, diverse visitor background, and variety of the visitor's experience combine to make research into the visit impact very difficult. Research methods need to take into account the different ways visitors respond to the same exhibit, for example, and whether or not they interact socially with others, because these things will affect the nature of any learning. A full research program would need to look for cognitive, affective, and skill outcomes, as well as unexpected outcomes, so data collecting techniques need careful consideration. On the one hand, we argue, the researcher needs to get "inside the visitors' heads" to find out about their thinking and ideas, but on the other hand, obtrusive measuring techniques vary the nature of the visit experience. Further, the element of time points to the methodological need for longitudinal research into the impact of museum visits, but there has been relatively little such research, partly because follow-up is so difficult with visitors dispersing in many directions. Nevertheless, longitudinal studies can reveal interesting outcomes. For example, parallel studies at a science museum and a science centre by Falk, Scott, Dierking, Rennie and Jones (2004), demonstrated the quite different impact experienced by visitors when measured immediately after the visit, compared to some months later. There is still much to learn, particularly about the nature of the science communicated during the visit.

Research Examples of Science Communication in Science Centres and Museums

As we have seen, each visitor's experience is unique. Apart from their individual backgrounds, visitors differ in terms of why they choose to visit, what selection of exhibits they might attend to, and the social circumstances of their visit. Research has established that visitors are able to learn some science knowledge (albeit different bits of knowledge, even when visitors attend together), and may also learn something about the process of science. What opportunities they have to learn are determined by the nature of the science stories offered by the exhibits in the informal institution. In the following sections, two examples of research into science centre and museum visits are overviewed to demonstrate the kind of science communication that can occur. These examples report on different contexts and use different research approaches. In Rennie and Williams (2006a) we explored visitors' learning about science in two different venues and reported on the findings from pre- and post-visit surveys and interviews designed to measure what visitors learned about science. Pedretti (2011) used interviews, observations, and comments in visitors' books to report on visitors' responses to a travelling exhibition, *Body Worlds*, built around plastinated human cadavers and located at a science centre.

Visitors' Ideas about the Nature and Communication of Science

Two similar studies were carried out into adults' learning about science from a visit to a science centre (Rennie & Williams, 2002) and a natural history museum (Rennie & Williams, 2006b). These studies focused on participants' thinking about science in their daily lives, the nature and use of scientific knowledge, its communication by scientists, and how this thinking might change as a result of their visit. The research was built around a pre- and post-visit survey developed and validated for use in the science centre (Rennie &

Williams, 2002) with small wording changes to fit the museum context (Rennie & Williams, 2006b). At each venue, 102 visitors completed the surveys. Of the science centre visitors, 75 were also interviewed, and 67 interviews were conducted with the museum visitors. Results from the interview data complemented and assisted interpretation of the survey findings. Staff members at both the science centre and museum were also surveyed and interviewed (see Rennie & Williams, 2002, 2006b), but here the focus is on visitors' perceptions about science and nature of science portrayed by the museum.

In Rennie and Williams (2006a), we reported remarkably similar patterns in the survey results from the science centre and museum. Not surprisingly, since they had chosen to attend their particular venues, all visitors were very positive about their interest in, and the value of, science, and these positive attitudes increased after the visit, although the increase (as measured by the survey) was not always statistically significant. Visitors also believed that the science at the venue was easy to understand, and most believed they had learned something new. All of the science centre visitors and 83% of museum visitors interviewed stated that exhibits at their respective venue helped visitors to understand more about science and technology. Visitors also believed that scientific research was beneficial, and there was little change in this view. At both venues, visitors arrived with moderately scientific views about the nature of scientific knowledge, but after the visit, there were statistically significant changes. Visitors reported more positive views about scientists' concern about the impact of their research findings and their ability to communicate about their work to the general public. But visitors also became more likely to think that science has the answers to all questions, that scientists always agree with each other, and that explanations in science are definite rather than uncertain. These changes were also statistically significant.

Visitors' "less scientific" views about nature of science expressed after the visit were unexpected, but indicate how difficult it is to communicate uncertainty and controversy in science. Visits to science centres and museums tend to be relatively short (in these studies visits averaged 156 min and 98 min at the science centre and museum, respectively), there is much to see and little time can be spent at most exhibits. Visitors show literally "passing interest" at many exhibits, so perhaps it is not surprising that they do not engage deeply with most of them. Exhibit designers are aware of this and also the variety of knowledge and interests of visitors, so they attempt to develop exhibits that are easy to understand. For science-related exhibits, this usually means that the science content is simplified and made straightforward; hence it is likely to be interpreted as uncontroversial. Further, science information in the form of factual knowledge is much easier to display than the processes of developing that knowledge. Telling stories about the twists and turns as scientific knowledge is produced, challenged, and refined, requires quite complex exhibits, none of which were present in either the museum or the science centre, nor were there any exhibits designed to show the human side of science. At the science centre, most exhibits exemplifying science concepts were presented as analogies, thus distancing them even further from real life. "Real" scientists, more plentiful at the museum than the science centre, were not visible to the usual visitor, because they worked "behind the scenes." The overall result was that exhibits and their physical environment seemed to succeed in glossing over any uncertainty and values in the science information they presented.

We concluded (Rennie & Williams, 2006a) that both institutions failed to challenge visitors to think beyond what they already knew. Interviews revealed positive ideas about the importance of science, but "in terms of their engagement with the nature of scientific knowledge, visitors ... seem to have had an enjoyable but unchallenging experience" (p. 890). Indeed, some visitors remarked that they saw what they expected to see, but did not express much excitement or wonder about it. We concluded that if these kinds of institutions wish to convey more about the real-world nature of science, they need to be more aggressive in their approach to displays in order to challenge visitors' thinking. This will require exhibits that tell science stories with an increased focus on social relevance, and attention to the diversity of views and values associated with controversial issues.

Visitors' Responses to Body Worlds and the Story of the Heart

Much more controversial, challenging, and definitely human exhibitions display the work of Gunther von Hagen, who has perfected the technique of plastination. In this process, human cadavers are dissected, then bodily fluids and soluble fats are extracted and replaced with resins and elastomers. The cadavers or body parts are posed in artistic ways for teaching purposes, and then become rigid and permanently preserved. *Body Worlds and the Story of the Heart* is a touring exhibition of over 200 specimens that display the human body and its muscles, organs and tissues in ways that illustrate form and function. Pedretti (2011) conducted research at a science centre where these plasticised, real human bodies—called plastinates—were displayed, together with separate organs and body slices relating to the cardiovascular system. Her purpose, in part, was to examine how visitors responded to and interacted with the plastinates, and how they made meaning of their experience. Unlike most science centre exhibits, plastinates are not interactive. Although visitors can walk around them, they are to be viewed, not touched. Visitors connect immediately with plastinates because they are essentially people just like themselves. In some ways, the plastinates are rather like the typical museum exhibit; the real object is displayed, but instead of needing to be interpreted, the story it tells depends on how visitors make personal connections with the object.

How did visitors respond? Pedretti's (2011) extensive data sources included visitors' written comments in two visitor books, 54 tape-recorded interviews with visitors, field notes describing visitors' interactions with one another, and content analysis of various media releases, pamphlets, etc., about the exhibition.

Her overall findings corroborated other reports describing people's strongly emotional reactions to the human cadavers, responses which ranged from amazement, fascination, and wonder at the intricacies of the human body, to disgust and aversion, with belief that the exhibition was disrespectful. Pedretti interpreted her findings in the context of McLuhan's (1964) assertion that "the medium is the message", that is, the message received is determined by the medium and not the content delivered. Pedretti analysed the meaning visitors made—the message—from their encounters with the plastinates—the medium.

There was no doubt the medium had effect. The plastinates evoked what Pedretti (2011) described as a celebration of the human body: "The very physicality and 'realness' of the exhibition inspired awe, fascination and aesthetic appreciation" (p. 49). Further, they elicited numerous personal narratives that visitors related to the researchers about themselves, their families, or acquaintances. The theme of universality, that we are all human regardless of colour or race, and the tendency of visitors to "transpose" features of the plastinates to their own bodies were powerful responses to *Body Worlds*. The lack of text panels or other information about the plastinates as previous living persons may have assisted this response, as it offered no distraction to visitors' ability to relate the human features to themselves and known others in their response to the medium. The messages visitors took from *Body Worlds* were intimately related to the medium; messages about health and life style, for example, an exhibit of diseased lungs sending a message about the need to quit smoking. Visitors also believed they gained knowledge and understanding about the human body by seeing the relationships between body parts that are hidden beneath skin and flesh in live bodies.

Pedretti (2011) wrote about the tensions such exhibitions bring to the fore. The medium—the plastinates—was at once both confronting and awe-inspiring. In this sense there was dissonance; "a kind of simultaneous fascination and disgust" (p. 55), which many visitors found difficult to negotiate. There was also tension relating to how the plastinates were displayed, with many in elaborate, athletic positions; poses that some visitors viewed as artistic, but others found disrespectful. Pedretti referred to this as the "(re)presentation of the human bodies" (p. 56), pointing out that such representations "unavoidably lead to questions about what is possible, and what is appropriate for museums to exhibit. … Should these bodies/objects be on display for public consumption (while generating large amounts of revenue?). Inevitably, one is led to ask if *Body Worlds* has commodified death?" (p. 57).

In her conclusions, Pedretti (2011) noted how visitors "overwhelmingly produced relationships between real bodies in and outside the exhibition. This speaks to a kind of symbiotic relationship between the message, the medium (see McLuhan, 1964), and the viewer" (p. 59). It was also clear that *Body Worlds* was a controversial exhibition. Technologies like von Hagen's plastination provide opportunities to exhibit objects that deal with very personal issues, prompting responses that many viewers find confronting, challenging their moral and ethical stances. As Pedretti (2004) has written elsewhere, such exhibitions personalize subject matter, evoke emotion, stimulate dialogue and debate, and promote reflexivity. They offer great stories about science and society that provide opportunities for learning as visitors grapple with the emotions and issues involved.

Discussion

This chapter presented the argument that effective science and technology communication in informal environments is dependent on a productive two-way interaction between the source of science information and the intended audience. The audience has considerable power in this interaction, because people can choose to engage, or not, with the science information that is available. It was pointed out that informal institutions and their visitors each have their own agendas and there needs to be a "coming together" of these agendas so that dialogue can develop and effective communication can occur. It was suggested that communicating science requires that the science that scientists deal with needs to be recast into "science stories" to make it accessible by its intended audience. The audience—the visitors to institutions or other members of our heterogeneous public—can engage with and interpret these stories according to their own needs and experience. Invariably, there will be as many interpretations as there are visitors and, almost always, communication will be differentially successful. The shortfall between what is intended and what is actually communicated about science can be explained, at least in part, by how the science is presented.

The two research examples described above were chosen to contrast different kinds of opportunities for science communication that visitors may experience in visits to science centres or museums. There were no special exhibits at the museum and science centre in the first example; the exhibitions were the kind of displays one might expect at such venues. Rennie and Williams (2002, 2006b) found visitors to be positive about their visit (an expected outcome, since they had chosen to come), they enjoyed it, found it interesting, and believed they had a learning experience, but in terms of learning about science, many came away with views that did not reflect how science works. We attributed this unexpected outcome to the comfortable and unprovocative nature of the exhibits. Exhibit developers endeavour to make exhibits understandable to the visitor. However, as the researchers pointed out, this often results in simplification of the science content, detachment from people and the social and political circumstances that attend the concept or phenomenon in everyday life. Science can appear to be certain and unproblematic.

In contrast, the visitors to the science centre in Pedretti's (2011) research experienced exhibits that were much more personally confronting and more thought-provoking than those of visitors to the science centre or museum described by Rennie and Williams. Visitors to *Body Worlds* expected to have a different and novel experience because the exhibition had travelled to the centre and was well advertised. Nearly all enjoyed it (98% of those interviewed said so), but there is no doubt that, because of the nature of the exhibits, some people chose not attend. The plastinates are once-living people, and their display raises questions about ethics, morals, and respect. Although all bodies are anonymous and were willingly donated, treating them as public spectacle is inherently controversial. There is risk of offending people according to their personal values and beliefs. As Pedretti pointed out, the choice of this medium "raises interesting questions about the nature of controversial exhibitions, relationships between objects, subjects (viewers) and our being in the world, and how museums might prepare for and scaffold the visitor experience, in light of the multiplicity of cultural, social and aesthetic norms" (p. 59).

What can we say about science communication in these examples? Clearly, the experiences were different and different stories were evoked. In both cases, there was a communicative dialogue between visitors and the exhibits but, not surprisingly, the outcomes were quite different. Unlike the typical objects in a museum, where interpretation is often required to provide the story of their significance, the plastinates were objects that provoked visitors to create their own stories, and these were personal. Very likely, visitors to *Body Worlds* had a more emotive, and perhaps more memorable, experience than did visitors to the traditional museum and science centre, but there is no basis for saying that one experience is "better" than the other, because the venues had different stories to tell. However, we can say that to promote effective communication, the challenge for exhibit developers is two-fold. First, they must design exhibits that motivate the visitor to engage thoughtfully, otherwise there can be no dialogue and no story told. Second, and in order to represent science as both process and product, they must strike a balance between clarity in explaining the science concept, phenomenon or issue, and conveying its complexity in terms of relevant social factors and the uncertainty inherent in many topical science issues.

Striking this balance is challenging. Exhibit developers are invariably constrained by their circumstances, including the allocation of time and funds. They must answer questions such as what information and or objects are available for display? What are the opinions of the scientists/curators, the educationists, and exhibit constructors about what is possible and what is desirable to exhibit? If the science issue to be displayed is contested, such as genetically modified foods, then presenting contested "facts" will involve multiple and potentially confusing interpretations. Should the science centre or museum risk confusing or discomforting visitors by displaying controversial issues? The answers to these questions will determine the kind of science stories told, but whether or not visitors choose to engage with them will determine what, if any, science is communicated. As museum consultant and visitor advocate Paulette McManus (1993) pointed out: "Visitors appreciate a simple story-line or a clearly expounded argument *prepared with them in mind*" (p. 62, emphasis added). The challenge is to get visitors to exercise their choice positively and engage in dialogue with the exhibit's story if there is to be effective science communication.

Notes

1   There are many other environments where lay persons seek information about science; Rennie and Stocklmayer (2003) provide examples and discussion.

References

Allen, S. (2004). Designs for learning: Studying science museum exhibits that do more than entertain. *Science Education, 88* (Suppl. 1), S17-S33.

Anderson, D., Lucas, K. B., Ginns, I. S, & Dierking, L. D. (2000). Development of knowledge about electricity and magnetism during a visit to a science museum and related post-visit activities. *Science Education, 84*, 658-79.

Bell, P., Lewenstein, B., Shouse, A. W., & Feder, M. A. (eds.) (2009). *Learning science in informal environments: People, places, and pursuits.* Washington DC: The National Academies Press.

Crowley, K., Callanan, M. A., Jipson, J. L., Galco, J., Topping, K., & Shrager, J. (2001). Shared scientific thinking in everyday parent-child activity. *Science Education, 85,* 712-32.

DeWitt, J., & Storksdierk, M. (2008). A short review of school field trips: Key findings from the past and implications for the future. *Visitor Studies, 11*(2), 181-97.

Ellenbogen, K. M., Luke, J. L., & Dierking, L. D. (2004). Family learning research in museums: An emerging disciplinary matrix? *Science Education, 88*(Suppl. 1), S48-S58.

Falk, J. H., & Dierking, L. D. (2000). *Learning from museums: Visitor experiences and the making of meaning.* Walnut Creek, CA: Altamira Press.

Falk, J. H., Moussouri, T., & Coulson, D. (1998). The effect of visitors' agendas on museum learning. *Curator, 41*, 107-20.

Falk, J. H., Scott, C., Dierking, L., Rennie, L., & Jones, M. C. (2004). Interactives and visitor learning. *Curator, 47,* 171-98.

Griffin, J. (2004). Research on students and museums: Looking more closely at the students in school groups. *Science Education, 88*(Suppl. 1), S59-S70.

Hodson, D. (1998). Science fiction: The continuing misrepresentation of science in the school curriculum. *Curriculum Studies, 6* (2), 191-216.

International Council of Museums (ICOM). (2007). ICOM Statutes (Article 03-3 Definition of terms, Section 1 Museum). Retrieved September 16, 2011 from http://icom.museum/who-we-are/the-organisation/icom-statutes/3-definition-of-terms.html#sommairecontent

Layton, D., Jenkins, E., Macgill, S., & Davey, A. (1993). *Inarticulate science? Perspectives on the public understanding of science and some implications for science education.* Nafferton, England: Studies in Education Ltd.

McLuhan, M. (1964). *Understanding media: The extensions of man.* New York: McGraw Hill.

McManus, P. M. (1993). Towards a general communication philosophy for the National Technical Museum. In J. Bradburne & I. Janousek (Eds.), *Planning science museums for the new Europe: Proceedings of a seminar held at the Národní Techniké Muzeum, Prague* (pp. 55-62). Prague: UNESCO/ Národní Techniké Muzeum, Prague.

——(1994). Families in museums. In R. Miles & L. Zavala (Eds.), *Towards the museum of the future: New European perspectives* (pp. 81-97). London: Routledge.

Milne, C. (1998). Philosophically correct science stories? Examining the implications of heroic science stories for school science. *Journal of Research in Science Teaching, 35,* 175-87.

Pedretti, E. (2004). Perspectives on learning through critical issued-based science center exhibits. *Science Education, 88*(Suppl. 1), S34-S47.

——(2011). The medium is the message: Unravelling visitors' views of body worlds and the story of the heart. In E. Davidsson & A. Jakobsson (Eds.). *Understanding interactions at science centers and museums—Approaching sociocultural perspectives* (pp. 45-61). Rotterdam, The Netherlands: Sense Publishers.

Perry, D. L. (1989). The creation and verification of a development model for the design of a museum exhibit. (Doctoral dissertation, Indiana University, 1989). *Dissertation Abstracts International, 50,* 3296.

Rennie, L. J. (2007a). Learning science outside of school. In S. K. Abell & N. G. Lederman (Eds.), *Handbook of research on science education* (pp. 125-67). Mahwah, NJ: Lawrence Erlbaum Associates.

——(2007b). Values in science portrayed in out-of-school contexts. In D. Corrigan, R. Gunstone, & J. Dillon (Eds.), *The re-emergence of values in science education* (pp. 197-212). Rotterdam, The Netherlands: Sense Publications.

Rennie, L. J., & Johnston, D. J. (2004). The nature of learning and its implications for research on learning from museums. *Science Education, 88*(Suppl. 1), S4-S16.

Rennie, L. J., & McClafferty, T. P. (1995). Using visits to interactive science and technology centers, museums, aquaria, and zoos to promote learning in science. *Journal of Science Teacher Education, 6,* 175-85.

Rennie, L. J., & Williams, G. F. (2002). Science centres and scientific literacy: Promoting a relationship with science. *Science Education, 86, 706-26.*

——(2006a). Adults' learning about science in free-choice settings. *International Journal of Science Education, 28,* 871-93.

——(2006b). Communication about science in a traditional museum: Visitors' and staff's perceptions. *Cultural Studies of Science Education, 1,* 791-820. (http://www.springerlink.com/content/b4k4082561696118/)

Rennie, L. J., & Stocklmayer, S. M. (2003). The communication of science and technology: Past, present and future agendas. *International Journal of Science Education, 25,* 759-73.

Rickinson, M., Dillon, J., Teamey, K., Morris, M., Choi, M. Y., Sanders, D., et al. (2004). *A review of research on outdoor learning: Executive summary.* Retrieved September 13, 2004 from http://www.field-studies-council.org/documents/general/NFER/NFER Exec Summary.pdf

Scott, C. (2003). Museums and impact. *Curator, 46,* 293-310.

Semper, R. J. (1990). Science museums as environments for learning. *Physics Today, 43*(11), 50-6.

Shields, C. J. (1993). Do science museums educate or just entertain? *The Education Digest, 58*(7), 69-72.

Stocklmayer, S. M., & Gilbert, J. K. (2002). New experiences and old knowledge: Towards a model for the public awareness of science. *International Journal of Science Education, 24*, 835-58.

Wellington, J. (1990). Formal and informal learning in science: The role of interactive science centres. *Physics Education, 25*, 247-52.

PART V
Communication of Contemporary Issues in Science and Society

13

COMMUNICATING GLOBAL CLIMATE CHANGE

Issues and Dilemmas

Justin Dillon and Marie Hobson

Introduction

A campaign is being launched across Australia to promote respect for science. The campaign comes after it was announced last week that climate scientists in the country had received death threats, with the Australian National University increasing security around nine climate scientists and administrative staff. With national debate over climate change becoming increasingly heated owing to the government's carbon tax, scientists are having to battle against what Anna-Maria Arabia, chief executive of the Federation of Australian Science and Technological Societies, called ''a noisy misinformation campaign by climate denialists''.

(Times Higher Education, June 30, 2011, p.16)

This chapter critically examines the relationship between science, scientists, and science communication through the lens of the major environmental issue facing society, climate change. Many, if not all, of the issues are relevant to science communicators wherever they work. The long-term nature of the challenges thrown up by climate change means that developments in the relationship can be identified and, as the subject will not 'go away' quickly, it allows opportunities for speculation about future trends and possibilities for science communication. Climate change is an issue that can provoke strong responses among experts as well as the lay public. Increasingly, as the story from the *Times Higher Education* above suggests, it is already an issue of life and death. A range of topics needs to be considered when discussing the communication of climate change. These include: the role of the media; public trust in scientists; public understanding of the science; and the nature of climate science; specifically, what counts as evidence.

So, this chapter will address the various conceptualizations of climate change/ global warming and look at how the relationship between science, scientists and the public has changed over recent years. We will also examine whether science communicators should promote the scientific consensus or encourage debate about the ways that science is carried out and at how scientists operate. Throughout the chapter, we will draw on the experience of the Science Museum, London and the planning that went into the design of a new gallery on climate change (*atmosphere ... exploring climate science*) that was opened in 2010.

The first thing to note is that, as with many issues, climate change has taken on a political dimension. Writing in *The Guardian* several years ago, the columnist Polly Toynbee described the tendency of the right-wing to line up against scientific evidence in issues concerned with public health and the environment:

Posing as hard-headed realists, those on the right are more prone to pit their ideology against the weight of science. Seat belts? Motorbike helmets? Chlorofluorocarbons and the ozone layer? Smoking bans? Advertising junk food to children? The science-based realos tend to be on the left, conviction funds on the right.

(Toynbee, 2006)

Five years on, Toynbee (2011), commenting on the influence of the US extreme right-wing on political debate in the UK, noted that 'a taste of the Tea Party arrives on these shores in the peculiar paranoia of the climate-change deniers'. Toynbee's position, which is mirrored in the public policies of the major UK political parties is that 'On matters of fact, those of us who are not scientists can only listen to what scientists say and trust such an overwhelming global consensus'. For many people, access to 'what scientists say' is moderated by the media and by science communicators on the Internet and in museums and science centres.

The question is, however, what does this moderation entail? Does it, for example, involve presenting the scientific consensus and looking at the range of predicted impacts of climate change? Or does it involve presenting climate change science as hotly contested? These, and other questions, are ones which faced the Science Museum when they decided to create a new gallery: *atmosphere ... exploring climate science.*

The gallery was designed to achieve the following goals:

- To deliver an immersive, enjoyable and memorable (life-enhancing) experience that increases interest, deepens understanding and is robust against deeply held convictions:

- To be recognised and admired as *the* UK destination for clear, accurate, up-to-date information on climate science for the non-specialist.

The exhibition was targeted at independent (non-specialist) adult visitors; families with children aged 8+; and high school science and geography teachers and their students (aged 11-16).

The development of the gallery was heavily influenced by research carried out by the second author and her colleagues in the Audience Research and Advocacy Department. This research involved consultation, desk-reviews and prototype testing. Initial consultations with the target audiences involved focus groups and in-depth interviews to find out their prior knowledge, opinions, and 'misconceptions' about climate change and climate science. The consultation also identified potential visitors' expectations of what the gallery would look like and tried to identify any barriers to attendance or engagement. The research team carried out desk-reviews of academic papers, public opinion polls, and other climate change exhibitions (both at the Science Museum and elsewhere). The final stage of the research involved prototype testing of the interactive exhibits to reduce and remove barriers to usability, comprehension, and motivation.

A key part of this work was a survey of 30 adult visitors' familiarity and understanding of terms and concepts, such as 'greenhouse gases', 'remediation', and 'carbon footprint' to develop a mental model of what visitors considered to be the causes of and possible responses to climate change. The findings were presented to the exhibition team as a two-page diagram, with areas of misunderstanding clearly highlighted, so it could be easily referred to when discussing content, to ensure that the team defined terms where necessary and avoided reinforcing misconceptions. This work showed that visitors have pockets of knowledge about climate change which they struggle to link together accurately and it demonstrated the need for the audience to have the science behind climate change clearly explained. The findings and implications of the research is described more fully below.

The Science and Terminology of Climate Change

Our understanding of climate change as a scientific phenomenon has developed rapidly since the 1980s. Climate change refers to long-term changes in weather patterns over a region or across the planet. Global warming refers to the process by which the average temperature of the Earth's near-surface air and oceans has increased relatively recently and continues to increase. The greenhouse effect refers to the process in which thermal radiation from the Earth is absorbed by gases in the atmosphere and then re-radiated. This re-radiation increases the temperature of the Earth.

The evolution in our understanding of climate change, and the issues which are currently open to debate, are outlined by Steve Jones, Emeritus Professor of Genetics at University College London, in his independent report to the BBC Trust, which was carried out in 2010:

In its early days, two decades ago, there was a genuine scientific debate about the reality of climate change (although that attracted rather little attention). Now, there is general agreement that warming is a fact even if there remain uncertainties about how fast, and how much, the temperature might rise. At present, the pessimists are in the ascendant and today's increase in floods and snow (as predicted for a warmer atmosphere which can take up more water) is on their side. A debate remains, and it deserves to be reported with as much objectivity as would any other unresolved issue.

(BBC Trust, 2011, p.68)

The scientific consensus on the key issues is outlined in the Third Assessment Report of the Intergovernmental Panel on Climate Change (IPCC 2001). The report's main conclusions are that the global average surface temperature has risen by 0.6 ± 0.2 °C since the late 19th century, and by 0.17 °C per decade since the 1970s, and that there is new and stronger evidence that most of the warming observed over the last 50 years is attributable to human activities. The Panel also concluded that if emissions of greenhouse gas continue, then the warming will also continue. Temperatures were projected to increase by 1.4 °C to 5.8 °C between 1990 and 2100. Accompanying this temperature increase will be increases in some types of extreme weather and a projected sea level rise of 9 cm to 88 cm.

The growing consensus about climate change science coalesces around findings such as these. Scientists, by their very nature, should be open to contrary evidence and to new ideas which might emerge with future research. Science has a habit of making new knowledge by falsifying old ideas. There is a debate about the degree to which climate change will impact on the planet—science can predict phenomena in the real world, but not always with 100 per cent accuracy—but the debate about what were once contested ideas has withered on the vine. The question is, though, to what extent have scientists and science communicators been successful in presenting these ideas and educating the public about some of the most important issues facing civilisation?

What Do the Public Know and Believe?

According to to a large-scale survey carried out early in this century, most of the UK's population had heard of the terms 'climate change', 'global warming' or the 'greenhouse effect' (DEFRA, 2002). At first, the public seemed more familiar with 'global warming' than 'climate change' (DEFRA, 2002; Whitmarsh, 2009). However, during the late 2000s, it is thought that people have become equally aware of both terms (Upham, et al., 2009).

An increased familiarity with both 'climate change' and 'global warming' suggests that the public regard the terms as synonymous. Only four per cent of Whitmarsh's respondents explicitly differentiated between 'global warming' and 'climate change' without being prompted to do so (Whitmarsh, 2009, p.410). This situation may have arisen because the terms are used interchangeably by journalists and scientists. Generally, however, the media prefers to use 'global warming', while scientists and policy writers prefer 'climate change' (Whitmarsh, 2009).

The Center for Research on Environmental Decisions (CRED) in the US (CRED, 2009) suggests that 'climate change' is a better choice than the term 'global warming' because 'climate change' better conveys broader changes in the earth's ecosystems and varying global temperatures from one year to the next (CRED, 2009, p.2). Unlike 'global warming', the term 'climate change' avoids 'the misleading implications that every region of the world is warming uniformly' and 'the idea that the only concern with increased greenhouse gases is higher temperatures' (ibid).

Surveys of the public suggest that CRED is right to make this recommend-ation. It does seem that the public associate 'climate change' with a broader range of impacts on the climate and weather (for example, hot summers, wetter winters, rainfall, and drought) and a broader range of causes (that is, both natural and human) than 'global warming' (Whitmarsh, 2009, p.410). Associated with the term 'global warming' are the notions of heat being 'trapped', increasing temperatures, human causes, and misconceptions surrounding the depletion of the ozone layer.

Atmospheric Science: The Controversy and Sources of Confusion

Atmospheric science entered the public consciousness some years before climate change grabbed the headlines. The discovery of a 'hole' in the ozone layer in 1985 attracted substantial media attention. Ozone depletion, a phenomenon that had been detected in the 1970s, refers to both the relatively steady decline in the total volume of ozone in the stratosphere, and to the seasonal decrease in the concentration of ozone above the polar regions. Ozone depletion has since been linked to the level of human use of substances, such as some refrigerants. The phenomenon persists and it has been associated with climate change although the nature of the link is unclear.

In the early 1990s, Boyes and Stanisstreet noted that global warming might present some significant challenges to educators. Their survey of undergraduate students' perceptions identified a range of incorrect ideas including a confusion between global warming and the depletion of the ozone layer (Boyes & Stanisstreet, 1992). Like many scientific phenomena, the key concepts are abstract and somewhat distant for students. It is difficult to carry out experiments or hands-on inquiries, so a lot of climate change education depends on data interpretation (with or without a computer), discussions, or watching videos.

Ironically, while the media coverage of the ozone hole was successful in raising public awareness of that issue, and prompting action to virtually wipe out the use of chlorofluorocarbons (CFCs), it is now presenting a barrier to the successful communication of climate change more broadly. In 2010, as part of the planning and preparation for the Science Museum's *atmosphere* gallery, the Audience Research and Advocacy team conducted a survey of visitors' knowledge of the science behind climate change (Science Museum, 2010a). One third of the participants thought that greenhouse gases caused the hole in the ozone layer. As one participant put it: 'They [greenhouse gases] are depleting our ozone layer – there are more rays coming through and we are heating up'.

The Science Museum adult visitor survey found that the analogy of a greenhouse fosters misconceptions, particularly relating to the ozone layer. Visitors in the survey thought that the greenhouse gases formed a physical layer in the atmosphere, like a ceiling, and that the ozone hole was in this layer. Consequently, in designing the gallery, Science Museum staff took great pains to avoid any graphical depictions of greenhouse gases forming 'ceilings', instead trying to depict varying concentrations of gases.

What and Who Do the Public Believe?

A person's level of understanding or mental model of climate change both affects, and is affected by, their level of belief. Altering either represents a huge challenge for climate change communicators as individuals are prone to 'confirmation bias', that is, a tendency to interpret information in such a way that it affirms their prior knowledge (CRED 2009). In the case of the ozone layer, sceptics may interpret the hole as letting hot air out to justify their belief that climate change is *not* happening, while a believer may interpret it as letting more hot air in to justify their belief that it *is* happening.

Public opinion about climate change varies significantly from country to country. It also varies within countries, according to various demographics. In the UK, believers in anthropogenic climate change tend to be children or adults aged between 18-40 and female, while sceptics/deniers are more commonly adults aged 40+ (particularly aged 60+) and male (Science Museum 2010b). Overall it seems most people believe climate change is happening, but not necessarily that it has anthropogenic causes. According to a survey conducted by the BBC in 2010, while 75 per cent believed climate change was occurring, only a third of them also thought it was caused by humans (BBC 2010). This finding resonates with Cardiff University research which found that 78 per cent of people surveyed believed in climate change, but only 31 per cent thought it was primarily the result of human activity (Spence, Venables, Pidgeon, Poortinga & Demski, 2010). In addition, the levels of belief in climate change appear to be decreasing over time. The Cardiff research revealed a swing of 13 per cent (down from 91 per cent to 78 per cent) from believers to disbelievers between their 2005 and 2010 surveys (Spence et al., 2010). The BBC identified a swing of eight per cent (down from 83 per cent to 75 per cent) between November 2009 and February 2010 (BBC 2010)—a period of a few months.

In 2010, George Mason University's Centre for Climate Change Communication (C4) published a report that compared US public opinion in November 2008 and June 2010 (Leiserowitz, Maibach, Roser-Renouf & Smith, 2010). The percentage of respondents classified as 'alarmed' or 'concerned' about climate change dropped from 51 per cent to 41 per cent during the period. The number of respondents classified as 'doubtful' or 'dismissive' rose from 18 per cent to 24 per cent. These data indicate the challenge facing climate change communicators, as well as the malleability of public opinion. The segmentation of the audience used in the analysis of the C4 report points to a need for different messages for different sectors of the public (or publics).

Issues of Trust

Earlier, we pointed to Polly Toynbee's statement that 'On matters of fact, those of us who are not scientists can only listen to what scientists say and trust such an overwhelming global consensus' (Toynbee, 2006). Trust in science and scientists underpins much of what passes for science education. At school, much of what is taught about science is the received wisdom of teachers and textbooks. What evidence do you have, for example, that the Earth spins on its axis? Trust, though, can be undermined, as recent events have all too clearly shown.

The 'Climategate' affair, in 2009, involving the release of 160 mb of emails and data from the University of East Anglia's Climate Research Unit, has changed the ways in which climate scientists operate and communicate with each other and with the public. The reporting of the affair may well be the major reason for the swing of eight per cent in the proportion of the population who believed that climate change was real, that the BBC identified as happening between November 2009 and February 2010.

Such was the potential damage to public trust in government climate change policy, that the UK political response to 'Climategate' involved then Prime Minister Gordon Brown being forced to comment that 'With only days to go before Copenhagen [Summit] we mustn't be distracted by the behind-the-times, anti-science, flat-earth climate skeptics'. Such language may not always help to build trust in the government's position as Leo Hickman (2011) noted in a *Guardian* blog. Hickman was concerned that a new term 'climate crank' had been added to the already long list of terms varying from relatively neutral to derogatory: 'sceptic, denier, contrarian, realist, dissenter, flat-earther, misinformer, and confusionist'. Characterising sections of society in such negative terms may backfire on those who use such strong terms.

The Public's Levels of Trust

In developing its exhibition, the Science Museum commissioned focus groups in April 2008, more than a year before Climategate, from TWResearch. The study revealed three factors which affect people's level of trust in sources of information about climate change: *hypocrisy* in that the public do not want to be told what to do without seeing any evidence of others taking action and practising what they preach'; *profit* in that there are two confusing dichotomies here: firstly, between the altruistic act of combating climate change being championed by businesses, such as energy companies encouraging home owners to switch to renewable energy sources, etc; secondly between 'independent' scientists being funded by the Government or businesses; and, *inconsistency* in that sources are expected to have a consistent stance, rather than, as the public perceive it, changing their views to suit the latest trend. As a result, the government, businesses, and the media were least trusted by the focus group participants, while scientists, charities, and non-profit public organisations (such as the Science Museum) were most trusted. Science communicators working in the media, then, would seem to be in a curious position—trusted because of their science credentials, but mistrusted because of their media employment.

The public's association of 'science' with 'truth' and 'facts' results in scientists being viewed primarily as independent truth-seekers (TWResearch, 2008). Confusion therefore arises over scientists' lack of agreement around climate change issues and this perceived 'inconsistency' is often cited as a reason for lack of belief in climate change. This observation suggests that, in order to communicate climate change effectively, more work needs to be done in communicating the nature of science to the public.

The Science Museum, as an educational, science-based institution, is considered a place to find reliable information based on evidence, rather than opinions. However, to overcome some of the potential barriers to trust in their exhibition, the Museum: employed a Sustainability Consultant; set up a Carbon Reduction Working Group, and succeeded in reducing its carbon footprint by 17 per cent between 2009 and 2010; retained editorial control from all sponsors; and focused on presenting the science behind climate change.

Teachers' Views of Climate Change

Teachers' understandings of global warming are critical in that many museum visits take place in the form of school visits. It is during the preparation and follow-up of visits that the messages of exhibitions can be enhanced and moderated. Science Museum focus groups with secondary school science and geography teachers revealed, however, that some of the science teachers were not convinced that climate change was caused by humans (TWResearch, 2010). Dove's (1996) survey of student teachers' understanding of the greenhouse effect, ozone layer depletion, and acid rain found similar confusions to those identified earlier by Boyes and Stanisstreet in university students. Dove, though, was puzzled as to why the prospective teachers understood the science behind the ozone layer but did not understand the greenhouse effect.

Following her survey of pre-service teachers, Dove hypothesized that while the link between CFCs and the depletion of the ozone layer was well-established, global warming was somewhat contentious. Another possible challenge might be that the science behind the greenhouse effect is more difficult than that behind ozone depletion. Dove noted that the difference in understandings of the different phenomena raised the question 'as to whether understanding would be improved by simply presenting the concepts involved, or if alternative teaching methods are needed to make the message clear' (p.99). Although Dove's research was carried out in the mid-1990s, its findings are likely to be relevant today.

Mason and Santi (1998) advocate that teachers should use constructivist approaches including discussions about different interpretations of evidence for global warming. The Science Museum focus groups revealed that teachers found dealing with conflicting and ever-changing evidence difficult as this sense of uncertainty conflicts with their perception of their role as teaching the 'truth' (TWResearch 2010). This is quite a paradox for teachers as they also need to teach about the nature of science, such as the tentative nature of some scientific knowledge and the value of disagreement over explanations of phenomena in the natural world.

Changing Attitudes/Behaviours

As the scientific consensus about the human causes of climate change has strengthened, and the potential consequences of global warming have become more immediate, some educators have advocated a shift in the type of education that is offered to students. Uzzell (1999) criticized much environmental education as being top-down and from the centre to the periphery, and argued that it did not have a good track record of changing the attitudes and values of children to the environment. More recently, educators have advocated new models and strategies for climate change education such as harnessing the power of community action (Moser & Dilling, 2004).

Cordero, Todd and Abellera (2008) reported that to be effective, climate change education should emphasize the personal connection between the student, energy, and climate change using methods such as environmental footprint calculations. Such strategies, they argue, can improve students' understanding of the links between personal energy use and global warming, a point echoed by Devine-Wright, Devine-Wright and Fleming (2004).

Barriers to Communicating Climate Change

Climate change communication is a challenging activity. In addition to the lack of understanding and lack of trust discussed above, the Science Museum focus groups (TWResearch 2008 and 2009) revealed a range of barriers for people engaging in climate change, regardless of their opinion as to whether it is happening, human caused or threatening:

- *Boredom:* climate change is constantly in the media and, for children, it is a topic they encounter in multiple subjects throughout their school career;
- *Irritation:* the public do not want to be told what to do and how to live their lives, particularly when it involves foregoing activities they enjoy, such as travelling abroad;
- *Powerlessness:* the public feel that individual actions are futile and have no sense of collective impact; they feel there has been little change and have a low awareness of international efforts, for example, the 2009 Copenhagen conference;
- *Fear:* the public do not know how bad the impacts will be, the effect it will have on themselves, or if it is even too late to act.

According to Roser-Renouf and Maibach (2010) action has been hampered by political partisanship and industry disinformation campaigns; principles of fairness in news coverage have given a far greater voice to the handful of skeptics than is merited by either their numbers or their evidence; and publication of their views has fostered a widespread perception in the public of scientific controversy, where none actually exists. Therefore, the issue remains a low policy priority for most people, and it is likely to remain so until the perception of controversy is overcome and people clearly understand both the dangers we face, and the actions we must take to avert these dangers.

The issue of impartiality is one that continues to vex science communicators. In his review of the BBC's science coverage, Steve Jones identified the issue as one that goes beyond climate change:

A belief in alternative medicine or in astrology and a fear of vaccines or of GM food are symptoms of a deep mistrust in conventional wisdom. Such scepticism should be part of every scientist's, every journalist's or every politician's, armoury. However, mistrust can harden into denial. That faces the media with a problem for, in their desire to give an objective account of what appears to be an emerging controversy, they face the danger of being trapped into false balance; into giving equal coverage to the views of a determined but deluded minority and to those of a united but less insistent majority. Nowhere is the struggle to find the correct position better seen than in the issue of global warming.

(BBC Trust, 2011, p.66)

Towards Overcoming Barriers

To communicate climate change effectively, communicators need to continue to engage and educate the public with the evidence that it is happening and that it is caused primarily by humans. Through audience consultation, the Science Museum (TWResearch 2008) identified the following strategies to *engage* the public with the issue of climate change:

- Focusing on humans: The public seem to be interested in the human stories, particularly those relating to:
  - the UK: these are emotive and can make the issue personally relevant;
  - the class war: the sense of injustice is motivating;
  - countries already experiencing the effects: that the effects are happening to people now helps make the issue seem less remote and more immediate;
- Personal relevance: Many visitors fail to relate to the global issue. They want to know how it will impact on them and in what time-frame or they can dismiss the issue as not relevant to them;
- Providing examples of possible adaptation and innovative solutions: Examples of action that could or have been taken can provide a message of hope in an otherwise gloomy picture. Visitors have low awareness of these broader solutions beyond re-using carrier bags and replacing incandescent light bulbs;
- Providing examples of solutions from other countries: Visitors are intrigued by what other countries have done and how it provides hope.

In order to educate, communicators need to present what the public considers to be evidence that anthropogenic climate change is happening. A crucial point to be borne in mind is that the public do not consider the effects of climate change (such as sea level rise), or statements about the consensus of scientific opinion, as evidence of climate change (even though they are frequently reported in the media)—they could be attributed to many different problems, not just climate change. Instead the public want to see why those impacts, responses, and opinions are related specifically to climate change. The Science Museum did this in the *atmosphere* exhibition by:

- Presenting graphs which demonstrate rising CO_2 or temperature levels, such as the Keeling Curve;
- Displaying objects which scientists use to work out how the climate has changed (for example, an ice core) or might change (for example, a weather balloon);
- Developing interactive exhibits which explain the carbon cycle and the greenhouse effect;

- Designing an exhibit which encourages visitors to compare natural and human causes of climate change to deduce which is the more likely cause of the current period of warming;
- Creating a whole zone of the exhibition on how scientists predict the future through climate modelling.

The C4 report noted that 'regardless of their beliefs about global warming, large numbers of Americans said they engage in energy conservation actions at home—turning off lights and electronics, reducing their use of heating and air conditioning, conserving water and replacing incandescent bulbs with compact fluorescents' (Leiserowitz et al., 2010, p.6). This finding suggests that a range of factors are influencing public patterns of consumption, including economics and perceptions of social responsibility. Education is just one of the influences on public behaviour. Changes in consumer behaviour are themselves likely to influence members of the public so that new patterns of behaviour emerge. One strategy for climate change communicators is to spread awareness of changes in public behaviour such as recycling or installing energy-saving lighting (actually, energy-saving is a misnomer, a more accurate term would be 'fuel saving').

The Science Museum survey and subsequent focus groups established that visitors participated in such 'energy-saving' activities although they did not know what else they could do beyond that. They wanted to know what else they could do, but they did not want to be dictated to. Neither did they want to make major lifestyle changes, such as giving up flying, as that would have too negative an impact on their lives. They had limited knowledge of mitigation or remediation, and they could not see how actions such as writing to an MP, or events such as the Copenhagen conference, would help.

Conclusions

Climate change presents significant opportunities and challenges to science communicators. The scientific consensus grows year by year, and there seems to be no justification nowadays for science communicators presenting the arguments of sceptics and deniers, except as an example of how hard it is to convince some sectors of society.

We know that public understanding of the issues is patchy and fragmented. The public, generally, trust scientists and their explanations. We suspect that part of the challenge facing science communicators is that the public do not fully appreciate the scale of the scientific consensus. Sometimes the uncertainty that scientists display, in terms of predicting what might happen in future decades, is seen as evidence that they disagree about whether or not climate change is caused by humans. Science communicators can help by showing people that uncertainty is part of how science works.

The London Science Museum's approach to researching public understanding and opinions about climate change provides a model for other institutions. The findings of the research provided a firm base for the exhibition developers. The realisation that visitors' often displayed serious misunderstandings, shifted the exhibition development team's focus towards explaining the science behind climate change, rather than, as was originally planned, on the broader impacts and issues raised by climate change.

What does the future hold? The sheer economic consequences of climate change will ensure that governments and industry takes action even if public opinion lags behind. Even in countries such as the US where there seems to be a perverse delight in ignoring scientific evidence, public opinion will eventually swing to accept that the human influence on the climate has been catastrophic. The challenge for science communicators is to maintain the public's interest in the topic while simultaneously showing that each individual can, and should, do their best to reduce their fuel consumption.

References

BBC (2010). *BBC Climate Change Poll*. Available at: http://news.bbc.co.uk/nol/shared/bsp/hi/pdfs/05_02_10climatechange.pdf. (Accessed on August 11, 2011).

BBC Trust (2011). *BBC Trust review of impartiality and accuracy of the BBC's coverage of science*. London: BBC Trust.

Boyes, E., & Stanisstreet, M. (1992). Students' perceptions of global warming. *International Journal of Environmental Studies*, 42(4), 287–300

Center for Research on Environmental Decisions (CRED). (2009). *The Psychology of Climate Change Communication: A Guide for Scientists, Journalists, Educators, Political Aides, and the Interested Public*. CRED: New York.

Cordero, E. C., Todd, A. M. & Abellera, D. (2008). Climate change education and the ecological footprint, *Bulletin of the American Meteorological Society*, 89, 865–72.

Department for Environment, Food and Rural Affairs (DEFRA) (2002). *Survey of public attitudes to quality of life and to the environment—2001*. London: DEFRA.

Devine-Wright, P., Devine-Wright, H. and Fleming, P. (2004). Situational influences upon children's beliefs about global warming and energy. *Environmental Education Research*, 10(4), 493–506.

Dove, J. (1996). Student teacher understanding of the greenhouse effect, ozone layer depletion, and acid rain. *Environmental Education Research*, 2(1), 89–100.

Hickman, L. (2011). The need for caution when 'calling out the climate cranks'. *The Guardian*. Available at: http://www.guardian.co.uk/environment/blog/2011/feb/14/climate-cranks-caution-sceptics-protest [accessed on August 8, 2011].

Intergovernmental Panel on Climate Change (IPCC 2001). *Climate Change 2001. IPCC Third Assessment Report*. Cambridge: Cambridge University Press.

Leiserowitz, A., Maibach, E., Roser-Renouf, C. and Smith, N. (2010). *Global Warming's Six Americas, June 2010*. Yale University and George Mason University. New Haven, CT: Yale Project on Climate Change.

Mason, L., & Santi, M. (1998). Discussing the greenhouse effect: Children's collaborative discourse reasoning and conceptual change. *Environmental Education Research*, 4(1), 67–85.

Moser, S., & Dilling, L. (2004). Making climate hot: Communicating the urgency and challenge of global climate change. *Environment*, 46, 32–46.

Roser-Renouf, C., & Maibach, E. (2010). Communicating climate change. In S. Priest (Ed.), *The Encyclopedia of Science and Technology Communication*, Sage Publications.

Science Museum (2010a). *Visitors' Mental Model of Climate Change* (unpublished).

——(2010b). *Audience's Attitudes Table*. (unpublished)

Spence, A., Venables, D., Pidgeon, N., Poortinga, W., & Demski, C. (2010) *Public Perceptions of Climate Change and Energy Futures in Britain: Summary Findings of a Survey Conducted in January-March 2010*. Cardiff: Cardiff University.

Times Higher Education (2011). Science is fair dinkum. *Times Higher Education*, June 30, 2011, p.16. Available at: http://www.timeshighereducation.co.uk/story.asp?story Code=416621&sectioncode=26 [accessed on August 10, 2011].

Toynbee, P. (2006). The climate-change deniers have now gone nuclear. *The Guardian*. Available at: http://www.guardian.co.uk/commentisfree/2006/jul/18/comment. politics3 [accessed on August 8, 2011].

——(2011). Britain must resist Tea Party thinking. *The Guardian*. Available at: http://www.guardian.co.uk/commentisfree/2011/aug/01/britain-resist-tea-party-thinking [accessed on August 8, 2011].

TWResearch (2008). *A Climate Change Gallery at the Science Museum* (unpublished).

——(2009). *Developing the Climate Change Exhibition* (unpublished).

——(2010). *A Climate Change Toolkit for Teachers* (unpublished).

Upham, P., Whitmarsh, L., Poortinga, W., Purdam, K., Darton, A., McLachlan, C., & Devine-Wright, P. (2009). *Public attitudes to environmental change: a selective review of theory and practice*. Swindon: Economic and Social Research Council. Available at: http://www.esrc.ac.uk/_images/LWEC-research-synthesis-full-report_tcm8-6384.pdf. Accessed on August 10, 2001.

Uzzell, D. L. (1999). Education for environmental action in the community: New roles and relationships. *Cambridge Journal of Education*, 29(3), 397–413.

Whitmarsh, L. (2009). What's in a name? Commonalities and differences in public understanding of "climate change" and "global warming". *Public Understanding of Society*, 18(4), 401–20.

14

SCIENCE COMMUNICATION DURING A SHORT-TERM CRISIS

The Case of Severe Acute Respiratory Syndrome (SARS)

Yeung Chung Lee

Introduction

Originating in southern China during late 2002, Severe Acute Respiratory Syndrome (SARS) is believed to have spread to Hong Kong via a medical doctor—a so-called "super-spreader"—who became infected after treating SARS patients. During his stay in a hotel, the doctor transmitted the disease to about a dozen residents, who in turn spread the disease within Hong Kong and to other countries through air travel (World Health Organization 2003a). During 2003, the disease affected people in 29 countries and claimed thousands of lives. The outbreak died down after July 2003 almost as rapidly as it emerged. The reason that SARS was successfully combated within such a relatively short period can be attributed to the utility of science as a way of seeking knowledge and solving human problems. However, behind this apparent success, there are numerous issues, dilemmas, tensions, and controversies relating to the operation and communication of science within the scientific community and society. Understanding the complexity of science communication and its underlying influences is important for resolving similar crises in the future. Because of space limitation, this chapter necessarily adopts a snapshot approach, by focusing on a number of critical incidents during the crisis to illustrate important issues at different levels of science communication. These issues are further reflected upon so as to throw light on the lessons learnt in this specific context. Before analyzing the case of the SARS crisis, it is important to revisit the roles that science plays in society, as these underpinned science communication during the crisis.

Roles of Science and Science Communication

There appears to be a consensus that the chief role of science is to solve problems, motivated either by curiosity or by a desire to have a positive impact on human life (Dawson, 1991). However, there are clear indications that modern scientific research is becoming increasingly applied in character (Layton, Jenkins, Macgill & Davey, 1993; Ziman, 2000). Layton et al. (1993) posit the following three categories that characterize the particular roles of science: *basic fundamental science* driven by a desire to understand natural phenomena; *strategic science*, which supports the development of technology; and *mandated science*, which assists in policy making or establishing standards. Additionally, it has been strongly argued that science is useful for all citizens, as it provides essential knowledge that enables individuals to make personal decisions (Bybee, 1997; Irwin, 1995) and, thereby, to participate in a democratic society (AAAS 1990). In light of these, science has at least four fundamental roles to perform—to enhance our understanding of nature, to support technological development, to assist public policy making, and to foster citizen action.

This chapter argues that the fulfillment of these four roles of science during the SARS crisis entailed the communication of science amongst scientists and various stakeholders. Any issues arising from this multilateral communication are likely to have had an impact on the utility of science in resolving the crisis. Hence, the analysis of these issues is instructive, and best situated within a multi-level framework that differentiates science communication into three levels. The first level is the science communication necessary for understanding and curing the disease, which mainly involved scientists and medical technologists. The relevant technology was highly dependent on scientific discoveries in the case of SARS, but there is limited space here for a detailed treatise of the highly specialized technology involved in the development of artifacts, such as test kits and drugs. Accordingly, this level of analysis focuses mainly on the scientific aspects of communication. The second level is science communication for policy making, which involved mainly government officials, scientists, and health professionals. The third level is scientific communication for citizen action, which needed to engage the public at large. Apart from the communication within each level, multilateral communication also occurred at the interface of the three levels, resulting in a complex pattern of interactions, which can be represented by a tripartite relationship, as depicted in Figure 14.1. As a further guide to unraveling this complex communication pattern, and to provide a focus for this study, the following questions are addressed throughout this chapter:

1. What are the roles of science communication at each level? How is science communicated within each level and across different levels?

2. How is the communication of science facilitated or impeded by factors related to science and the social context?

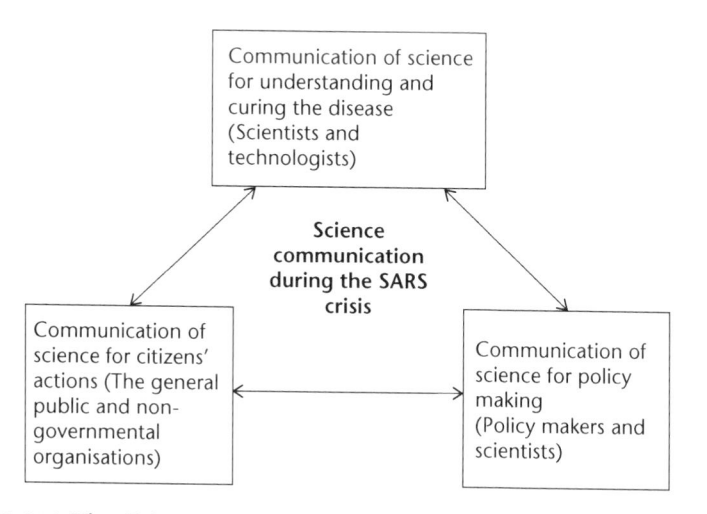

FIGURE 14.1 The Tripartite Relationship among the Three Levels of Science Communication during the SARS Crisis

3. What are the issues, tensions, or dilemmas arising from the interactions among these factors?
4. What impact did the communication of science have on the crisis? How well has science communication fulfilled its various roles?
5. What lessons can be learned about science communication from this crisis?
6. What are the challenges for science communication in resolving similar crises in the future?

Level 1: Science Communication for Understanding and Curing the Disease

During the SARS crisis, communication among scientists was essential for discovering many aspects of the disease such as the causative agent, the transmission modes, and the source of the pathogen. As the generation of scientific knowledge is ultimately governed by the nature of science, a brief detour through the nature of science is essential to understand the dynamics and issues involved in communication at this level.

The nature of science can be defined as a way of knowing, or the values and beliefs that are inherent to the development of scientific knowledge (Lederman & Zeidler, 1987). It is also portrayed as "a rich description of what science is, how it works, how scientists operate as a social group and how society itself both directs and reacts to scientific endeavours" (McComas, Clough & Almazroa, 1998, p.4). McComas et al. identified fourteen tenets of the nature of science that can be categorized into philosophical, psychological, historical, and sociological aspects. From a philosophical perspective, science is an attempt to explain natural phenomena. The search for scientific knowledge relies on observation, experimental evidence, rational argument, and skepticism. Scientific knowledge, although durable, is tentative. From a psychological point of view, scientists are creative and must be open to new ideas, and their observations are theory-laden. From a historical perspective, science reflects social and cultural traditions, and scientific ideas are affected by their social and historical milieus. From a sociological perspective, all cultures are seen to contribute to science, new knowledge must be reported clearly and openly, and scientific work requires peer review, and replication of results. How these important aspects of the nature of science are manifested in communication within the scientific community leading to the unveiling of the mysteries of SARS can be illustrated by two episodes, to which we now turn.

Episode 1—The Search for the Perpetrator of SARS

In the early stages of the epidemic, evidence was contradictory as to which pathogen caused SARS. Various microorganisms were implicated including the avian flu virus, the paramyxovirus, chlamydia, and the human metapneumovirus (hMPV) (WHO 2003a). The real breakthrough came when researchers at Hong Kong University (HKU) observed, using an electron microscope, a coronavirus growing in monkey kidney cells that had been inoculated with clinical samples from two SARS patients (Peiris, et al. 2003). The viruses were capable of inducing antigen-antibody reactions in the serum from the SARS patients. The results of random genetic screening of infected cells then lent support to the hypothesis that the disease was caused by a new coronavirus. Despite this, hMPV continued to appear in samples from SARS patients in various countries, suggesting the possibility of co-infection by two viruses.

All these findings were uploaded onto the Internet continuously for peer review and confirmation by other researchers. While other hypotheses received mixed results, HKU's findings were confirmed by other laboratories. As specified by Koch's postulates, the "acid test" for the coronavirus hypothesis was whether the same disease could be produced by inoculating the virus into the original host species or a related species (Fouchier, et al. 2003). This experiment was carried out at the Erasmus University of Rotterdam. Two macaques were inoculated with a coronavirus sample, and both showed symptoms of SARS. In contrast, subsequent inoculations of hMPV did not result in a more serious version of the disease (Fouchier et al. 2003). With this piece of evidence, scientists concluded that the coronavirus was the single true cause of the disease. This discovery offered not only a brilliant showcase of the utility of science in explaining natural phenomena, but also a vivid demonstration of how science, when communicated through an iterative review process based on the rules of logic, led to the confirmation of one hypothesis and the falsification of its rivals.

This communication process may be even more important for its potential to build consensus within the scientific community than for its role in seeking an absolute answer, especially in cases where the evidence is not as unequivocal as in the previous example. The next scenario illustrates this contention.

Episode 2—The Amoy Gardens Mystery

Towards the end of March 2003, an unusual cluster of 320 cases of SARS broke out in Amoy Gardens, a multi-block housing estate in Hong Kong. There was also an unusually large cluster found in Block E of this eight-block estate. These patients, who invariably exhibited diarrhea, prompted researchers to focus their investigation on possible transmission by way of the sewage system.

The sewage system in Amoy Gardens was so designed that the effluent from all the floors in the same wing drained into a common vertical soil stack (Whaley 2006). Scientists in the Hong Kong Department of Health postulated that as the bathroom in each apartment was fitted with a powerful exhaust fan, the negative pressure created by the fan could cause a reflux of air, contaminated with droplets from the soil stack, into the bathroom through the floor drain. It was further suggested that contaminated droplets could have been transmitted to the light wells between adjacent wings of the blocks, thus entering other units through open windows. An oil droplet test showed that the aerodynamics of the light well could produce a "chimney effect" that caused the droplets to rise (Hong Kong Department of Health 2003). However, the evidence was mainly circumstantial, as no trace of the coronavirus was found, presumably due to the long time lapse.

There was no unanimous agreement regarding the "sewage system" hypothesis. Ng (2004) contended that it would be difficult for contaminated droplets to reach the top floors through the sewage system, and that the theory could not explain the wide distribution of cases in blocks other than Block E. He argued that rats, which heavily infested the estate, might have acted as the dynamic source (or "vector") of the virus. The World Health Organization carried out an independent investigation into the incident and arrived at the same conclusion as the Hong Kong Department of Health (WHO 2003b). Although the inference drawn by these two official bodies was considered more acceptable at that time, scientists later conceded that Ng's theory remained a possibility (Ellis, Guan & Miranda, 2006).

Judging from these two episodes, science communication at this level ideally involves the search for a seamless connection between the evidence and a hypothesis, through interrogation of the evidence, argument, and inferences from peers. However, in reality, it does not necessarily lead to a single solution acceptable to all. When an unequivocal conclusion cannot be reached, consensus has to be sought to guide further actions, with the understanding that communication will be re-invoked when new evidence emerges that suggests the contrary.

Influences on Communication Within the Scientific Community

Cooperation and Competition

Communication between scientists is characterized not only by logic or rationality, but also by the nuanced interactions among scientists in pursuing the same goals. In stark contrast to the AIDS virus, which took three years to discover, the identification of the SARS virus took less than three months. This remarkable rate of knowledge generation was made possible by an unprecedented scale of collaboration, which occurred not only among the scientists in individual research teams, but also among different research laboratories and institutes around the world. Under the coordination of the World Health Organization (WHO MCNSD 2003) whose specific roles in this crisis will be further explored in a later section, a network of the world's most renowned research laboratories specializing in epidemiology was formed and connected by a shared website, which enabled daily communication in real time. Chemicals and specimen cultures were rapidly transported between laboratories within hours, and discoveries published online within days, thereby expediting the generation of knowledge to an incredible extent. The following description of a single day's work of this network illustrates vividly the nature of the collaboration that was typical during this period of time:

> On March 20, human metapnenmonvirus primers were tested in four additional laboratories. The Chinese University of Hong Kong found parmyxovirus-like particles in respiratory samples … The Rotterdam laboratory sent test kits for human metapneumonvirus to Singapore and Hong Kong laboratories and shared, via the website, the phylogenetic tree of the isolated paramyxovirus. The laboratory in Canada shipped convalescent sera to Rotterdam for further testing of isolates. During the daily conference call, the Singapore laboratory reported round pleomorphic structures in samples, and scientists in Germany and Hong Kong described similar findings.
>
> *(WHO MCNSD 2003, p. 1730)*

Paradoxically, amidst this extensive collaboration, different research teams were covertly competing with one another. A prominent example is the sequencing of the coronavirus genome. At least four research teams were working on this separately at the same time. A comment by Leung, the HKU team leader, exemplifies the intense atmosphere of the race: "The pressure was intense, but that's where the road to discovery leads. It is always competitive" (HKU Genetic Sequencing Team 2003, p. 9).

Eventually, the team from Canada came first, with their paper being published in the journal *Science* (Marra, et al., 2003). Despite this, Leung contended that fairness is an issue underlying this competition. What is contentious are the rules of the game, rather than the inequality of resources available to individual teams. Leung lamented that after his team had uploaded their findings to the world wide web, the winners had to issue corrections to the sequences they reported earlier. This led Leung to believe that his team deserved to be the first even though they were denied the reward of publishing in such a reputed journal (HKU Genetic Sequencing Team, 2003).

Another exemplification of the intense competition among scientists is the report by Abraham (2004) that WHO wished to published a single paper on the identification of the SARS virus in the names of all collaborating laboratories, but the HKU, the US, and Germany all raced to publish their findings in separate papers. Abraham commented on this atmosphere of competition bluntly: "While WHO paid tribute to the selflessness of the various participating laboratories, in reality the old competitive habits die hard." (Abraham, 2004, p. 95). He attributed these old habits to a combination of "human egos and the realities of scientific funding" (p. 93).

Competition among scientists has been escalated to an overt form, as some outcomes now offer more significant gains than those associated with publishing papers or securing research funding. A case in point is the fight to patent the SARS genome, in which the University of Hong Kong, which was the first to identify the SARS virus, competed with the US Center for Disease Control and Prevention and the British Columbia Cancer Agency, which was the first to sequence the virus genome. The disputes reportedly stemmed from concerns as to whether the genomic information could be managed and used most effectively for the public good, which, paradoxically, all three teams—the CDC, BCCA, and HK's Versitech—claimed to be able to do better than the others (Gold, 2003; Tsui, 2003).

It may be argued that competition amongst scientists can accelerate discovery, and hence is a necessary evil. However, excessive competition is likely to impede communication and information flow among scientists, thereby delaying scientific discoveries of a more complicated nature. Clearly, how to encourage scientists to collaborate in a culture of competition is an important tension that needs to be resolved.

Influence of Social and Cultural Milieus

The literature on the nature of science shows that science operates in particular social and cultural milieus. Two critical incidents illustrate the covert and overt influences of the social environment on science communication in the case of SARS, which had very different results.

The first incident again arose from the genome sequencing process undertaken by the HKU Genetic Sequencing Team. Wong et al. (2008) reported in an interview with the team leader, that the data obtained by his team conflicted with those obtained by the US Centers for Disease Control and Prevention (CDC). The following interview excerpt illustrates the dilemma faced by the team in deciding which data to report:

In the US sequence, the first base pair of the head was an "A", but ours started with a "T"... my students asked me if we should put an "A" or a "T"... Over and over again, we found that it was a "T"..." he (one student) said the "...CDC could not be wrong." And I said, "...our research confirmed over and over again that we started with a "T" instead of an "A", so we should enter a "T" instead of an "A".

(*Wong et al. 2008, p. 113*)

Apart from demonstrating the importance of honest reporting of data as a fundamental tenet of the nature of science, this episode also shows the tacit influences to which scientists, especially less experienced ones, are susceptible in communicating with their peers. The second incident reflects a similar dilemma, but the outcomes were sadly the opposite, leading to potentially fatal consequences.

After conducting extensive interviews with Chinese scientists, Enserink (2003) reported the political, social, and cultural factors that influenced scientific operations and impeded the progress of science during the SARS crisis in a paper published in *Science*, entitled "China's Missed Chance." Enserink found that a team from the Academy of Military Medical Sciences (AMMS) in Beijing had discovered the SARS virus weeks before it was identified by the team at HKU. However, the team was not confident enough to challenge the chlamydia hypothesis that was dominant in the Chinese mainland at that time. The chlamydia hypothesis had been promoted by Hong, a highly regarded senior microbiologist, and "it would not have been respectful" to challenge it (Enserink, 2003, p. 294). This hypothesis was so influential in China that even after the international community had reached the consensus that a new coronavirus was the culprit, the Chinese Ministry of Health continued to adhere to their hypothesis and even controlled any publicity about it. Had the AMMS team been able to inform WHO of their discovery, the SARS virus might have been identified weeks earlier and control measures put in place much sooner. However, at that time, the Chinese mainland was not part of WHO research network. The Chinese scientists attributed their failure to "systemic problems in Chinese science: a lack of coordination and collaboration, stifling political influence, hesitation to challenge authorities, and isolation from the rest of the world" (Enserink, 2003, p. 294).

These socio-cultural influences reflect another tension at this level of communication. To what extent can scientists exercise independent scientific rationality in communicating their findings without being unduly influenced by their socio-cultural or sociopolitical contexts? This raises a more general question: how can scientists remain impartial amidst social, economic, and political pressures that have potential implications for the resource allocations on which their research funding or even their livelihoods depend?

Level 2: Science Communication for Policy Making

While science provided evidence-based guidance for policy makers during the SARS crisis, the need for effective policy making also drove scientific research by generating questions for investigation. It is often difficult to determine the direction of influence between scientists and policy makers, but suffice it to say that science and policy-making mutually influence each other. To facilitate communication between scientists and health officials in resolving the crisis, scientists were often engaged in high-powered coordination and steering committees established by the state. This bilateral communication proved to be fruitful in many instances. A prominent example is the critical finding that there were no reports of SARS being transmitted before the appearance of symptoms (Merianos & Plant, 2006), leading to the adoption of the policy of isolating patients immediately after the onset of symptoms. Another example is the finding that the disease has an incubation period of around two to ten days (Doberstyn 2006), which provided a scientific basis for governments to adopt a ten-day quarantine period for any person coming into close contact with a SARS patient. A third example is that as soon as civets were proven to be carriers of a virus nearly identical to the coronavirus, the rearing and selling of civets was banned in Guangdong province, and thousands of civets were exterminated (Normile 2004).

As scientific knowledge is tentative, there is always the possibility that public policies that hinge on science will backfire. Based on the knowledge that SARS is transmitted through droplets or close personal contact, it was concluded that isolating the residents of Amoy Gardens in their own flats should minimize transmission. However, the number of infections in Block E continued to mount. New evidence emerged that SARS virus was relatively stable in faeces, and that the sewage system of the building might be a channel of transmission. In light of this, the isolation order appeared to have put the residents in an even more dangerous position. The policy then shifted from isolation to evacuation, with residents being transferred at midnight to holiday camps for quarantine purposes (Whaley, 2006).

This incident reflects the constant tension in emergent crises between the meticulous and time-consuming process of collecting scientific evidence, and the pressure on governments to make timely decisions in order not to put more lives at stake. However, given the tentativeness of science, particularly front-line research, the decisions made, albeit evidence-based, often fall short of the solid grounding required for policy making in matters as important as public health.

A second source of tension is the conflict between the state's concerns for public health and its adherence to societal practices for either cultural or economic reasons. The trading of wild animals, which are sold for consumption, is related to both economic incentives and Chinese culinary habits based on the belief that the consumption of these animals can "enhance the vitality of the body" (Zhong & Zeng, 2007, p. 32). However, with strong evidence indicating a link between exotic animals and the SARS coronavirus, the sale of these animals was temporarily banned in Guangdong province. However, this was not imposed in other provinces (Loh, 2004; WHO, 2006), and the ban in Guangdong was lifted by the end of July of the same year (Abraham, 2004), reflecting the Chinese government's decision to take a midway course to address this tension.

There are other more fundamental sociopolitical influences on science communication for policy making regarding SARS. A comparison of the policy-making processes in three areas that were affected by SARS—the Chinese mainland, Singapore, and Hong Kong—reveals the intricacies of these influences. The Chinese mainland was where SARS originated. However, the disease spread for at least three months before its existence was even acknowledged. This procrastination has been attributed to the Chinese government's "customary obsession with secrecy" (Loh, 2004, p. 140), particularly as the event tied in with the meeting of the National People's Congress that so preoccupied the government at that time. Only after the issue of a travel advisory notice for Hong Kong and Guangdong did the Chinese government begin to take more serious action to reverse the situation (Balasegaram & Schnur, 2006).

In contrast, the Singapore government responded much more rapidly when the first case was identified in early March, with measures that included the rapid isolation of cases, the tracing of contacts and sources, and the placement of suspected cases in mandatory home quarantine. The government even went so far as to hire a commercial security firm to serve Home Quarantine Orders and install cameras in homes to ensure compliance (Whaley & Mansoor, 2006). In Hong Kong, the government was criticized for its slow reaction to the outbreak. Yeoh, the then Secretary for Health, Welfare and Food, openly denied that SARS had spread to the community through a private doctor who had been infected after being consulted by a SARS patient, and "accused" WHO of spreading panic" following its global alert on Hong Kong (Chiu & Galbraith, 2004). Although the outbreak of SARS started earlier in Hong Kong than in Singapore, it was made a statutorily notifiable disease—the legal basis for imposing quarantine—ten days later than in Singapore (Whaley & Mansoor, 2006). This was due to concerns about driving patients into hiding, and issues such as civil liberties and public acceptability (Benitez, 2006; SARS Expert Committee, 2003).

This cross-cultural comparison reveals at least two other tensions in science communication for policy making. The first is the dilemma between transparency and restricting the flow of information. This crisis revealed that how this dilemma is resolved depends largely on the political culture and philosophy of governance of individual states. The second tension is where the actions warranted by a crisis are incompatible with core societal values. Apart from the insensitivity of the Hong Kong health officials (Ma, 2003), the government's laissez faire policy, and the high regard for personal freedoms within Hong Kong society, are obviously at odds with the stern action required for health crisis management.

Level 3: Science for Citizen Action

In the context of the SARS crisis, the role of science communication in helping citizens to gain a better understanding of the nature of the disease and to take informed action was mediated either by public policies with which citizens were expected to comply, or through communication directly with scientists. This occurred through the mass media, such as press conferences and phone-in programmes. Most citizens in SARS-affected areas heeded the advice of scientists by wearing face masks in public areas, using household bleach to disinfect their homes, and checking their body temperature.

However, a number of politico-economic, socio-cultural, and personal factors impeded this communication process. First, restrictions on the flow of information inhibited the public's engagement with scientific information, as was the case in the Chinese mainland. When communication of the relevant science was absent at the early stage of SARS, Chinese citizens turned to folk medicines to inform their actions. For instance, many people boiled vinegar at home, believing the vinegar vapor had an antiseptic effect on air-borne pathogens. Ironically, this practice led to at least two deaths due to carbon monoxide poisoning (Kong, 2003).

Second, when a public decision based on scientific understanding was at odds with perceived interests, people's interests often won out, an obvious example being the breaking of quarantine by residents of Amoy Gardens (Beech, 2003). This raised the question of "the balance of freedom of action of the individual and freedom from infection of the community at large" (Weiss & McLean, 2007, p. 112). Third, during the crisis people tended to let their instinctive feelings and subjective reasoning steer their actions, which often resulted in stigmatization and discrimination. In Toronto, for example, some people viewed the local Chinatown "with suspicion, and outright accusation" (Abraham, 2006, p. 131). The same kind of thinking was also manifested by group action. A US university asked Hong Kong students and their relatives to stay away from graduation ceremonies (Lee & Gibb, 2003), and Hong Kong athletes were banned from participating in the Special Olympics in Dublin (Doran & Maitland, 2003).

These incidents reflect a major problem in communicating science to and among citizens. It seems that even if science communication was not cut off, many citizens still failed to adopt scientific views to guide their actions. This may have stemmed from the perceived scientific uncertainty surrounding various issues, the ineffectiveness of science communication, the public's inability to understand scientific evidence, or disagreements between scientists and citizens over the interpretation of evidence. Many people got around this problem by resorting to subjective judgments, folk science or pseudoscience, and communicating their judgments to others, culminating in mass actions. Some of these citizens' judgments have in turn influenced state policy making. For example, parents called for the closure of schools in fear that their children would be infected through close contact with other children, even though scientists contended that school closures were not necessary, and that children would actually be safer in schools where hygienic measures were routinely practiced (Chiu & Galbraith, 2004). Some citizens established a website to release information based on public reports about areas and buildings in which residents were infected with SARS (Loh & Welker, 2004), which the government had been careful to avoid. In both cases, the government eventually succumbed to citizens' demands. Yet other citizens' actions had resulted in various degrees of social turbulence, for example, the panic buying of vinegar or anything erroneously believed to be effective in curing or preventing SARS.

The Role of the World Health Organization (WHO)

The SARS crisis saw the emergence of a new supra-level of science communication, orchestrated by WHO, which is essential to coordinate the communication of science at all the three levels. At the level of scientists and technologists, WHO functions as a leader and coordinator, networking scientists from different parts of the world to capitalize on their expertise. For state policy makers, it acts as a consultant to the state, providing support in the form of expert advice and resources, which prove to be particularly important for developing countries. At the citizen level, WHO serves as an information provider and adviser that provides the public with updated information and advice concerning disease prevention.

Fidler (2005) highlighted three breakthroughs in WHO's global governance of health during the SARS crisis. The first is the global surveillance of non-governmental sources of information through the Global Public Health Intelligence Network (GPHIN), which was established after the outbreak of Ebola disease in the 1990's (Heymann, Kindhauser & Rodier, 2006). This network allows WHO to scan for rumors or informal reports circulating on the Internet, thus empowering WHO to circumvent government health authorities reluctant to disclose national information due to economic concerns. The second breakthrough was the setting up of global networks to draw expertise from institutes with substantial research capacity. In addition to the Global Outbreak Alert and Response Network (GOARN) established in 2000 to coordinate 120 networks and institutes, a virtual network of scientists in 13 laboratories was established during the SARS crisis to facilitate collaboration and communication among scientists in real time, thereby accelerating the pace of discovery.

The third breakthrough is the issuance of travel advisory notices, which allows WHO to bypass the machinery of the state and directly communicate with non-state actors. This direct engagement with citizens assists in translating scientific understanding into citizen action, thereby forcing the state to become more transparent. However, the exercising of such acts without consulting individual states has created tensions between WHO and nation states as to how

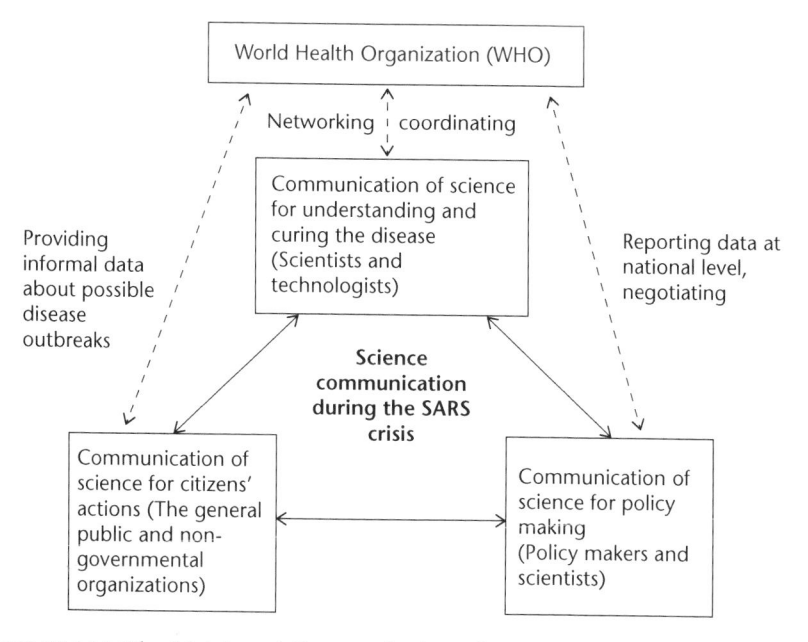

FIGURE 14.2 The Multilateral Communication of Science During the SARS Crisis under the Governance of the WHO

to balance national and international interests in controlling a global crisis. In light of the key role played by WHO, Figure 14.1 is modified to incorporate the various kinds of communication occurring between WHO and other stakeholders, as illustrated in Figure 14.2.

The Impact of Information Communication Technology and the Mass Media

The role played during the SARS crisis by information communication technologies (ICT), an essential part of the infrastructure of a globalized society, is undeniable. The global networks coordinated by WHO could not have functioned without high-tech facilities, such as websites, teleconferencing and videoconferencing, and email. In the context of Hong Kong, ICT has played a crucial role in "policing" the disease by tracking its spread using high-tech devices developed originally for crime investigation (Kong, Tsang, Liu & Chan, 2006). The expediting of information flow by ICT also greatly enhanced the role of the mass media. Unlike the secure websites for communicating within the scientific community, communication through the mass media is open and accessible to all and, hence, facilitates communication between the various stakeholders. The SARS crisis has shown that the mass media serves as a channel for scientists to communicate their findings to the public, and for governments or WHO to keep citizens up to date on policies or advisories, to the extent of mobilizing the whole nation through the state propaganda, as in the case of China (McNally, 2003). Besides, the media provided a platform for citizens to air their views to health officials, politicians, and lay citizens. Paradoxically, scientists also utilized the media during the crisis to bypass the state to communicate "scientific truths" to the public. An example is the open declaration by Chung, Dean of the Medical Faculty of the Chinese University of Hong Kong, that SARS had spread to the local community, which contradicted Yeoh's assertion (Loh, Galbraith & Chiu, 2004). Another example is the revelation by Jiang, a retired doctor in Beijing, to foreign media, that the Health Minister had covered up cases occurring in the city (Chiu & Galbraith, 2004). Both incidents had a snowball effect on policy making, leading to more active measures being put in place to control SARS, and more open communication between the state and citizens regarding science.

Some of the media exerted a substantial influence on policy making by advocating their own stances or policies, which is tantamount to "advocacy journalism" and is highly controversial (Loh, Galbraith & Chiu, 2004, p. 213). Some media organizations were also accused of capitalizing on public sentiment, thereby aggravating social turbulence, although it is not easy to draw a clear line between sensationalism and responsible journalism in a messy situation such as a major health scare. With scientists still grappling to determine the "correct" information about the crisis, it may be unrealistic to expect the media to present information without some distortion.

Lessons Learnt and Challenges Ahead

This study provides a range of insights regarding future crises. The control of SARS reinforces the vital role of science in solving short-term crises, all the more so when there is an extensive degree of collaboration among scientists. However, the communication of scientific knowledge does not automatically lead to problem resolution, unless it is translated into effective public policies and citizen action. The SARS case reveals that science communication is fraught with difficulties at all levels, partly because of the limits of science, and partly because of the social, political, and cultural contexts in which science is communicated. The latter is inextricably related to the nature of policy making, political ideology, and the level of scientific literacy of citizens, which interact with each other and with other factors such as human nature, indigenous culture, and globalization. The tensions that arose from these influences in relation to science communication presented formidable challenges to all stakeholders. Three major lessons or challenges distilled from these tensions may throw light on the resolution of similar crises in the future.

First and foremost, honesty and transparency in communication are the best policies in dealing with a crisis like SARS, which is a life-or-death situation (Doberstyn, 2006). In the absence of scientific evidence, especially at the early stage of the crisis, or of sensible engagement with the evidence, the public's reliance on subjective reasoning and pseudoscientific ideas proved to be detrimental to personal and public health, and led to the victimization of particular groups of citizens. The deliberate attempt to restrict public access to scientific evidence posed even greater problems. However, it is not a simple matter of lying or not lying, or disclosing or not disclosing *per se*, that underlies the problem, but rather a matter of whether the evidence is communicated promptly and used sensibly to make informed and timely decisions within the limits of science. For scientists, it is a matter of being honest and letting the data speak the truth. For policy-makers, it is a matter of being transparent, utilizing evidence and making policies responsibly, incorporating sufficient checks and balances within the system of health governance (McNally, 2003), and appreciating the implications of their actions for the wider community (SARS Expert Committee, 2003). As for lay citizens, it is a matter of viewing the nature and utility of science in perspective, differentiating claims and evidence based on science, pseudoscience or mere hearsay, and adapting their actions to emergent understanding of the crisis.

The second lesson is that the ability of science communication to resolve crises will be limited if the science is unable to promote deep reflection on the roots of the problem in relation to several aspects. These include the political and social culture, principles of governance, prevailing values of society in balancing the pursuit of health and economy (Plant, 2003), attitudes toward, and respect for, nature (Tang, 2003), and social responsibility—both in terms of responsible citizenship in the local and global sense, and honesty of the state across national borders (Lee, 2003). The third challenge concerns the promotion of public understanding of science in the context of a new crisis. Such scientific knowledge may not be readily available to lay people and is likely to be different from unequivocal textbook knowledge. The purpose of science communication should extend beyond the mere comprehension of scientific knowledge to the understanding of the nature of scientific inquiry and evidence. Only with this understanding will citizens' decisions not be overshadowed by subjective reasoning, instinctive reactions, and mob psychology. The responsibility for addressing this challenge should not only rest with science educators, but also with front-line scientists who are in the best position to communicate to citizens how scientific problems are being resolved. When all these challenges are met with sufficient confidence by the major stakeholders, we can be sure that we will be able to deal more effectively with similar crises in the future.

*Note that an earlier version of this paper was published in the *International Journal of Science Education, 30*, 515-541.

References

American Association for the Advancement of Science [AAAS] (1990). *Science for all Americans*. New York: Oxford University Press.

Abraham, C. (2006). Canada: how a hospital coped. In World Health Organization, Western Pacific Region (ed.) *SARS: How a Global Epidemic was Stopped* (pp. 126-132). Geneva: World Health Organization Press.

Abraham, T. (2004). *Twenty-first century plague: The story of SARS*. Hong Kong: Hong Kong University Press.

Balasegaram, M., & Schnur, A. (2006). China: from denial to mass mobilization. In World Health Organization, Western Pacific Region (ed.) *SARS: How a Global Epidemic was Stopped* (pp. 73-85). Geneva: World Health Organization Press.

Beech, H. (2003). Doing battle with the bug. *Time*, April 14, 45-48.

Benitez, M. A. (2006). Hong Kong (China): hospitals under siege. In World Health Organization, Western Pacific Region (ed.) *SARS: How a Global Epidemic was Stopped* (pp. 86-93). Geneva: World Health Organization Press.

Bybee, R. W. (1997). *Achieving Scientific Literacy: From Purposes to Practices* Portsmouth, NH: Heinemann.

Chiu, W., & Galbraith, V. (2004). Calendar of events. In C. Loh and Civic Exchange (Eds.) *At the epicenter: Hong Kong and the SARS outbreak* (pp. xv-xxvii). Hong Kong: Hong Kong University Press.

Dawson, C. (1991). *Beginning science teaching*. Melbourne, Australia: Longman Cheshire.

Doberstyn, B. (2006). What did we learn from SARS? In World Health Organization, Western Pacific Region (Ed.) *SARS: How a Global Epidemic was Stopped* (pp. 243-254). Geneva: World Health Organization Press.

Doran, H., & Maitland, T. (2003). Athletes face Special Olympics ban. *South China Morning Post*, 17 May.

Ellis, A., Guan, Y., & Miranda, E. (2006). The animal connection. In World Health Organization, Western Pacific Region (Ed.) *SARS: How a Global Epidemic was Stopped* (pp. 243-254). Geneva: World Health Organization Press.

Enserink, M. (2003). SARS in China: China's missed chance. *Science, 301*, 294-296.

Fidler, D. P. (2005). Health, globalization and governance: an introduction to public health's 'new world order'. In K. Lee and J. Collin (Eds.) Global change and health (pp. 161-177). Berkshire, England: Open University Press.

Fouchier, R., Kuiken, T., Schutten, M., van Amerongen, G., van Doornum. G. J. J., van den Hoogen, B. G., & Osterhaus, A. D. (2003). Koch's postulates fulfilled for SARS virus. *Nature, 423*, 240.

Gold, E. R. (2003). SARS genome patent: symptom or disease? *The Lancet, 361*(9374), 2002.

Heymann, D. L., Kindhauser, M. K., & Rodier, G. (2006). Coordinating the global response. In World Health Organization, Western Pacific Region (ed.) *SARS: How a Global Epidemic was Stopped* (pp. 49-55). Geneva: World Health Organization Press.

Hong Kong Department of Health (2003). *Outbreak of Severe Acute Respiratory Syndrome (SARS) at Amoy Gardens, Kowloon Bay, Hong Kong: Main Findings of the Investigation*. Hong Kong: Department of Health.

Hong Kong University (HKU) Genetic Sequencing Team, (2003). A race against time: cracking the genetic code of SARS coronavirus [electronic version]. *The University of Hong Kong Medical Faculty News, 8*(1), 9.

Irwin, A. R. (1995). *Citizen Science: A Study of People, Expertise, and Sustainable Development*. London: Routledge.

Kong, E. (2003). Two dead after boiling white vinegar. *South China Morning Post*, 13 February.

Kong, J. H. B., Tsang, T. H. F., Liu, S. H., & Chan, A. L. (2006). Policing a communicable disease with IT innovations: a fresh paradigm. In J. C. K. Chan, and V. C. W. T. Wong (eds.) Challenges of Severe Acute Respiratory Syndrome (pp. 59-79). Singapore: Elsevier.

Layton, D., Jenkins, E., Macgill, S., & Davey, A. (1993). *Inarticulate Science*. East Yorkshire: Studies in Education Ltd.

Lederman, N. G., & Zeidler, D. L. (1987). Science teachers' conceptions of the nature of science: Do they really influence teacher behavior? *Science Education, 71*(5), 721-734.

Lee, E., & Gibb, M. (2003). WHO rebukes US universities for SARS ban. *South China Morning Post*, 10 May.

Lee, P. S. (2003). SARS—Lessons on the role of social responsibility in containing an epidemic. In T. Koh, A. Plant, and E. H. Lee (Eds.) *The new global threat: Severe Acute Respiratory Syndrome and its impacts.* (pp. 273-282). Singapore: World Scientific Publishing Co.

Loh, C. (2004). The politics of SARS: WHO, Hong Kong and mainland China. In C. Loh and Civic Exchange (eds.) *At the Epicentre: Hong Kong and the SARS Outbreak* (pp. 139-162). Hong Kong: Hong Kong University Press.

Loh, C., Galbraith, V., & Chiu, W. (2004). The media and SARS. In C. Loh and Civic Exchange (Eds.) *At the Epicentre: Hong Kong and the SARS Outbreak* (pp. 195-214). Hong Kong: Hong Kong University Press.

Loh, C., & Welker, J. (2004). SARS and the Hong Kong community. In C. Loh and Civic Exchange (Eds.) *At the Epicentre: Hong Kong and the SARS Outbreak* (pp. 215-234). Hong Kong: Hong Kong University Press.

Ma, N. (2003). SARS and the HKSAR governing crisis. In T. Koh, A. Plant, and E. H. Lee (eds.) *The new global threat: Severe Acute Respiratory Syndrome and its impacts* (pp. 107-122). Singapore: World Scientific Publishing Co.

Marra, M. A., Jones, S. K .M., Astell, C. R., Holt, R. A., Brooks-Wilson, A., Yarron, S. N., & Roper, R. L. (2003) The genome sequence of the SARS-associated coronavirus. *Science, 300*(5624), 1399-1404.

McComas, W. F., Clough, M. P., & Almazroa, H. (1998) The role and character of the nature of science in science education. In W. F. McComas (Ed.) *The Nature of Science in Science Education: Rationales and Strategies* (pp. 3-40). Dordrecht: Kluwer Academic Publishers.

McNally, C. A. (2003). Baptism by storm: The SARS crisis' imprint on China's new leadership. In T. Koh, A. Plant, and E. H. Lee (Eds.) *The new global threat: Severe Acute Respiratory Syndrome and its impacts* (Singapore: World Scientific Publishing Co.) (pp. 69-89).

Merianos, A., & Plant, A. (2006) Epidemiology. In World Health Organization (ed.) *SARS: How a Global Epidemic was Stopped* (Geneva: World Health Organization Press) (pp. 185-198).

Ng, S. (2004) The mystery of Amoy Gardens. In C. Loh and Civic Exchange (Eds.) *At the Epicentre: Hong Kong and the SARS Outbreak* (pp. 95-116). Hong Kong: Hong Kong University Press.

Normile, D. (2004). Viral DNA match spurs China's civet roundup. *Science*, 303(5656), 292.

Peiris, J. S. M. et al. (2003). Coronavirus as a possible cause of severe acute respiratory syndrome. *The Lancet*, April 8, 2003. Retrieved from http://image.thelancet.com/extras/03art3477web.pdf.

Plant, A. J. (2003). Editorial. In T. Koh, A. Plant, & E. H. Lee (Eds.) *The new global threat: Severe Acute Respiratory Syndrome and its impacts* (pp. xiii–xxiv). Singapore: World Scientific Publishing Co.

SARS Expert Committee, HKSAR (2003). *SARS in Hong Kong: from experience to action, Report of the SARS Expert Committee*. Hong Kong: SARS Expert Committee.

Tang, S. P. (2003). Fighting infectious diseases: One mission, many agents. In T. Koh, A. Plant, and E. H. Lee (eds.) *The new global threat: Severe Acute Respiratory Syndrome and its impacts* (pp. 17–29). Singapore: World Scientific Publishing Co.

Tsui, L. C. (2003). SARS genome patent: to manage and to share. *The Lancet*, 362(9381), 406.

Weiss, R. A., & McLean, A. R. (2007). What have we learnt from SARS. In A. McLean, R. May, J. Pattison and R. Weiss (Eds.) *SARS: A case study in emerging infections* [electronic version] (Oxford Scholarship Online), pp. 112–116.

Whaley, F. (2006). Lockdown at Amoy Gardens. In World Health Organization (ed.) *SARS: How a Global Epidemic was Stopped* (pp. 155–162). Geneva: World Health Organization Press.

Whaley, F., & Mansoor, O. D. (2006). SARS chronology. In World Health Organization, Western Pacific Region (Ed.) *SARS: How a Global Epidemic was Stopped* (pp. 3–48). Geneva: World Health Organization Press.

Wong, S. L., Hodson, D., Kwan, J., & Yung, B. H. W. (2008). Turning crisis into opportunity: enhancing student-teachers' understanding of nature of science and scientific inquiry through a case study of the scientific research in severe acute respiratory syndrome. *International Journal of Science Education*, 30(11), 1417–1439.

World Health Organization (2003a). *Severe Acute Respiratory Syndrome (SARS): Status of the outbreak and lessons for the immediate future* [electronic version] (Geneva: World Health Organization).

——(2003b). *Inadequate Plumbing Systems Likely Contributed to SARS Transmission*. Retrieved February 2004, from http://www.who.int/mediacentre/releases/2003/pr70/en/print.html.

World Health Organization, Western Pacific Region (2006). *SARS: How a Global Epidemic was Stopped*. Geneva: World Health Organization.

(WHO MCNSD) World Health Organization Multicentre Collaborative Network For Sars Diagnosis (2003). A multicentre collaboration to investigate the cause of severe acute respiratory syndrome [electronic version]. *The Lancet*, 261, 1730–33.

Ziman, J. (2000). *Real Science: What It Is, and What It Means*. Cambridge: Cambridge University Press.

Zhong, N. S., & Zeng, G. Q. (2007). Management and prevention of SARS in China. In A. McLean, R. May, J. Pattison, & R. Weiss (Eds.) *SARS: A case study in emerging infections* [electronic version] (Oxford Scholarhip Online), pp. 31–34.

15
COMMUNICATION CHALLENGES FOR SUSTAINABILITY

Julia B. Corbett

Introduction

The concept of "sustainability" dates back to ancient Greece, but recently the word has found new popularity and cachet. At times, it appears that "sustainability" is everywhere. Walmart has a Sustainability Index and BP has a Sustainability Review. DuPont and Coca-Cola employ a Chief Sustainability Officer, and the US golf industry embraces sustainability. More than 2,500 local governments from across the European Union have signed the charter of the Sustainable Cities and Towns Campaign. Sustainability offices operate on many college campuses, and numerous academic journals have "sustainability" in their titles. Some people call sustainability the new and improved environmentalism, while others claim that "sustainable" is attached to so many things that it's no more meaningful than the word "green."

Yet regardless of the word and how it's used, there is good reason to examine "sustainability" more closely. There is abundant scientific evidence that the way humanity currently lives on the Earth cannot be sustained. Phytoplankton, which account for about half the production of organic matter on Earth, are in great decline in the oceans, oceans that are becoming increasingly acidic. Greenhouse gas emissions increased by a record amount in 2010 to the highest carbon output in history: 30.6 gigatons. The Arctic has warmed over 4 degrees, melting vast expanses of ice. The planet has air pollution, species extinction, chemical contamination, topsoil erosion, deforestation, and a global population of almost 7 billion that increases by a quarter-million each day. Decades of scientific data tell us that the myth of unlimited resource use and consumption needs to be retired. Obviously, "sustainability," in some iteration is extremely important to our lives on this planet.

This chapter investigates "sustainability" from several angles. First, it examines the term itself. Because it has been defined in dozens of ways, "sustainability" can be strategically exploited and used to communicate very different things. Second, it examines how sustainability has been conceptualized and practiced by different types of organizations. Third, the chapter discusses "sustainability science," an umbrella term for an emerging "transdisciplinary" science that addresses coupled human-environment problems and practices. Finally, the chapter explores communication strategies and pitfalls for sustainability. Given the term's many uses, there is also a variety of positive and negative aspects to communicating about sustainability.

The Lexicon of Sustainability and its Rhetorical Uses

An instructive place to begin defining sustainability is breaking the word down into "sustain" and "ability."

The first dictionary definition for the verb "sustain" is to keep going, to continue or maintain. Therefore, something is sustained if it endures and carries on. A second definition has to do with providing sustenance, as in food sustaining life and providing service in the form of nourishment. A third definition is to bear or endure, such as in sustaining an injury. And a fourth definition gets its meaning from law and science: to corroborate, prove, or affirm. For example, when the judge "sustains" the lawyer's objection, she affirms it.

The first two definitions seem good fits for environmental protection and human survival: we want to endure as a species, which requires protecting all that sustains life on this planet. Interestingly, the fourth definition resonates with how popular language surrounding "sustainability" can be used to "affirm" or legitimate certain worldviews and ideologies and make the concept self-affirming and -reinforcing.

For example, in *Hot, Flat, and Crowded* (2008), *New York Times* columnist Thomas Friedman said the current approach to sustainability is one in which "everybody gets to play, everybody's a winner, nobody gets hurt, and nobody has to do anything hard...[and] that's not the definition of a revolution. That's the definition of a party" (p. 252). Friedman and others question whether modest changes in things like energy efficiency or purchasing products labeled "sustainable" serve the *image* of sustainability more than they bring about the enormous changes needed for true sustainability. A variety of scholars has argued that sustainability readily accommodates (rather than challenges or changes) the existing languages of capitalism, colonialism, consumerism, and growth (Kendall, 2011; Peterson, 1997; Corbett, 2006).

The word "ability" can be defined two ways: one, being able to do something and having the capacity for it, and two, having the skill or talent (innate or learned) to do something. This too provides an interesting insight: humans may have the "capacity" to live sustainability, but our current "skills" are to live *un*sustainably, and these skills may blind us (individually and collectively) to a different way of being on Earth. What many of us now possess (especially in the more developed world) is sustain-*in*ability.

For example, a recent advertisement for the US restaurant chain Red Lobster proclaimed "Endless Shrimp." Many of us in affluent, developed countries live according to this mentality, treating resources as though they are indeed endless—if not naturally, then with a bit of technology and engineering. As ocean stocks of shrimp are severely depleted, shrimp farming in southeast Asia has exploded, transforming biologically diverse mangrove swamps into mono-culture shrimp farms. This technology causes fouled water supplies, intrusion of salinity, decline of marine environments, human rights abuses, and a host of other impacts. Examples in the US of our skills at "living large" are increasingly large houses and large rented storage units.

A popular concept to measure and picture sustainability is an ecological or carbon footprint. A footprint is a tally of resources used by an individual (or organization, city, or country) that estimates the land area needed to support that person's lifestyle. For some web-based calculators, even if you live modestly and use little energy, your country's resource use determines your footprint. If you live in the US, it takes almost 24 acres of productive land to support you. The footprint in the UK is 14 acres; India's is 0.2 acres. Carbon footprints are helpful for awareness regarding lifestyle impacts on the environment, but the individual-level suggestions for reducing one's impact are insufficient without macro-level changes to make significant impacts on sustainability. According to Global Footprint Network researchers, we are living so far beyond our planet's means that September 27 in 2011 was Earth Overshoot Day, the day when "humanity's demand for ecological resources and services in a given year exceeds what the Earth can regenerate in that year" (www.footprintnetwork.org).

The most widely-cited definition of sustainability is not from the dictionary, but from a 1987 report by the United Nations World Commission on Environment and Development, commonly called the Brundtland Report. This commission was interested in global "sustainable development" which they defined as: "development that meets the needs of the present without compromising the ability of future generations to meet their own needs" (WCED, 1987, p. 41).

This definition more than any other has shaped our current understanding and discussion of sustainability and deserves a closer look. What key words stand out? To whom does it refer? And, what importantly is left out?

The UN Commission wanted to move discussions of global development away from the tensions between the world's developed North and less-developed South. They recognized that environmental quality is of common interest to all countries, insofar as it greatly affects (if not exacerbates) conditions such as poverty, pollution, population growth, and industrial development. Therefore, they concluded, there are three pillars underlying future "sustainable" development that needed to be reconciled: environmental, social, and financial.

The Brundtland definition has a decidedly anthropocentric (human-centered) slant in discussing human "needs," and it is ambiguous as to what those "needs" are. Many less-developed countries do not now meet the needs of their citizens, while many developed countries not only meet most needs, but also a great many "wants." This definition also could be interpreted to mean that in meeting human needs, a variety of habitats, species, and ecosystems could be sacrificed.

But even for ecologists and environmental scientists, sustainability is a contested term. Generally, a sustainable ecosystem condition is one in which biodiversity, renewability, and resource productivity are maintained over time. The science of ecology has provided rich data about the elements and processes of healthy, functioning ecosystems, as well as the points beyond which those systems are no longer sustainable, in other words when they have reached their optimal or maximum "sustained yield." However, relying on existing ecological knowledge (which is still extremely limited) can legitimize exploitation: for example, setting a "sustained yield" harvest for one species of fish to the detriment of other important substocks of fish.

The Brundtland definition focuses on humans (and their future generations), which positions humans as somehow separate from the ecosystems they inhabit, and to which they are fundamentally and inextricably welded. This dualism— believing we are separate from the natural world—is a key stumbling block not only to defining sustainability, but also to being able to live it and to learn the skills to move beyond our well-honed sustain-inability.

The lack of a universal and unifying definition of the "human environment" makes "sustainability" open to interpretation and appropriation to serve certain ends. Communication scholars claim that the rhetorical strength of "sustainability" lies in its philosophical ambiguity and range (Peterson, 1997). The many contradictions encountered in the use of the word stem from the kinds of claims people make with the concept and in its interest. Depending on what people want sustainability to mean, they are able to strategically exploit its inherent flexibility.

Sustainability—as word and concept—requires us to integrate human-environment issues not just in the biological and physical sciences but also in the social sciences, humanities, and other fields. Understanding sustainability means considering all the contexts in which the term is used—justice, freedom, ethics, environmental policy—and at levels from the personal to the organizational to global. When any organization or individual talks about sustainability, it is important to ask, "Sustainability with regard to what? And in whose interest?"

Practicing Sustainability

If we need to consider so many disciplines and contexts for sustainability, how on earth are we able to *be* sustainable and truly practice it? It just seems so, well, BIG. One way is to think of the various levels of sustainability. Though a common focus is on individual actions (Paavola, 2001), another place to practice sustainability is at the organizational level.

Virtually every corporation, a fair number of municipal or country governments, as well as non-profit and trade associations have something on their websites about sustainability or "green" commitments. This widespread attention demonstrates the popularity and "mileage" of buying into sustainability, but does it truly demonstrate a commitment to sustainability?

Again, it depends on "sustainability in terms of what?" It also depends on how this broad concept is measured. Some organizations undertake self-reports and ratings that they claim represent actions towards sustainability. Increasingly, independent rating systems are springing up, some driven by socially responsible investing firms that believe sustainability scores may affect the price of stocks or bond ratings. Measurement systems vary widely but often include resource indicators like waste minimization, water conservation, paper reduction, and decreased greenhouse gas emissions and energy use. Some rating systems also include a variety of non-environmental indicators like taxes paid, safety, leadership, and ratio of CEO-to-employee pay.

Many sustainability accounting systems refer to a *triple bottom line*, such as "social, environmental, and financial." If you envision those elements as three overlapping circles in a Venn diagram, the part where the social and financial circles intersect could be called "social equity." Where the financial and environmental circles intersect is "eco-efficiency," and where social and environmental overlap, "sustainable environments." Where all three of those intersections overlap is where true "sustainability" resides.

A 2011 research report by MIT Sloan Management Review (Haanes, 2011) states that most companies defined sustainability in terms of the long-term viability of the company, its employees and customers, not in terms of climate change or environmental concerns, which ranked near the bottom. This fits the "social equity" overlap, and matches the "enduring" and "stable" definitions of the word "sustain," but it still manages to separate humans from the natural world and not consider them fundamentally a part of it.

The MIT report also found that "embracers," whose companies are shifting their business models toward sustainability, believe sustainability will open new markets, attract employees, and improve their public reputations. If you consider the conditions under which businesses operate, this strong survival instinct makes sense. Most businesses worldwide operate in capitalistic economies based on continual growth. The stock market and stockholders consider a company successful only when each year, more money is made (and more resources consumed, more products produced). This is not sustainable.

Nevertheless, the planet's people still need to eat, stay warm, and live somewhere (we won't even consider "wants" for now). One crucial step to using the earth's resources more sustainably is accurately "valuing" environmental resources. If you value something, you tend to treat it well and not waste it; you can do that individually, and the economic market system can also attach a value. Some economists believe that "free-market environmentalism" (which depends solely on the marketplace to price the Earth's resources) is sufficient for valuing the environment. Many economists point out, however, that while the "invisible hand" of the marketplace is able to price a ream of paper or a bushel of corn, it is unable to account for (that is, attach a price to) other significant environmental costs.

One such cost is "externalities," or costs of production that are not paid by the producer, and that have an impact on others who were not involved in the economic transaction. For example, when a natural gas producer sells its product, the cost reflects the drilling, transportation, and delivery; it does not include attendant environmental impacts, such as air pollution near drilling sites and the health, social, and environmental costs that the product and its use create. Likewise, if a factory releases effluent into a river, the costs of that pollution are borne by others (human and non-human) downstream, not by the factory owner or even those who bought the factory's products. Pollution represents inefficiency, and wastes of all kinds are externalities that are not priced (and not valued) by the market.

Externalized costs can be significant and are crucial for determining sustainability. The Center for Investigative Reporting (cironline.org/reports/price-gas-2447) determined that a gallon of gas (3.8 litres of petrol) would cost upwards of US$15 if all external costs were included (instead of 2012 prices of about $3 per gallon). Another cost not priced by the marketplace is what scientists call "ecosystem services," discussed in the next section.

What some sustainability experts (and engineers and scientists) recommend to reduce externalities is "closing the circle," i.e., eliminating waste and inefficiencies, and even designing products and practices to more closely match processes in the natural world. "Biomimicry" studies nature's designs and functions and then imitates them to solve human problems. For example, studying how a leaf absorbs sunlight could help engineers design a better solar cell.

Some businesses have made great strides in reducing waste and improving efficiency. But to make true progress toward sustainability, significant changes are required in environmental accounting, policy, and governance. For example, sustainable societies cannot be built around or operate on nonrenewable fossil fuels (ancient solar energy, if you will). Yet our industries, cities, transportation, and food systems all developed around cheap, abundant supplies of fossil fuels. Part of their affordability is due to those externalities just discussed. Another reason they are cheap is large, long-standing government subsidies and tax credits, a marketplace perturbation that some economists question, particularly in a world with significant air pollution and climate change. To revamp fossil fuel-based systems toward more sustainable ones would require enormous amounts of political will, investment, and cultural cooperation.

Agriculture is another sector in need of a vast overhaul, according to sustainability experts (McConnell & Abel, 2008). Like energy, agriculture receives large, select government subsidies that slow progress toward sustainability and are often harmful to the environment. To become more sustainable, substantial food supplies must be produced locally and ultimately be based on organic methods (McConnell & Abel, 2008; Montague, 2009). The growing worldwide demand for meat also must be addressed. Industrial-style meat production bears significant environmental costs (waste, pollution, hormones, and synthetic chemicals), and the vast acreage required includes biologically diverse habitats (such as rainforests) that have been converted to meat and grain production.

Sustainability in Higher Education

As an example, I will explore sustainability efforts in one type of organization: institutions of higher education. The history of sustainability on campuses and universities is brief, enthusiastic, and stretches around the globe. Three-quarters of sustainability officer positions at US campuses were created between 2003 and 2007. The Association for the Advancement of Sustainability in Higher Education was founded in 2006 and now has 800 colleges and universities as members, primarily in Canada, the US, and Mexico, though members also hail from beyond those borders. In the UK, the Environmental Association for Universities and Colleges has over 300 institutional members seeking to "embed" sustainability on campuses. In 2009, the International Alliance of Research Universities began a sustainability initiative, which includes universities in Australia, China, Japan, Indonesia and Europe.

Some attribute this growth to awareness of sustainability issues and to global inaction on climate change. Others say that sustainability offers college students a strong sense of personal involvement and self-efficacy: it focuses on humanity's use of natural resources, and it puts climate at the center of the discussion. Also, in the 1990s, more and more universities were promoting engaged learning and connection to community in several forms, through service learning classes, civically engaged research and service projects, and co-curricular activities in residence-life offices; sustainability efforts were a natural fit. One commentary in the *Chronicle of Higher Education* said sustainability on campus shifts the focus from environmentalism to the imagined future and needs of Earth itself, and it replaces the old focus on pollution dangers with the idea that Western society itself is profoundly at odds with the Earth (Wood, 2010).

A university functions much like a small city in terms of autonomy and resource use; in fact, my university calculated its footprint as the equivalent of 24,000 average US homes. Thus, a campus office of sustainability institutionalizes this practice (for both good and bad) amongst thousands of students, faculty, and staff, potentially affecting every facet of university operation. Sustainability offices work with campus stakeholders to identify operational and institutional changes, from transportation to food service, and may even develop plans for making their university's operations "carbon-neutral." On one campus, students (working with faculty members) submit proposals to improve campus sustainability, from installing motion-sensors on hallway lights, to campus gardens, to retrofitting fans in labs. The competitive proposals are funded by a small levy on student tuition.

Only recently has there been much agreement on auditing and accounting methods for campus sustainability. Early rating systems and self-report surveys did not use particularly relevant nor valid assessments and contributed primarily to 'image'. In 2010, STARS was launched: the Sustainability Tracking, Assessment and Rating System is the first comprehensive, third-party-verified auditing and reporting system specifically for higher education. By 2012, over 300 universities had enrolled in STARS.

For all their growth and popularity, there is a lack of critical scholarly attention to campus sustainability efforts. One academic publication, the *International Journal of Sustainability in Higher Education*, has partially filled the gap, though it focuses largely on practice-oriented lessons and case studies.

Like all large organizations, universities face entrenched habits and routines when it comes to changing resource use and human relationships to the environment. They also encounter meager budgets to make needed upgrades and changes, particular at government-funded schools. Nevertheless, the enthusiasm of large groups of campus "sustainatopians," seeking to preserve the integrity and quality of human and natural environments, are converting thousands of young people to the hope and promise of sustainability.

Sustainability Science

If one searches a university library for journals with "sustainable" in the title, one finds plenty: *Journal of Sustainable Agriculture, Renewable and Sustainable Energy Reviews, Journal of Sustainable Forestry*, and *Journal of Sustainable Tourism*, to name a few. Sustainability is also discussed in many other journals, such as *Ecological Economics, Ecological Indicators*, and *Environmental Ethics*. A growing number of journal titles include "sustainability science," a subfield so robust that the US *Proceedings of the National Academy of Science* created a new section for it.

Sustainability science explores the provisioning (cf. the definition of "sustain") of humankind in a way that does not threaten the earth's support systems (Kates et al., 2001; Kates & Dasgupta, 2007). It views human and environmental subsystems as intimately linked, and it recognizes that the environment provides services required to maintain humankind, regardless of our awareness of them or the lack of economic value placed on them. This research community is closely aligned, as you might imagine, with global climate and environmental change research (Kastenhofer, Bechtold & Wilfing, 2011).

Sustainability science is particularly interested in questions of *vulnerability* and *resilience*, terms registered by Working Group II of the Intergovernmental Panel on Climate Change (IPCC). The concept of *vulnerability*, developed largely in the social sciences, is connected to environmental risks and hazards; within sustainability science, vulnerability refers to the degree to which a coupled human-environment system (CHES) is likely to experience harm from exposure to a hazard. *Resilience* emerged from the ecological sciences and was expanded from solely ecosystem changes to include changes in the larger CHES—in other words, how much disturbance a system can absorb and still remain the same or function well (the "endure" definition of sustain). CHES is one of three "pivot points" to which sustainability science research pays attention (Turner, 2010). A second pivot point is environmental services, and the third is the tradeoffs between those services and human outcomes and uses.

The second pivot—environmental services (sometimes referred to as "natural capital")—is the direct benefits and life-supporting processes that come from the natural environment. Some services operate at the global level (like an atmosphere) and some are local or regional (like an aquifer or watershed, or a place to recreate or be renewed). Humans have long taken environmental services for granted, particularly those that are regulating (like climate and flood regulation) and supporting (like soil formation and nutrient recycling) (Daily et al., 2000). These crucial functions are not explicitly valued in most economic and socio-political systems.

The tradeoffs pivot point is relevant every time humans utilize earth materials for some particular outcome, whether to grow food or generate heat. Oftentimes, we endeavor to expand the limits or reduce the uncertainties of natural systems to produce material well-being for people, such as applying synthetic fertilizer to increase food production or damming rivers to produce electricity and irrigation water. However, the consequences of one activity cascade through the entire system and produce tradeoffs (Batterham, 2006). Creating a reservoir may improve crop production with a steady supply of water, but it will alter the original streamside habitat and environmental services provided by it.

Tradeoffs are generally calculated according to economic value or by comparing a physical measure, such as amount of crops produced. However, the economic measure falls short because, as mentioned above, many environmental services have no market value.

Vulnerability and resilience can be considered different but complementary approaches. When vulnerability focuses on a small number of human outcomes (such as hunger or housing) for which the environmental subsystem serves merely as a backdrop, the concept gives minimal attention to tradeoffs in environmental services. But when vulnerability and resilience are both fully applied to CHES, vulnerability serves to identify the weakest parts, and resilience identifies which characteristics will make systems more robust to disturbances.

Because sustainability science treats the human and environment subsystems as coupled, it endeavours to remain at the core of those three overlapping circles comprising sustainability. Though sustainability science can be considered a young discipline with much more research ahead, indicators and indices developed from its findings hold promise as valid and accurate ways to judge whether a practice labeled "sustainable" truly qualifies as such.

There are numerous challenges ahead for this discipline, not the least of which is the complexity of its coupled approach (Turner, 2010). In addition, research is needed in virtually all locations where human-environment activity takes place: though lessons from one region may apply to others, unique characteristics (such as habitat, species, climate) may differently affect the vulnerability and resilience of a particular area. Another challenge is that environmental change is already occurring so quickly in some regions (such as the Arctic) that researchers must continually revisit the drawing board.

Communication Challenges for Sustainability

Communicating a large, complex human-environmental issue—whether sustainability or global climate change—is an enormous challenge. Worldwide, only 45 percent of people in 111 countries surveyed by Gallup in April 2011 see global warming as a threat to themselves and their families. "Sustainability" does not even grab the attention of pollsters (Pugliese & Ray, 2011).

Though the tendency is to view climate change as solely about greenhouse gas emissions, it is very much about living unsustainably. It is important for communicators of both climate change and sustainability to convey that humans are not immune or separate from the ecosystems they inhabit and alter, and that ensuring the resilience of human-environment systems is necessary for our very survival (Meppem & Bourke, 1999). This is extremely difficult when urbanizing populations become more and more disconnected from, and lack knowledge about, the non-human world.

For some people, sustainability may sound more acceptable and "doable," more distant from climate change and environmentalism, and therefore palatable. That may be good news for communicators needing to reframe some well-worn messages. However, sustainability's inherent ambiguity and broad scope also make it an easy target for appropriation (Peterson, 1997) to a wide variety of projects, products, and organizations that are at best "sustainability-lite."

A similar fate befell the word "green." It was so overused and misused to denote "good for the environment" that the word lost its meaning and accuracy. The result was "green-washing," when an individual or organization boasted of some "green" action that was not an accurate portrayal or representation of the entire story or record. A well-known example was the Ford company boasting of a green "living roof" at one of its US auto plants that manufactured some of the worst gas-guzzlers in the industry. The roof was part of Ford's campaign "Greening the Blue Oval" to which the group CorpWatch gave a greenwashing award. Sustainability is likewise susceptible to having proponents over-advertise small or cosmetic actions and thus create "sustainable-washing."

The enormous scope of sustainability can be seen as both a positive and a negative for communicators. The immenseness of the challenge can be paralyzing, causing individuals to tune out, ignore, and lack the self-efficacy to act. When other issues top the list—wars and unrest, economic woes, political change—it is hard to keep individuals (and countries) focused and concerned about ongoing issues that seem to have more to do with distant places than their daily lives. Making an issue like sustainability more integral and linked to daily life is a constant communication challenge.

A negative consequence of a sustainability "craze" is that marketing and consumption have been increasingly tied to environmentally responsible behavior. Consumers are told they can shop their way to guilt-free sustainable lives by purchasing cloth shopping bags, fluorescent lightbulbs, and hybrid cars. While it is true that individuals are key in any significant social change, this rather "upperclass" approach puts an inappropriate burden—and blame—on consumers. Ethical shopping and green consumerism are strategies that are not scaled to the size of the challenge of sustainability, and consumerism comes with exceptionally high social and environmental costs. According to some scholars, pushing unsustainable consumption (as key to sustainability or to "help" the economy) has contributed to the global financial collapse, the "export" of pollution to less developed producer countries, and large trade deficits (Cohen, 2010).

Social change scholars know that achieving meaningful and substantive change (at any level, let alone global) is very difficult to achieve. One reason is that those who seek to maintain the current unsustainable system (because they benefit from it) will exert significant pressure to thwart fundamental change, a process sociologists call "social control" (Corbett, 1998). Scholars point to the widespread environmental activism of the 1960s and '70s to show how environmental challengers were often co-opted and accommodated, and many of the changes they desired never fully materialized (Corbett, 2006; DeLuca, 2005).

Another positive reframing of sustainability tells us that its very enormity means it must be a matter of shared responsibility. We cannot rely only on personal actions, nor just policy and legal solutions, nor solely actions by business. Action is needed at all levels, and all levels need accurate and persuasive communication.

Another positive is the possibility that sustainability will truly help humans think about themselves and their lives in terms of global systems, both social and ecological. Also, the very ambiguity of the word "sustainability" means it *can* be defined and tailored in ways that best fit different world cultures, customs, and practices according to their particular human-environment subsystems. There is no "one-size-fits-all" answer that will effect sustainability the world over.

Scholars maintain that sustainability science is a good match for the goals of "civic science," which include increasing public participation in the production and use of scientific knowledge (Bäckstrand, 2003). Scientists can identify tradeoffs in human-environment subsystems, but they need active participation from citizens, policy-makers, and stakeholders across disciplines to reach just, sustainable decisions. A civic science approach also is amenable to bringing in place-based and indigenous knowledge as grounded perspectives to supplement traditional science.

A final positive note: there is currently a variety of enterprises around the globe working to make human lives more sustainable. In the US, there has been an explosion in backyard farms and poultry, participation in Community Supported Agriculture and farmers' markets. In the EU, local governments in over 40 countries have pledged to "local sustainability," which seeks to "mainstream" sustainability at the level of cities and towns (www. localsustainability.eu). The Sustainable Cities and Towns Campaign targets topics such as governance and management, natural common goods, traffic, health, and social justice (www.localsustainability.eu).

Another positive example is the Transition Initiative, a movement begun in the UK in 2006 that has spread to Australia, South Africa, Canada, and beyond (Griffiths, 2009). Like the EU campaigns, it seeks change at the middle level of *community*, where people feel connected and more empowered. Each transition unfolds at the direction of its community, as it trains citizens to self-organize and create initiatives that rebuild resilience and reduce CO_2 emissions. Transition communities have examined their food production, waste, energy use, and transport for ways to become more self-reliant.

At present, there are little survey data and limited communication research specific to sustainability, with the majority being case studies and rhetorical analyses of the term's use. However, there is now a solid body of theoretical research about climate change communication to add to an already robust literature about attitude and behavior change specific to science, environment, and health, as well as risk communication, media coverage, social marketing, and other areas. A sustainability communicator has many solid paths to follow.

For students and citizens of sustainability, perhaps one of the most constructive first steps is seeing ourselves as indeed connected to the extra-human world. As sustainable science scholars maintain, human-environment coupling and interaction take place every moment of every day. In a classroom or at the grocers, we are surrounded by nature: paper, minerals, cloth, embodied energy. We continually breathe an atmosphere, drink from a watershed, and participate in a climate. And for every bite of food and every product purchased, there is a tradeoff, a cascade of interactions and effects. This is a basic science lesson that is key to understanding, and living, sustainably.

References

Bäckstrand, K. (2003). Civic science for sustainability: Reframing the role of experts, policy-makers and citizens in environmental governance. *Global Environmental Change*, 3(4), 24-41.

Batterham, R. J. (2006). Sustainability—The next chapter. *Chemical Engineering Science*, 61, 4188-4193.

Cohen, M. J. (2010). The international political economy of (un)sustainable consumption and the global financial collapse. *Environmental Politics*, 19(1), 107-126.

Corbett, J. B. (1998). Media, bureaucracy, and the success of social protest: ewspaper coverage of environmental movement groups. *Mass Communication & Society*, 1(1-2), 41-61.

——(2006). *Communicating nature: How we create and understand environmental messages.* Washington, D.C.: Island Press.

DeLuca, K. M. (2005). Thinking with Heidegger: Rethinking environmental theory and practice. *Ethics & Environment, 10*, 67-87.

Daily, G. C., Soderqvist, T., Aniyar, S., Arrow, K., Dasgupta, P., Ehrlich, P. R.,…D., Walker, B. (2000). The value of nature and the nature of value. *Science* 289, 395-396.

Friedman, T. L. (2008). *Hot, flat, and crowded: Why we need a green revolution-and how it can renew America.* New York: Farrar, Strauss and Giroux.

Griffiths, J. (2009, July-August). The transition initiative: Changing the scale of change. *Orion*, 40-45.

Haanes, K. (2011). *Sustainability: The "embracers" seize advantage.* North Hollywood, CA: MIT Sloan Management Review.

Kastenhofer, K., Bechtold, U., & Wilfing, H. (2011). Sustaining sustainability science: The role of established inter-disciplines. *Ecological Economics*, 70, 835-843.

Kates, R. W., Clark, W. C., Corell, R., Hall, J., Jaeger, C., Lowe, I.,…Svedin, U. (2001). Sustainability science. *Science* 292, 641-642.

Kates, R.W., & Dasgupta, P. (2007). African poverty: A grand challenge for sustainability science. *Proceedings of the National Academy of Sciences* 104, 16747-16750.

Kendall, B. E. (2011). *Lay theory, communication, and organizing: A study of a university's office of sustainability.* (Unpublished doctoral dissertation). University of Utah, Salt Lake City, USA. (Direct link: http://gradworks.umi.com/34/60/3460340.html)

McConnell, R. L., & Abel, D. C. (2008). *Environmental issues: An introduction to sustainability.* Upper Saddle River, NJ: Pearson.

Montague, F. (2009). Gardening: An ecological approach to individual, community, and global health. Wanship, Utah: Mountain Bear Ink.

Meppem, T., & Bourke, S. (1999). Different ways of knowing: A communicative turn toward sustainability. *Ecological Economics*, 30, 389-404.

Paavola, J. (2001). Towards sustainable consumption: Economics and ethical concerns for the environment in consumer choices. *Review of Social Economy*, LIX(2), 227-248.

Peterson, T. R. (1997). *Sharing the earth: The rhetoric of sustainable development.* Columbia, SC: University of South Carolina Press.

Pugliese, A., & Ray, J. (2011). Fewer Americans, Europeans view global warming as a threat. Washington, D.C.: Gallup. (http://www.gallup.com/poll/147203/fewer-americans-europeans-view-global-warming-threat.aspx)

Turner, B. L. II. (2010). Vulnerability and resilience: Coalescing or paralleling approaches for sustainability science? *Global Environmental Change, 20*, 570-576.

WCED (World Commission on Environment and Development) (1987). Our Common Future. Oxford: Oxford University Press.

Wood, P. (2010, Oct. 3). From diversity to sustainability: How campus ideology is born. *The Chronicle of Higher Education,* (www.chronicle.com/article/From-Diversity-to/124773).

16

THE VALUE OF INDIGENOUS KNOWLEDGE SYSTEMS IN THE 21st CENTURY[1]

Yonah Seleti

Introduction

For the past 500 years, indigenous knowledge systems (IKS) have been viewed and expressed through the lens of Western thought, language and perception. As Europe spread its influence across the globe, it also projected itself as the source of the only knowledge that counted in the world. The Western world has hence come to see other cultural traditions and sources of knowledge through the filters of the modern view of the world, their own view of the world. Colonization deliberately created paradigmatic opposites: traditional vs. modern; oral vs. written and printed; illiterate vs. literate; rural and agrarian vs. urban and industrialized. In this classification, the traditional, oral, illiterate, rural and agrarian were considered to be inferior and backward, and so were their knowledge and their theories of knowledge. The role of the colonized was consigned to that of consumers of western knowledge. Technology and science, as perceived in the western sense, was used as a measure of the level of civilization of the societies that were brought under colonization. This resulted in the trivialization of the entire mode of life and spiritual framework of millions of people.

In recent times, however, in Africa and elsewhere, the quest for self-determination in British, French, Portuguese, and Spanish colonies has focused on political emancipation. The attainment of political liberation was accompanied by a cultural renaissance that saw a growth in education and the creation of new universities as symbols of independence. However, while these countries invested in education, they did not seek to overthrow the disenfranchisement that colonialism had imposed. They were happy to seek modernization with its accompanying underpinnings that placed western knowledge systems as drivers of this development. For over fifty years, therefore, most African countries have continued to pursue their dreams of development within the confines of western theories which have discounted knowledge from the periphery as of no value.

The IKS dilemma has been characterized by a narrow definition of knowledge. Western taxonomies continue to dominate the definition of what constitutes knowledge, much to the chagrin of indigenous knowledge systems. A significant challenge is the current focus of the knowledge sector on the distribution and coordination of *external* knowledge at the expense of the production of *relevant* knowledge. In Africa, this situation has resulted in an intellectual dependency of African universities—and hence of their scientific research and science communication—on western knowledge institutions. There is no doubt that the national systems of information rest on weak linkages between the creators of knowledge products and the consumers.

Defining Indigenous Knowledge

To see the value of IKS, one must be able to see it through the indigenous peoples' lens and hear their stories in their voices and through their experience. The definition of indigenous knowledge proffered in this chapter is a working definition which should be understood holistically. For the purpose of this chapter:

> Indigenous Knowledge Systems are the sum total of the knowledge and skills which people possess and use to identify themselves as indigenous people of a particular geographic place. They are based on a combination of cultural distinctiveness and prior territorial occupancy relative to a more recently arrived population with its own distinctive and subsequently dominant culture. Knowledge and skills are passed on orally from generation to generation.
>
> *(defined in Hoppers, 2002; ILO, 1989)*

Many scholars, especially in science, continue to regard IKS as "mumbo jumbo," which has no space in the knowledge agenda for societal transformation. To scholars of IKS, it delineates a cognitive structure in which theories and perceptions of nature and culture are conceptualized by indigenous people (Hoppers, 2002). The cognitive structure includes definitions, classifications, concepts of physical, natural, social, economic, and ideational environments. IKS is a knowledge system that has distinguishable knowledge domains. The knowledge domains include and are not limited to the following: agriculture, meteorology, ecology, governance, social welfare, peace building and conflict resolution, medicine and pharmacology, legal and jurisprudential matters, music, architecture, textile manufacture, metallurgy, food technology, and many more. The indigenous knowledge domain also includes technologies ranging from garment and weaving design, medical (pharmacology and obstetrics), food preservation, agricultural practices, fisheries, metallurgy, astronomy, and many more. The indigenous knowledge domains are surrounded, bounded, and embedded by cultural contexts and practices such as songs, rituals, dances, and fashion (Hoppers, 2002). The rituals and taboos have been used as a mechanism for knowledge protection and also for the protection of the society from harmful knowledge. The production, exploitation, and dissemination of knowledge are regulated through communities of practice. In this knowledge system, knowledge production is not for profit, but for the well being of the whole society. Knowledge holders and practitioners are allowed to eke out a living and gain rewards from their professions, but not to use the knowledge for personal aggrandizement.

Why Study Indigenous Knowledge Systems?

Interest in the role that indigenous knowledge can play in knowledge production has increased dramatically in recent years. Its role has occupied a centre stage in political, social, and even economic discourses. This has led to a rapid expansion of IKS as fields of investigation primarily focused on how they should locate themselves within the context of global discourse and context. This interest is reflected in the myriad of activities generated within communities that are recording their own knowledge, for use in their educational systems where IKS is now regarded as an invaluable national resource that builds upon and strengthens community-level knowledge systems and organization.

What is an IKS Epistemology?

Epistemological concerns are necessarily central to the study of indigenous knowledge. In the quest for redeeming IKS from its consigned limbo as a marginalised knowledge, IKS scholars have been at pains to show that both IKS and science are socially constructed and context dependent at the level of explanation, evidence, and truth. The process of questioning the normative definitions of what constitutes knowledge is the purview of epistemology, the study of the nature of knowledge. 'Epistemology' is defined in English as the "theory or science of the method or grounds of knowledge" (OED, 2012). Marie-Joëlle Broweys (2004) writes that, without epistemology, there is no possible scientific reflection. Broweys observed, however, that the modern concept of epistemology builds on two traditions, the Anglo-Saxon and the French. In contemporary Anglo-Saxon countries, "epistemology is a philosophical term meaning 'theory of knowledge'." On the other hand, "epistemology" is defined in the French philosophical tradition as ''the philosophy of science in general'' (p.20). In this tradition, epistemology tries to understand the foundations of knowledge, its development, its object, its purposes, and its objectives (Dortier, 1998).

Notwithstanding these different emphases, Browaeys (2004) concludes that there is no real incompatibility between both definitions, except that the French definition emphasizes scientific thinking per se. This 'thinking' is central in the works of Bachelard (1971, p. 15) who said that 'thinking is a force, it is not a substance.'' It forms the foundation of ''complexity thinking'' discussed by Morin (1994). Morin sees complexity ''as a challenge and as an instigation to think.'' One of the principles of complexity thinking is the dialogic principle described elsewhere in this book (see, for example, Chapter 2). Dialogue gathers up both epistemological approaches because it offers the opportunity to maintain duality, while at the same time transcending that duality and creating a unity from the whole. It takes into account both the Anglo-Saxon and French approaches, which thus join in an inseparable way.

The ascendance of complexity theories of knowledge has also stretched the original concept of epistemology. Complexity theory has grown out of systems theory and chaos theory in an attempt to demonstrate why the whole universe is greater than the sum of its parts, and how all its components come together to produce overarching patterns as the system learns, evolves, and adapts. Dann and Barclay (2006) point out that:

Scientific thinking has fundamentally shifted from the practices of the Scientific Revolution of Newton, Descartes and Galileo. Their models were mechanistic, study was restricted purely to measurable and quantifiable characteristics and the 'world as a machine' became the metaphor for the era. Here, all that happened had a definite cause and gave rise to a definite effect, thus making outcomes and relationships predictable. Scientific analysis was dominated by Descartes' reductionist approach which asserts that phenomena can only be understood by breaking them into increasingly smaller parts and only once the part properties are known can the whole be determined (Capra, 1996). Centuries later, following a series of backlashes and resurgences of the mechanistic view, the paradigm has given way to the new sciences of 'complexity' (p.21).

Systems thinkers (Dann & Barclay, p.22) contradict traditional reductionist thinking by asserting that:

• A system is an integrated whole, whose properties cannot be reduced to the sum of its parts;

• All phenomena are interrelated yet independent. Thus each system forms part of a larger system, yet each has its own individual properties. This is the idea of systems being nested or arranged in a hierarchy;

• Each system exhibits properties that do not exist at lower levels within the hierarchy. These are so-called ''emergent properties'' e.g., life and consciousness;

- The observer influences the determination of the system boundary i.e., what is part of the system, what is excluded, and the purpose of the system, thus making the definition of the system and its constituent parts critical;
- Systems are subject to "feedback," i.e., the influence of one element on another within the system. The nature of this feedback can vary, being either positive (amplifying and providing 'gain' in the system), or negative (declining, 'damping' of the system). This is so called 'non-linear' behaviour.

According to Dann and Barclay,

> Chaos theory relates to this latter point that systems behave in a non-linear fashion, particularly that small changes in initial conditions can have a large impact on outcomes. Rosenhead (1998) highlights that chaotic systems do not exhibit pure random behaviour as common use of the term 'chaos' might suggest. These systems are neither stable nor unstable; rather they operate at a boundary between the two zones in what has been termed 'the edge of chaos'. Chaotic systems refer to systems that display behaviour that, though it has certain regularities, defies prediction. (p.22)

Building Blocks of an IKS Epistemology

The concept of things always in a state of change leads to a change being perceived as the driving concept in knowledge production. In most indigenous societies the world is perceived to be in constant motion and flux. Things are constantly undergoing processes of transformation, forestation and deforestation, and restoration. In this worldview, the essence of life and being is movement. Without movement, there is no life and there is no being. Movement is the expression and manifestation of being. This constant movement, flux, and motion give rise to a "spider web" network of relations between beings and things. In this view of the world everything is interrelated and interconnected. In most African societies, it is assumed that everything in creation consists of a unique combination of energy waves. From this point of view, the existence of all animate things in space or place consists of energy waves and interrelationships. All things are imbued with energy waves or the spirit. In Bantu societies this is referred to as the *ntu*. The *ntu* is a life force that gives effect to movement and change. Every living being is thus imbued with this life force.

Relationships between people hold pride of place in society and are best captured by the African concept of *Ubuntu*. *Ubuntu* is the container of the *ntu;* it carries the life force of *ntu*. It carries, depicts, gives, and imparts the life force of *ntu*. To have *Ubuntu* is to be caring, generous, hospitable, and compassionate to other humans. It is always about how you relate to others. From the perspective of *Ubuntu*, harmony, friendliness, and community are the greatest social good and are celebrated and rewarded in society. On the other hand, anger, resentment, revenge, and success through aggressive competition corrode this social good and are to be discouraged, as they impart negative energy (Hoppers, 2002). *Ubuntu* does not seek to conquer or debilitate nature, but rather stresses the interrelatedness and interdependence of all phenomena. In this context, knowledge production, exploitation, and dissemination is about *building* the collective and communal well-being. Anything that contradicts and works against the well-being of society may be considered a negative force, probably the same as witchcraft. *Ubuntu* does not seek mastery, certainty, and hegemony but rather seeks harmony, consensus, and dignity for all. From this point of view, the outcomes of knowledge production, exploitation, and dissemination should therefore result in a harmonious society based on consensus and providing dignity for all.

The world view associated with most southern African people is clearly captured in the citation below from a Zulu personal declaration (quoted in Asante & Aubrey, 1996):

> My neighbour and I have the same origins
> We have the same life-experience and a common destiny
> We are the obverse and reverse sides of one entity
> We are unchanging equals;
> We are the faces which see themselves in each other
> We are mutually fulfilling complements
> We are simultaneously legitimate values
> My neighbour's sorrow is my sorrow;
> His joy is my joy
> He and I are mutually fulfilled when we stand by each other in moments of need
>
> His survival is a precondition of my survival.
> I am sovereign of my life
> My neighbour is sovereign of his life
> Society is a collective sovereignty
> It exists to ensure that my neighbour and I realise the promise of being human
> I have no right to anything I deny my neighbour
> I am all; and all are me.
> I come from eternity
> The present is a moment in eternity
> I belong to the future.
> I can commit no greater crime than to frustrate life's purpose for my neighbour
> Consensus is our guarantee for survival
> I define myself in what I do to my neighbour

In the indigenous knowledge systems paradigm, renewal is an important aspect. Out of the constant flux, indigenous peoples have observed certain regular patterns such as seasons, migrations of animals, or cosmic movements. In these knowledge systems, human beings are intrinsically bound in relationship with these regular patterns. Through regular ceremonies, rituals, and economic activities they maintain a life rhythm that calls for renewal and maintenance of certain regularities that are foundational to their continuing existence. Failure to maintain the patterns could result in extinction of livelihoods. Hence, communities have evolved systems of knowledge production, exploitation, management, and dissemination focused on maintaining the ecology of knowledge that ensures their existence. Some of this knowledge is managed through ceremonies to mark the change of seasons, and it is often coded in rituals performed publicly and privately by the custodians of such sacred knowledge. From the perspective of outsiders, these ceremonies are seen as static and ritualistic, but to the communities, the yearly routines speak to them about the complex interconnectedness, interdependencies, and interrelatedness between human beings and the environment. The resultant knowledge from such a system reflects a theory of knowledge embedded in a complex netwok of interrelationships and interdependencies that are renewed constantly.

In this paradigm land is an important reference point. Over centuries, indigenous people have observed patterns, cycles, and happenings from the land. Land plays a significant role in spirituality, as well as the socio-economic well being of communities. The concepts of land and space are interlinked in multiple directions and interactions with the animal migrations, cycles of plant life, seasons, and cosmic movements. These events and activities are detected from particular spatial locations. IKS holds that there are sacred places that have to be avoided and conserved. Basically, indigenous communities map the interconnectedness of events to the spatial allocations and attributes. Natural phenomena like rivers and mountains play a significant role in the psyche and constitution of communities bringing with it responsibilities and obligations resulting in ecological conservation practices. The knowledge generated from such a close relationship with land and location is contextualized local knowledge.

Gregory Cajete (2000) provides a useful discussion of the process of indigenous knowledge that he calls "native science." He has described methodological elements and tools that underpin indigenous knowledge (pp. 66-71). These methodologies for knowledge production, exploitation, and distribution include: observation, experimentation, attribution of meaning and understanding, objectivity, unity, models, causality, instrumentation, appropriate technology, spirit, interpretation, explanation, authority, cosmology, representations, paths, and more.

To illustrate the similarities and differences between indigenous knowledge and modern western methodologies, a few key issues are discussed. In both cases, observation is a central methodology in pursuit of knowledge. In indigenous knowledge, careful observation of plants, animals, weather, celestial events, healing processes, the structure of natural entities, and the ecologies of nature constitute its essence. Unlike western modern science, where laboratories are used for controlled experimentation, indigenous people have applied practical experimentation to find ways in which to maximize the benefits to their society. In indigenous knowledge the question of *objectivity* is perceived to be founded on *subjectivity* arising from a personal and collective closeness to nature, the object of study. In the area of causality, indigenous knowledge accepts that causes can be both physical and spiritual, but both would have the transformative energies to cause change. The aim of indigenous knowledge experimentation is to seek meaningful relationships and an understanding of one's responsibility to the entities on which they depend for their livelihood. The aim of indigenous knowledge is not to seek control of nature. It will not be surprising to note then that indigenous knowledge explanations of events in nature embody a multiplicity of metaphoric stories, symbols, and images.

Indigenous knowledge is associated with terms "local knowledge" or "ethnoscience," indicating knowledge systems that are specific to cultures or groups in particular social or historical context. The ascription of "indigenous" as a prefix to this kind of knowledge is an attempt to separate what is considered to be universal, value-free, static truth, from the situated, value-laden, changing cultural beliefs of indigenous knowledge. However, since the 1980s, social scientists have viewed indigenous knowledge as a coherent system of belief and explanations that are homologous with scientific thought and that can offer better accounts of local phenomena than those theorized by universal science.

Western Modern Science and Its Critiques

Many scholars have presented scathing critiques of Western modern science and its role in marginalizing and denigrating IKS. Scholars of indigenous knowledge trace the subjugation of diverse knowledges, through systematic misinterpretation, misrepresentation, and marginalization of IKS, to the colonial encounter with indigenous people. The combination of colonialism, capitalism, and modern western science has been highlighted as a lethal combination of historical processes in the expansion of Europe that not only culminated in the colonization and subjugation of the entire world, but also led to the trivialisation of the ways of life of the colonized. This section provides a few examples as illustrations, rather than a repetition of the well-argued critiques of modernity.

In *We Have Never Been Modern* (1991), Bruno Latour[2] highlights four premises of modernity that have shaped the making of the modern world, namely:

1. The separation of nature from society, by making nature's laws "transcendent," such that we can do nothing about them;

2. The separation of society from nature by making society "immanent," *i.e.*, artificially constructed and so totally free, with no unlimited possibilities;
3. The separation of powers between nature and society; nature will remain without relationship to society;
4. A "crossed out God," *i.e.*, an arbitration mechanism without a deity.

Latour must be handled with care, in that his critique of modernism and also post-modernism is not a wholesale rejection of these ideas, because he recognizes values that must be retained. His caution to antimodern philosophy must be read carefully, lest the Indigenous Knowledge Systems paradigm be lumped in with the antimodern. It is not the same. IKS scholars should avoid the tradition of the Africanist school that succeeded in creating mirror images of the colonial historiography. The Africanist historiography rejected the colonial history that celebrated Europeans as actors in history and replaced them with African centred activities with Africans as the heroes. The interfacing of IKS with other knowledge system seeks a mutual respect and reciprocal recognition of knowledge systems.

The aim of the critique by Latour was to find a third space, where the separation between human and machines, society and nature comes back together in a new paradigm.

In her rigorous critique of modernism, Odora Hoppers (2002) points to the effects it wrought among the subjugated:

*   Epistemological disenfranchisement;
*   Imposition of a Western paradigm that is cruel, blind, and has no place for defeated knowledges or alternative theories of knowledge;
*   Absence of cognizance and moral sensitivity to local conditions;
*   Cognitive injustice;
*   Condescending and paternalist attitudes;
*   Production of paradigmatic opposites: traditional vs. modern; oral vs. written and printed; illiterate vs. literate; rural and agrarian vs. urban and industrialized;
*   Trivialization of the entire mode of life and the spiritual framework of millions of people.

The Quest for IKS and democratic politics

At the global level of knowledge production, the work of Michel Foucault and Jacques Derrida ignited a movement that allowed for the emergence of a new moral and cognitive space within which constructive dialogue between people and between knowledge systems could occur. For Thomas Kuhn, science is not an evolutionary, progressive march towards greater and greater truth, but rather "a series of peaceful interludes punctuated by intellectually violent revolutions," in which one point of view is replaced by another (Kuhn, 1962, quoted in Van Gelder, 1996).

The quest for the recognition of the role of indigenous knowledge in the knowledge sector in South Africa did not arise from the universities or science councils. In fact, the articulation of the African Renaissance and New Partnership for Africa's Development created a space to search for an African rationale for an African-centred initiative. In South Africa, the birth of IKS emerged outside the leading academic institutions. It was a product of intellectual activism and political elites in Parliament. Initially, it was the historically disadvantaged universities—North West University (Mahikeng Campus) and the University of Venda—who took an active interest in the promotion of indigenous knowledge.

Another contributing factor to the ascendance of IKS in recent years is the positioning of knowledge as an intrinsic part of democratic politics. In the liberation struggles against colonialism and apartheid, Africans relied on knowledge from within their societies to wage their campaigns. Indigenous knowledge played a critical role in challenging the knowledge injustices of the past. Knowledge production, access to knowledge and information, and its dissemination and distribution have become contested issues in modern democracies. Further, the role of society in raising issues on governance of knowledge—knowledge politics—is on the rise, as civil societies challenge the role of research institutions in setting the research agenda. Knowledge generators are being challenged to orient their contributions to the welfare of local communities.

The increasing involvement of society in requesting accountability from knowledge producing entities should contribute to the creation of a holistic framework for societal development. The need for new ethics of knowledge production that is cognisant of indigenous concepts of justice should be stressed in this chapter; otherwise the rising momentum and pressure from below for change could lead to rendering knowledge-producing institutions, such as scientific research organisations, less relevant to society.

In South Africa, the knowledge agenda, as reflected by science curricula and qualifications, continues to mirror western knowledge systems at the cost of excluding and marginalizing local or indigenous knowledge. For an inclusive and democratic knowledge network to emerge within Africa, it will require a significant mind-shift to be in a position to recognize the contribution of other knowledge systems. Africa's knowledge needs must be determined by those who produce it. As Africans, we are seeking our own programmes of renewal and rebirth by using indigenous knowledge systems for physical, spiritual, psychological, socio-economic, and political well-being. We should guard against the risk of dividing society that could emanate from the enunciation of African views based on IKS, but should strive to indicate how IKS can enrich and inform Western science to the benefit of both.

Enriching Science Knowledge Systems

This chapter argues that the integration and interfacing of indigenous knowledge systems into modern knowledge systems will lead to recognition that science, hitherto dominated by western knowledge systems, has been premised on simplistic theories of knowledge. Indigenous knowledge systems, by virtue of their methodology and processes of knowledge production, bring the benefit of a holistic approach that transcends the disciplinary boundaries that have characterized modern western knowledge. IKS is projected as a knowledge system that goes beyond the over-simplifications and over-regularisation of knowledge that is a dominant feature of modern western knowledge. What the chapter calls for is a reexamination of epistemology that is built on empiricism, reductionism, and positivism.

This chapter holds that indigenous knowledge systems proffer an epistemology of hope to humankind that acknowledges cognitive flexibility. The chapter maintains that the value of IKS to the modern western knowledge systems is that IKS promotes the acceptance of multiple truths or knowledges, and rejects the projection of western modern knowledge as the unique and universal truth. It urges the use of multiple mental and pedagogical representations that certainly do not culminate in one truth, but multiple truths. At the centre of this approach is the promotion of multiple alternative systems of linkages among knowledge elements that reflect the multiply interconnected nature of knowledge. This paper has presented IKS as a humanizing agent that complements the cosmopolitanism of universities and science centres. The value of IKS is that it is cognizant of moral sensitivity to local conditions. It also calls for a rethinking and enlargement of modern science and technology, and their communication, by integrating IKS into them in the twenty-first century.

In contrast to reductionist thinking, this chapter has identified IKS with complexity theories of knowledge, holding that indigenous knowledge as a system is an integrated whole, whose properties cannot be reduced to the sum of its parts. The chapter also noted that IKS perceives all phenomena as interrelated yet independent and that each system forms part of a larger system, yet each has its own individual properties. This is the idea of systems being nested or arranged in a hierarchy. IKS is thus embedded interdisciplinary knowledge, that celebrates the interrelatedness of all knowledge, including things living and inanimate. This chapter thus calls for reciprocity and co-existence between knowledge systems. It is hoped that it has highlighted the need to build fraternity between knowledge systems at the cognitive level. The value of this approach will culminate in an inclusive citizenship of knowledge characterized by cognitive justice—the right of different knowledges to survive to the benefit of humankind.

Notes

1   The views expressed in this chapter are solely those of its author.
2   Latour attempts to reconnect the social and natural worlds by arguing that the modernist distinction between nature and culture never existed. He claims we must rework our thinking to conceive of a "Parliament of Things" wherein natural phenomena, social phenomena, and the discourse about them are not seen as separate objects to be studied by specialists, but as hybrids made and scrutinized by the public interaction of people, things, and concepts.

References

Asante, M. K., & Aubrey A. S. (eds.) (1996). *African Intellectual Heritage*. Philadelphia: Temple University Press

Bachelard, G. (1971). *Epistémologie*. Paris: PUF.

Browaeys, M-J. (2004). *Complexity of epistemology: Theory of knowledge or philosophy of science?* Presented at the Fourth Annual Meeting of the European Chaos and Complexity in Organisations Network (ECCON), 22-23 October 2004, Driebergen, NL.

Cajete, G. (2000). *Native Science: Natural Laws of Interdependence*. Santa Fe: Clear Light Publishers.

Capra, F. (1996). *The web of life: A new scientific understanding of living systems*. New York: Anchor Books.

Dann, Z., & Barclay, I. (2006). *Complexity Theory and Knowledge Management Application. The Electronic Journal of Knowledge Management, 4*, 1, pp. 21-30, available online at www.ejkm.com

Dortier, J-F. (1998). *Sciences Humaines*, Paris: PUF (cited in Marie-Joëlle Browaeys, *Complexity of epistemology: Theory of knowledge or philosophy of science?* Presented at the Fourth Annual Meeting of the European Chaos and Complexity in Organisations Network (ECCON), 22-23 October 2004, Driebergen, NL).

Hoppers, O. (2002). *Culture, Indigenous Knowledge and Development: The Role of Universities*. Pretoria: Centre for Education Policy Development.

ILO. (1989). International Labour Organization Convention on Indigenous and Tribal peoples in independent Countries, Geneva, Article 1 C.

Kuhn, T. (1962). *The structure of scientific revolutions*. Chicago: University of Chicago Press.

Latour, B. (1991). *We have never been modern*. Cambridge, Massachusetts: Harvard University Press, (Translation in 1993).

Morin, E. (1994). *La complexité humaine*. Paris : Flammarion

OED (Oxford English Dictionary) (2012). Definition retrieved 6.4.2012 from http://www.oed.com/view/Entry/63546?redirectedFrom=epistemology#eid.

Rosenhead, J. (1998). *Complexity Theory and Management Practice (Working Paper Series, LSEOR 98.25),* London: London School of Economics and Political Science

The Zulu Personal Declaration (1825). in M.K. Asante and A. S. Aubrey (eds.) (1996). *African Intellectual Heritage*. Philadelphia:Temple University Press: pp. 371-378.

Van Gelder, L. (1996). Obituary: Thomas S. Kuhn, scholar who altered the paradigm of scientific change, dies at 73. *New York Times*, June 19, 1996, p. B7

17

SCIENCE COMMUNICATION

The Consequences of Being Human

Chris Bryant

Introduction

In the past, many scientists chose their profession because they were more adept in dealing with ideas and facts than with the unpredictable ferment of human relations. Communication, other than with their scientific peers, did not have a high priority: indeed, it was often actively avoided. The resulting lack of empathy with the general public led them into many false assumptions—'they're not interested', 'they wouldn't understand' and, by way of contrast and in the words of one eminent scientist passionately advocating immunisation, 'Show them the figures—how can they *fail* to understand? How, indeed? It is the role of the science communicator to explain to the scientist how 'they' can fail to understand and, to the community, what the scientist meant.

Humans are remarkable for their consciousness; for their ability to reason, their self-awareness, their capacity for empathy, for remembering the past, awareness of their environment and, above all, to make choices from among many possible 'futures'. In the 21st century, the availability of information from myriad sources to inform those choices is unsurpassed. Increasingly, people are becoming connected to friends, neighbours, and workmates all over the world. They access information about government, the stock market and multinational corporations. They access scientific and medical information, not only from web pages designed for general use but directly, from scientific and medical journals. They establish specialist websites to facilitate the exchange of information and opinion. They Google, they blog, they tweet, they subscribe to Facebook and YouTube. They have mobile phones and other devices so that they can communicate on the way to work, or during lunchbreak, or on the beach, or climbing mountains. Never before have so many people become connected by the invisible strings of modern technology. This connectivity and the consequent complexity that it creates is a critical attribute of modern communication, accelerating the development of higher organisational structures (Westwood & Clegg, 2003).

Here is a definition. *Connectivity* is the state, property or degree of being interconnected (Frazer, 2003). One important consequence of connectivity is the emergence of order, beautifully illustrated by Kauffman (1995). He describes a thought-experiment.

Imagine a scatter of similar buttons on a flat surface, distributed so that no button is touching another. Select randomly two buttons and join them with a thread. Repeat the process. For the first few tries, it is unlikely that you will pick a pair of buttons that have already been joined. As more and more buttons are joined it becomes increasingly likely that at the next try you will pick up an unjoined button and one that has already been joined to another. By connecting the unjoined button to one of the two already connected you create a string of three buttons. Continue this experiment and eventually each button you select will be attached to several other buttons. Networks form. If you plot the size of the largest cluster of buttons against the ratio of the number of threads to the number of buttons at each step you obtain a sigmoid curve (Figure 17.1). The scattering of buttons has moved, quite rapidly, from one level of organization to another. The process of stringing clusters of buttons together, and clusters of clusters leads, given an inexhaustible supply of buttons and string, to higher and higher levels of organization.

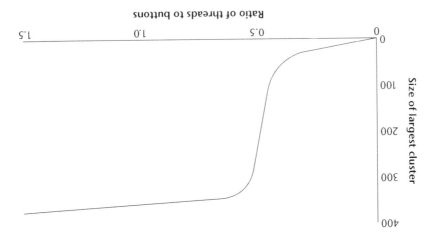

FIGURE 17.1 Clustering of Connected Buttons (*Source*: Kauffman, 1995, p.57)

Emergent Phenomena

Kauffman (1995), and others (eg. Waldrop, 1992; Boal, Hunt & Jaros, 2003) extend this analogy to a wide range of disciplines. The idea of 'order for free' is that repetition of simple processes creates complexity. This is an especially important idea for biologists. For example, the cells of an embryo responding over and over again to a simple repetitive series of 'commands' create a fully organised foetus.

Another and even better example is that of organic evolution itself (Chamberlin, 2009). Over the course of three billion years matter has organised itself into a system of the utmost complexity by the repetition of the simple imperatives of natural selection. Morphological and physiological innovations arise and each time there are consequences—emergent properties explicable with hindsight but not predictable - that enable the innovation to extend its geographical and physiological ranges. Then another innovation is selected and the process recurs. This chapter explores that process and examines some of the consequences for humanity, and the degree to which they influence understanding of natural phenomena (see, for example, Corning, 2008).

This process of simple, self-repeating events indefinitely continued—iteration—is at the heart of evolution by natural selection (Bird, 2003). Assembly of macromolecules according to simple rules led to the emergence of protocells. Protocells became the first prokaryotic cells, represented today by bacteria and blue green algae. Prokaryotes entered into associations to give nucleated 'eukaryotic' cells (Margulis, 1970). Each eukaryotic cell is thus the product of symbiosis, of connectivity between at least three prokaryotes and sometimes more. This is the type of cell from which all animals, plants and fungi are constructed. In the next stage, eukaryotic cells aggregated to become coordinated colonies of identical cells acting in unison (as in the alga *Volvox*). Assigning different tasks to cells within the colonies permitted higher levels of organisation. This stage is represented in modern sponges, which can be separated into individual cells of different types that subsequently recombine if left alone. Such division of labour became more or less permanent, as in *Trichoplax,* a small multicellular cellular mat with only three different sorts of cooperating cells. Finally, the permanent differentiation of cells into tissues occurred, as in all modern animals, plants and fungi—and humans.

This magnificently inadequate description of the origin and evolution of life shows only a few of the major events but illustrates another important phenomenon. Each successful step, each increase in connectivity gives rise to a population explosion followed by adaptive radiation into all available ecological niches, creating possibilities of life unavailable to precursors. Connectivity between units brings with it unforeseeable consequences. These 'emergent properties' are characteristic of life at each level of organization.

The idea that 'the whole is more than the sum of the parts' is widely attributed to Aristotle (384–322BC). The attribution is often made with reference to Greek theatre, to make the point that the whole play has integrity and makes an impact on the audience that is independent of the contribution of any one of its characters. This metaphor illustrates the concept of emergent properties. They are the properties of complex systems, whose complexity is the consequence of many simple interactions, but such that the behaviour of the whole cannot be predicted on the basis of the behaviours of the individual components.

When the three or more prokaryote precursors came together to produce a eukaryote cell, they created an entity whose potential (emergent properties) far exceeded the potential of the precursors, even taken together (Macklem, 2008). These emergent properties include the rest of the biosphere. They include human beings who form families, villages, towns, cities and nations. Humans cooperate in technology and together are on the verge of producing artificial intelligence, an evolutionary leap that changes the very nature of evolution. This has been made possible because, in order to survive, they shared resources and cooperated in buying and selling goods. From that emerged an entity called a 'market' that, in the whole, has properties and consequences that are unpredictable on the basis of the behaviour of the individuals who participate in it.

The evolution of neural systems underwent the same sort of process. All members of the 'single-celled' Protista show responsiveness to stimuli. In the multicellular Cnidaria, to which *Hydra*, jellyfish and sea anemones belong, responsiveness has, however, become largely the role of specialised cells called neurons. *Hydra* has only a simple neural network, yet is capable of such complicated behaviours as somersaulting. As Cnidaria become larger and more complex, many more neurons are needed to form large numbers of interneuron connections. The neural network of jellyfish is organised into a conducting nerve ring around the rim of the bell to control contractions. The next evolutionary step is represented in the flatworm, *Planaria*. It possesses two eyespots, each of which is associated with a ganglion, a knot of interconnected nerve cells, and from each of which a nerve runs backwards to the tail. This simple type of nervous system is greatly developed in all higher animals; many sense organs are concentrated at the front end because it is an advantage to have advance information about what you are getting into, and their associated clusters of ganglions multiply and become brains.

The vertebrate brain is the product of vastly increased connectivity. It culminates in the giant brain of modern humans, an enlargement that started with the modest brain of an ape a mere two million years ago. Connectivity is hugely extensive in the adult brain, one estimate putting the number of neurons at 100 billion, implying interneuron connections well into the trillions. As remarked earlier, the outstanding emergent property of the human brain is consciousness. This is not the place for a philosophical discussion of consciousness (for this, see Chalmers, 1996, 2010a) but it is, however, appropriate to consider some of its attributes.

Some Attributes of Consciousness

Awareness of 'self' is amongst the most important characteristics in the suite of adaptations to the environment provided by consciousness. Awareness of 'self', and knowing the boundaries where 'self' ends and the environment begins, implies awareness of other components of the environment and the ability to deal with its challenges.

Awareness of the environment is of obvious adaptive advantage. Unfortunately, the environment is potentially dangerous, so the downside is fear. An extension of the awareness of self creates an environment comprising potentially dangerous things that are 'not self'. Rocks and trees and bushes are perceived to have their own spirits: spirits that must be appeased, made harmless. The human capacity for visual patterning, to see faces in the fire, in clouds, or in foliage, leads first to animism, then deism, and finally structured religion. This basic human response to the environment creates problems for science communicators, who may find themselves confronted by religious sects claiming that meddling with nature is against God's laws.

There is a downside to the awareness of self that we call selfishness. Once the concept of 'my environment' becomes 'this territory is mine', there follows an instinctive desire to protect it. Another example of selfish behaviour is the reliance of selfish people on 'herd immunity' for keeping themselves and their families disease-free. They argue that others can incur the minimal risks involved in vaccination. This works in the short term as long as the great majority of people accept vaccination. As more people become complacent and refuse vaccination, the strategy fails, as the reappearance of formerly common infectious diseases in the First World has shown.

The ability to imagine what another person is feeling and adjust decisions and behaviour accordingly is known as empathy. It is an extension of the awareness of self to include an awareness of others. Of extraordinary survival value in small societies, where future options are limited, unfortunately it becomes qualified in large societies. Thus, an instinct to form a cooperative pack or herd or family group, which is of obvious survival value, leads to a more sinister outcome as populations become larger. It leads to attitudes such as 'it's my family against that one in the next valley' bolstered by spurious reasons, such as, 'they are not like us—their skin is the wrong colour' or 'they have the wrong shaped eyes'. This is the road to doing away with all those who are 'not like us', to cruelty, sadism, racism, nationalism, war, and genocide.

In communities with large numbers of people, the empathic faculty becomes attenuated, strong for family and friends, but progressively weaker for people who live in the next street, or town or another country. If this were not so, ordinary men and women who would balk at hitting another person over the head with a club could not bring themselves to drop a bomb they know may kill a hundred people.

Consciousness offers a way of 'seeing into the future' that confers enormous survival value. Every organism can 'see' a little way into the future. Enzymes in bacteria are produced when they detect a few molecules of useful food substrate. The bacteria then move up the concentration gradient to the source. The roots of plants actively grow towards water and nutrients, 'confident' of a good outcome. These are statistically valid responses, encoded in their genes. They are the end product of many trials in which the many bad outcomes did not lead to survival.

Most organisms on Earth manage very well without consciousness. Plants have a very narrow range of options and do not need consciousness to enable them to 'choose'; this is provided by various physiological responses, such as geo- or photo taxis. Animals, possessing the capacity to move from place to place, have a larger range of options; if they pick the wrong one, survival is compromised. While humans spend more than half their time self-aware, other vertebrates seem to run on 'automatic pilot', with brief flashes of self-awareness in circumstances of, say, danger. This, in humans, is like travelling along a very familiar path. For example, one's conscious mind might be grappling with a knotty problem and becomes so engrossed in it that it is with some surprise that one finds one has arrived at one's destination. Lower order brain function has taken care of what might have been quite a complicated journey.

Humans, however, can visualise numerous possibilities, options that offer a large range of outcomes that vary from 'very good' to 'disastrous'. Setting aside potential winners of the Darwin Award whose disastrous choices are likely to remove them from the human gene pool, most people select from a range of choices from 'good' to 'not so good'. There are far more 'not so good' options than 'good ones', and far more 'not so good' choosers than 'good' choosers. 'Not so good' outcomes therefore, are more likely to be chosen. This has important consequences for science communication. The sheer complexity of modern society created by billions of people has also created an enormous range of future options of the 'not so good but not disastrous' category.

Ability to make choices about the future follows the same rules as other biological phenomena, and conforms to a normal distribution curve. Those in the population who are able to make 'good' decisions (and for science communicators 'good' means based on sound scientific understanding) will occupy one small tail, while the 'not so good' deciders will occupy the huge hump clustering around the mean. The other tail is occupied by the 'disastrous' choosers. Given that the population of Earth is 7 billion, that hump will comprise huge numbers of people, most of whom will be intelligent, and many of whom will be influential and may have an interest in promulgating 'not so good' scenarios.

If one accepts the evidence for evolution, one must also accept the fact that humans are adapted to living on Earth. 'Adaptedness' may be defined as the best available fit to the environment—it is what is possible but is not necessarily either maximal or optimal. Rabbits did better when they got to Australia than in their native Spain. A camel is better adapted to living in the desert than to life above the Arctic Circle. It presumably feels more comfortable in the desert, whereas prolonged exposure on an ice shelf will lead to the supreme discomfort of dying. It is, however, possible that there is an ideal camel environment somewhere on Earth that is empty of camels because camels have never been in a geographical position to discover it; but, almost certainly, something else has.

Animals—and humans—do the best they can with what they've got. Life is a series of approximate fits, and they are always chasing 'adaptedness' because, inevitably, the environment changes around them. This is called 'Running the Red Queen's Race' (Van Valen, 1973) because, as the Red Queen said to Alice in *Through the Looking Glass*, "… it takes all the running you can do to keep in the same place." Beauty, comfort, and contentment can be considered to be personal measures of adaptation in humans, a sense that things are 'right'. But things continually change (both within and without) so 'nostalgia', for things as they used to be, especially for childhood things, is a common emotion that engenders resistance to change.

As the brain develops, and it gradually acquires self consciousness, it begins to observe first, its internal, corporeal environment and then its external environment. It needs this information to build a picture of itself and its place in the world. With it, it begins to assemble information about what is needed for survival. As it continues to develop, the brain makes hypotheses about the environment and tests them against what actually happens. It often takes the form of 'if I do this, such and such will happen.' If such and such does not happen the hypothesis will be weakened. If such and such happens, the hypothesis will be strengthened; successful repetition establishes it as a 'fact', *even if it is untrue*. It is an attempt to make sense of the environment in a way that will have a predictive quality that enhances survival. It can, of course, go badly wrong and end up in the form of meaningless ritual. 'I slew a black cockerel yesterday evening and the sun rose this morning' becomes 'I have to slay a black cockerel in the evening, otherwise the sun will not rise in the morning.' Cause and effect become confused; thus, belief systems are established as mixtures of truths and untruths. They may lead to conflict between communities, as in 'my belief system is truer than yours.' With the growth of larger and larger communities, belief systems progress from the personal to the collective.

Belief Systems and Science Communication

Belief systems are thus part of everyone's heritage. As we have seen in Chapter 9, 'belief' may refer to a principle accepted as true without supporting objective evidence, or it may mean a conviction based on such evidence. In the statement 'many believe that black cats bring good luck', the word takes the former meaning. In 'having considered all the evidence, the scientist believes evolution to be a fact' it is the latter meaning that is implied. Without carefully defining 'belief' it is impossible to have a useful discussion. In any event, the term describes an internal process that is extremely difficult for an outsider, such as a science communicator, to alter. At a lower level there are superstitions to which everyone is subject. In some cases they too can be unshakeable, especially in times of stress. Highly educated fighter pilots carry good-luck mascots; travellers carry St Christopher medallions for safety; unfortunates who descend into neurological disorder become enmeshed in a personal web of rituals that they must perform in a certain way; remnants like these weave a rich, dense fabric that is not easily penetrated by reason.

Entrenched beliefs abound in medicine and science. The placebo effect—the beneficial effect, say, of a sugar pill on someone who believes it to be a medicine – and its opposite, the nocebo effect, are now well established and the power of suggestion has to be reckoned with in modern pharmacopoeias. People are notoriously hard to shift away from their favourite nostrums, many of which can have no possible therapeutic effect but some of which occupy the uneasy hinterland of the marginally effective. There are nonsense palliatives such as rhinoceros horn, bear bile, and bat dung. Some years ago, however, I had occasion to go through an English translation of *A Barefoot Doctor's Manual* (translated by Fogarty, 1977) looking for specifics against parasitic disease and found no fewer than forty, amongst some hundreds of prescriptions, that would probably have some desirable effect. A great vindication of one Chinese medicine has been a new antimalarial derived from a daisy, *Artinisia*, concoctions of which have been used to treat the disease since at least 340AD (Klayman, 1985). Folk medicines characteristically have this mix of the effective and the magical. Science communicators find it notoriously hard to persuade people to jettison the magical. People tend to draw their own conclusions about modern medicine as well. Thus, the conclusion that, because antibiotics are brilliantly successful against bacterial infections, they must work against the common cold, is compounded of misunderstandings both about antibiotics and the common cold.

It is worth pointing out, however, that given a sufficiently large target audience, which will necessarily include people of many different beliefs, some groups will form coherent foci of opposition to attempts at science communication and will have an adverse effect on any measure of success that might be attempted.

Science Communication: Success or Failure?

One is justified in asking whether the possession of foresight in humans is a trait that confers biological success. It has certainly enabled us to pioneer and colonise more environments than any other biological species, though we tend to bring along our pets, pests, and parasites. It has allowed us to reduce infant mortality and to increase life expectancy to variable extents in different parts of the world. As a result, we have a population explosion; we are a monoculture that has become susceptible to epidemic diseases. Water is becoming scarce. We are using up other non-renewable resources at a rapid rate. We are beginning to compete for them and, as we seem to be addicted to wars of varying ferocity, including nuclear, there are several bleak scenarios ahead of us. Science surely has a role in helping to avert potential disasters. Science communicators have a role in helping the world community to understand the problems and the possible solutions but only, it seems, with limited success at present. But how does one measure success?

There are two measures that might be used to determine the success of science communication. The first is the time taken for the public to embrace a particular issue: the shorter the time, the more effective the communication. The second is the proportion of the target population that accepts a scientific idea: the greater the proportion, the greater the effectiveness.

Take, for example, the acceptance of evolution by natural selection. By any standards this has been a successful piece of science communication, one that has had a huge impact on society. Darwin published *The Origin of Species* in 1859. A majority of scientists and the educated elite had accepted it by the turn of the century, but there were still professional enclaves who remained sceptical. Even in the social revolution that took place after WW1, there were still scientists who rejected it.

After Huxley published *The Modern Synthesis* in 1942, it was possible to say that the *vast* majority of scientists were evolutionists, as were most non-scientists who were familiar with the idea. But even today, when much of the underpinning of evolutionary theory is in place, evolution is not universally accepted. A 2006 survey of the USA, Japan, and 32 European countries administered 152 years after *The Origin of Species* was published (National Geographic News, 30 Oct. 2010) shows that only about a third of Americans accept the evolution of humans from lower animals. The UK and Japan each had about 80% acceptance; only Turkey ranked lower than the USA, with about 25%. About two thirds of all people surveyed agreed that evolution had taken place. Is this a success of science communication or a failure? Is the glass two-thirds full or one-third empty? The implication is that even the best science communication can never be 100% effective. Science communicators do not have a monopoly of intellect.

History is full of stories of the promulgation of scientific knowledge. Unfortunately, a clear demonstration that the idea is correct does not guarantee the immediate success of its communication. Aristotle (c384-322BC) deduced that the Earth was round on the basis of empirical data derived from the observation of eclipses. Eratosthenes (c276-195BC) determined its circumference with an accuracy of between 1% and 16%, depending on whether Egyptian or Attic stadia were used as yardstick. A practical demonstration by Columbus, 1700 years later, should have put the matter to rest, but there are still Flat Earthers.

When the Royal Society was established in 1660, the communication of its findings was considered to be of high priority by the members. They created two journals to serve this purpose—*Philosophical Transactions* and *Proceedings*. The target audience was a well-defined and educated elite, mainly members of the Society itself. As an exercise in the communication of science it was astonishingly successful, ushering in the Age of Enlightenment.

Individual discoveries—or re-discoveries—also shared this success: Edward Jenner (1796; vaccination), Ignaz Semmelweiss (1847; prevention of puerperal fever), Louis Pasteur (1862; germ theory), Joseph Lister (1867; antisepsis), Florence Nightingale (c1859; medical statistics and epidemiology) are some names from the nineteenth century that resound in the history of medicine.

More recently, the banning of smoking in Australia to prevent heart disease and lung cancer has engendered legislation to make smoking in public places illegal. The widely advertised *Slip, Slop, Slap* campaign (*Slip* on a shirt, *Slop* on sun blockout cream, and *Slap* on a hat) to prevent melanoma is a triumph, as a quick glance into the schoolyards at the pupils in broad-brimmed hats will testify. Unfortunately, a downside is emerging, as cases of vitamin D deficiency are on the rise. The discovery of a fly in the scientific ointment is commonplace and is seized on by those who wish to oppose any particular measure. The greater good becomes, for them, obscured by the lesser evil. The debate over the MMR vaccine (against measles, mumps and rubella) is a recent example, in which the antivaccination case was helped along by a fraudulent paper linking the vaccination with autism. More recently, unfortunately, a flawed flu vaccine has caused some morbidity amongst small children and has reinforced the antivaccinators (Godlee, 2011).

Medicine is clearly a special case. The examples in the previous paragraphs are about health, a subject in which any audience will have a vested and immediate interest. Nevertheless, the messages have been accepted by the overwhelming majority of people, surely a great measure of success. A significant minority, however, still reject them. There are still those who cluster in draughty doorways to smoke, others who are unwilling to accept the risks of vaccination for their children, still others who continue to tan themselves on beaches without effective sun-screens.

New Problems for Science Communicators

At the beginning of the twenty-first century, humans are enjoying yet another layer of connectivity, linking to one another via IT media, to add to their biological heritage. Kauffman's (1995) button thought-experiment is reenacted over and over again as unforeseen political consequences emerge, such as the fall of the Berlin Wall and the turmoil in the Arab states (the 'Arab Spring'). People have greater access to information than at any time in the past, and that includes access to scientific information. They have access to the words and thoughts of scientists and to scientific journals. It is, for the most part, a good thing. The disadvantage, one that concerns scientists and science communicators, is that the conclusions that are drawn by people who do not have the benefit of a background in science may be based on misunderstandings—or rather, they may interpret science in a way that does not conform with the view of the science communicator. It is difficult to argue with people who have as good, or greater, command of the facts, even if their understandings are flawed. The medical profession has always found this to be a nuisance.

Paramount among the misunderstandings is the concept of risk. Science communicators have great difficulty communicating the concept of risk. Risk for scientists, on the other hand, has a very firm basis in statistical theory and probability.

As we have seen in Chapter 6, failure of the general public to evaluate risk properly is compounded by their failure to understand the function of doubt and certainty in science. Science can progress by the gradual accretion of information until eventually the case for a given hypothesis is so strong that the great majority of scientists accept it. The theory of evolution discussed above is one such case. Another is plate tectonics. When I was an undergraduate in the 1950s, continental drift was an interesting but way-out idea that appeared to make sense of some knotty problems in zoogeography. Few scientists now would deny Wegener's (1912) hypothesis. But it was a hard-fought battle that polarized scientific opinion at the time.

Another way that science progresses is in leaps, in which the existing paradigm is rejected and a new one takes its place (Kuhn, 1962). For example, for twenty years biochemists sought a mythical seven-carbon acid that the then-current paradigm suggested was central to energy synthesis in cells. One biochemist, Peter Mitchell, rejected this idea and suggested instead that the mechanism involved pumping protons across cell membranes (Mitchell, 1966). Immediately there was internecine warfare among scientists but, within a surprisingly short time, Mitchell's view, for which he was awarded the Nobel prize, prevailed.

The public finds it very difficult to follow these sorts of arguments, where scientist is pitted against scientist, especially if they find themselves being invited to take sides. Realistically, nobody except the scientist is going to care much about proton pumping, or even plate tectonics. The current issue of climate change, however, provides a good example of the public bewilderment that can ensue when the current paradigm is hotly contested, as has been described in Chapter 13.

When there is debate about a hypothesis, the public feels that it is necessarily flawed. Vigorous debate is considered by the scientific community to be a strength of the scientific method, but is perceived by a section of the general public as weakening the position. In any large community, such as Australia, that section may also be large and vocal. The problem is compounded when the debate becomes politicized, thus entering an arena where majority opinions prevail as a matter of course, irrespective of the intrinsic merits of the issue. In such cases, the response from some sections of the public may be, 'if you can't defeat them, demonise them.'

At bottom, the climate change debate is about preserving for posterity the climate that humans have enjoyed for the last few hundred years. It is an example of a type of debate that has been going on among ecologists for decades. Are those bits of the environment that are sequestered for national parks then to be preserved in a pristine condition for posterity, or are they to be conserved, managed, manipulated within well-defined guide-lines for public use? It is a problem that is exacerbated by the short human life span. The ecology of the Earth is a story that is three billion years long: that of humanity is a niggardly two million years, while the luckiest of us will only see about one hundred. The Earth is about halfway through its life cycle and even a conservative estimate suggests that life has a further expectancy of about one billion years (Conway Morris, 2003). Life began more than two billion years ago. Cambrian organisms, including many that we can recognize as ancestral to modern forms, flourished about half a billion years ago. So there is plenty of time left for innovation and, unless humans sterilize the Earth, life will not have to start from scratch after a wholesale extinction. There is enough DNA in a handful of estuarine mud to populate a planet. Repopulation will not have to start from single-celled organisms either, as myriad multicellular beings live their lives in the refuge of sub-aqueous and sub–marine anoxic slime.

'Preservation of the environment' has become major catch-cry of modern times. Humans as individuals and collectively are anxious to preserve that one tiny cross-section of their present when they see that it is about to change. Sadly, change is inevitable. They also tend to put a human value on the environment, as if it is an actively nurturing entity. But as Lynn Margulis remarked:

> Gaia is a tough bitch—a system that has worked for over three billion years without people. This planet's surface and its atmosphere and its environment will continue to evolve long after people and prejudice are gone.
>
> *Quoted in Brockman (1995)*

Identifying with nature is not necessarily a bad thing but we should be clear about what it implies and about our motives. The Earth will continue to exist. It will not be 'our' Earth, any more than the Cambrian or Devonian Earths were ours. There has been an Earth without free oxygen, there has been an Earth with far higher concentrations of atmospheric carbon dioxide than any that humans have added or will add, and there has been a 'snowball Earth' with glaciation at the equator. There have been Earths flourishing in the aftermath of meteor strikes. Just as humans can look forward and see a profusion of possible futures, they can look back and see 'the one true past'—and if they don't like what they see they can edit it to conform to a more desirable narrative. The need for preservation and conservation has a basis in this nostalgia, a longing for things, persons, or situations of the past. It is an important human attribute, a by-product or consequence, in a conscious being, of adaptation to environment. A second attribute is a sense of posterity, the idea that the sacrosanctity of children extends beyond the immediate family to grandchildren and future generations. Both contribute to the perceived need to preserve the environment of the present, even though common sense tells us that the environment is continually changing.

It is a debate that has led to many notable confrontations within the community. It is an area of much dispute, even among scientists—and science communicators—who might legitimately take different positions because there are good cases to be made for both sides. In the end, it is a matter of personal choice, of opinion, and of emotion fuelled by different concepts of posterity. It is therefore a minefield for science communicators who themselves might come into conflict over the issues and contribute to the problem.

The biological implications of the ethical codes of religions resonate with the opponents of science. 'God made man in his own image' suggests that that image is holy and should not be tampered with. Issues like blood transfusion, organ transplants, and especially transplants between animals and humans engender considerable suspicion. Reproductive science is a particularly sensitive area. Positions on abortion are as much about religion as about conflicting ethical philosophies. *In vitro* fertilisation, implantation of embryos, in natural or in surrogate mothers, is both accepted and vilified. Science communicators can only make small gains against entrenched attitudes, as in all these cases the mood of public may descend into outrage. They may have to await generational change but it is important to remember that success achieved glacially is still success.

Genetic manipulation is an area fraught with perils for the science communicator. At bottom, the issue is about the usurping of god's prerogative. Mary Shelley's Frankenstein has reverberated among the anti-science lobby for almost 200 years, but builds on a substantial earlier history of the alchemist and 'mad scientist'. Frankenstein's monster escaped from the laboratory; so too could transgenic organisms. Inserting genes from one species into another is seen as violating the species barrier—notwithstanding the fact that it frequently occurs in nature—and possibly unleashing some future plague on humanity. Questions about the point at which a genetically manipulated tomato, say, becomes something else are important to the general public. 'How many animal genes' asks the vegetarian, 'do you insert into a tomato before it violates my convictions so I may not eat it?' That position does not change even when the vegetarian is informed that tomatoes and animals already have in common at least 30% of their genes.

'Unleashing the monster' is an opprobrium that also attaches to the peaceful use of atomic energy. As well as the tragic loss of life, measured in hundreds, due to reactor accidents, it is feared that radiation leakage into the environment may cause long-term pollution that threatens unborn children and posterity, thereby violating one of the most significant biological imperatives. Radiation exposure is thus another example of a malignant escape from control whose risks are magnified, in the minds of the community, well beyond reality. Conventional fuels have, since the Industrial Revolution, caused millions of deaths and even today, thousands die every year.

Stem cell research, in spite of the promises that are now gradually being fulfilled, got off on the wrong foot, because initially it used cells derived from discarded early human embryos and, later, from umbilical blood. This contravened two of the great biological imperatives, 'going against Nature' and, more important, the sanctity of children. Arguments that an early embryo is not, in fact, a child, are not considered relevant by many who point out that, at the very least, it is potentially a child. The sanctity of children extends to the perception of risk. Adults will accept a much higher personal risk than they will permit for their children.

This is particularly well illustrated by the vaccination of children. Vaccination is a difficult area where the sense of violation is as immediate as the needle-prick. A mother has to exercise willpower to allow this to happen. Instinctively, she wants to protect her child from the immediate discomfort, while, at the same time, intellectually, she understands that she is protecting it from a future and more dangerous peril. Unfortunately, when instinct and intellect come into conflict, instinct sometimes wins. It is therefore not surprising when a measure such as the MMR vaccine was promoted for the common good that some 20% of people rejected it. This has been either a very successful science communi-cation phenomenon (80% vaccinated) or a failure (20% unvaccinated). It illustrates the very important need to identify the boundaries of success before embarking on any science communication program.

Failure to understand the range of opinions and abilities in a very large target audience has led many into error. Frazer (2003), in a study to determine why parents would not immunise their children in the face of abundant evidence of its benefits, adduced a remarkable fact. The non-immunisers were not, as supposed, ignorant and unschooled. Rather, they were highly intelligent with a high level of education, in some cases in the field of immunology. They were competent to understand figures and read graphs and yet they made, in the view of the scientists and science communicators, the 'not so good' decision. Arguments about maintaining 'herd immunity' had no impact. Their foresight, their view of alternative futures, led them to choose the much greater risk of subjecting their children to a potentially fatal childhood disease, rather than a vanishingly small one of side effects from immunisation.

New media have enabled such views to be propagated across countries and around the world, further compounding the problem.

Science Communication and the Internet

If you put the title of this section into your web browser you will get (January 2012) about 113 million hits, and 2 million from Google Scholar. Adding the word 'blogs' attracts 68 million hits, and 'twitter' more than 21 million. About a year earlier, the figures were, respectively, 7 million, 2 million, 2 million, and 1 million. This, if illustration were needed, the extraordinary and continued growth of the internet, and its use as a medium for science communication.

A far-from-exhaustive scan of science blogs shows that they are often informative, interesting, but without the scientific rigour or credentials that one might find even in the more chatty science-based journals. Quite frequently one encounters writings that are at best journeyman pieces, earlier essays by young scientists who, presumably do not want this evidence of effort to be lost to posterity. In this it contributes to a general flaw of the internet. Much of the material has not been scrutinised by experts, so its quality varies. This is in contrast to one of the great strengths of science: peer review. Peer review has its shortcomings, but generally validates the integrity of the material that is published in scientific media.

Moon (2011) used *Twitter* as a device to highlight scientific issues that air on the internet. She found that she was able to correlate the immediacy of a scientific issue with the rate at which numbers of 'tweets' increased or decreased. Others are creating Twitter sites to encourage debate about science communication issues. For example, Kirstin Alford launched a discussion in April 2011 entitled #onsci: *telling better science stories*. Over about an hour there were 395 tweets from 34 tweeters, and it became a 'trending topic' in Australia. The results were meagre; the consensus was that science stories should have strong narrative lines, rich in metaphor and symbolism. Any first-year text on science communication would make these points in the first chapter. Imagine, however, that this process was multiplied a million times, each iteration building on what had gone before …

And so, with *Twitter* and *Facebook* and *Youtube* in mind, we return to Kaufmann's button experiment and 'order for free'; for threads, substitute the Internet; for buttons, substitute people with personal computers.

Conclusion

This chapter has considered the roles that five attributes of consciousness—awareness of self, of others, of the future, of the past and of the environment—play in modifying the responses of the public to attempts at science communication. Consciousness is itself an emergent phenomenon, a result of increasing biological complexity, that is now adding another layer of complex interaction in the form of the developments of information technology.

The notion that organic evolution leads to complexity and that at each evolutionary stage there are emergent consequences, is of some antiquity (see, for example, Morgan, 1923). This is true for evolution of our own species. It has also been commonly believed that, as all natural processes are limited by thermodynamic principles, evolution is moving towards some end point. One endpoint was the 'sphere of human thought' or 'noosphere' of Teilhard de Chardin (1956):

> In the case of man … it is with … the appearance of the ability to reflect that we may link the whole cluster of the neo-properties that determine the formation of the noosphere.

Divested of its spiritual overtones, the notion that increases in biological complexity and connectivity are approaching some thermodynamically limiting endpoint has been embraced by many information scientists, who refer to it as a 'singularity' in evolution (Chalmers, 2010b). Observing that technological growth is exponential, Kurzweil (2005) adduced a law of accelerated returns that suggests that, in about thirty years, we will have arrived at a point where it exceeds that of the sum of human intelligence. According to James Martin (2007), a renowned futurologist, there will be a break in human evolution that will be caused by the staggering speed of technological advance. Human contribution to science will cease as technological development is taken over by artificial intelligence.

At that point, apparently, there will be no role for science communicators.

References

Bird R. J. (2003). *Chaos and life: Complexity and order in evolution and thought*. New York: Columbia University Press.

Boal K. B., Hunt, J. G., & Jaros S. J. (2003). Order is free: on the ontological status of organizations. In R. Westwood & S. Clegg. (Eds). *Debating organisation*. (pp. 84-97). Oxford: Blackwell.

Brockman J. (1995) *The third culture: Beyond the scientific revolution*. New York: Simon & Schuster.

Chalmers, D. J. (1996). *The conscious mind*. Oxford: Oxford University Press.

——(2010a). *The character of consciousness*. Oxford: Oxford University Press.

Chalmers, D. (2010b). The singularity: a philosophical analysis. *Journal of Consciousness Studies, 17*, 7-65.

Chamberlin, W. (2009). Networks, emergence, iteration and evolution. *Emergence: Complexity & Organization, 11*, 91-98.

Conway Morris, S. (2003). *Life's solution. Inevitable humans in a lonely universe*. Cambridge: Cambridge University Press.

Corning, P. A. (2008). Holistic Darwinism. *Politics and the Life Sciences, 27*, 22-54.

Darwin, C. (1859). *The Origin of Species*. London: John Murray.

Fogarty, J.E. (Translator) (1977). *A barefoot doctor's manual*. Philadelphia: Running Press.

Frazer, C. (2003). Bridging the gap between the science of childhood immunisation and parents. Vols. 1 & 2. PhD thesis, Australian National University.

Godlee, F. (2011). Wakefield's article linking MMR vaccine and autism was fraudulent. *BMJ, 342*, c7452.

Huxley, J. (1942). *Evolution: The modern synthesis*. London: Allen and Unwin.

Kauffman, S. (1995). *At home in the universe: The search for laws of self-organization and complexity. London:* Penguin Books.

Klayman D. (1985). Qinghaosu (Artemisinin): Antimalarial drug from China. *Science, 238,*1049.

Kuhn, T. S. (1962) *The structure of scientific revolutions.* Chicago: University of Chicago Press.

Kurzweil, R. (2005). *The Singularity is near.* London: Viking Press, Penguin Books.

Margulis, L. (1970). *Origin of eukaryotic cells.* Hartford: Yale University Press.

Macklem, P. T. (2008). Emergent phenomena and the secrets of life. *Journal of Applied Physiology, 104,* 1844-1846.

Martin, J. (2007). *The meaning of the 21st Century.* New York: Riverhead Penguin.

Mitchell, P. (1966). "Chemiosmotic coupling in oxidative and photosynthetic phosphorylation". *Biological Reviews, 41,* (3), 445–502.

Moon, B (2012). PhD Thesis, in preparation. Canberra: Australian National University.

Morgan C. L. (1923). *Emergent evolution.* London: Henry Holt and Co.

Teilhard de Chardin, P. (1966). *Man's place in nature* (transl. René Hague). London: Collins Fontana Books.

Van Valen, L. M. (1973). A new evolutionary law. *Evolutionary Theory 1,* 1-30.

Waldrop, M. M. (1992). *Complexity.* London: Viking Press, Penguin Books.

Watts, D. J. (2004). *Six degrees: The science of the connected age. London:* Vintage Press.

Wegener, Alfred (1912). Die Herausbildung der Grossformen der Erdrinde (Kontinente und Ozeane), auf geophysikalischer Grundlage. *Petermanns Geographische Mitteilungen, 63,* 185-309.

Westwood, R., & Clegg, S. (Eds). (2003). *Debating organisation.* Oxford, Blackwell.

PART VI

Further Exploration

FURTHER EXPLORATION

In this section, we suggest further readings for each of the chapters in this book. We also suggest some points suitable for group discussion or tutorial work, and some extended projects or assignments.

Generic Activities and Projects: Expanding Horizons

The following activities are suitable for any of the chapters in this book.

Further Reading

Many sources provide stimulating further reading on the communication of science. Three general categories are useful here: academic literature in leading academic journals; governmental reports and discussion pieces; and opinion writing targeted at discourse about science.

For Discussion

1. From a daily newspaper, select three recent articles on scientific themes. Are the articles positive, negative, or neutral in their presentation? What specific technique is being used to engage the reader? Are there examples of scientific jargon or complicated ideas?
2. Think about some of the controversies over scientific practice that are important to members of **your** community, such as those listed below. What is the basis of the conflicting opinions? What are the contextual reasons why people think and feel the way they do? How might such responses affect their attitudes towards science in general?

a. animal experimentation
b. chemotherapy
c. mobile phones
d. smoking
e. wind-powered electricity
f. HIV/AIDS
g. space shuttle missions
h. suburban expansion or urban infill

Projects

1. Choose three papers from the references in the relevant chapter, with publication dates later than 2000. Read and summarise these papers in the form of a critical review.
2. Choose an area of science with which you are very familiar. Design a 5-minute presentation on this topic for a specified audience. Explain why you believe this audience will be interested in this topic.

Activities for Each Chapter

Part I: Models of Science Communication—Theory into Practice

Chapter 1: The place of public engagement with science

Further Reading

Cheng, D., Claessens, M., Gascoigne, T., Metcalfe, J., Schiele, B., & Shi, S. (2010). *Communicating Science in Social Contexts: New Models, New Practices.* Springer Science.

Clark, T. W., & Kellert, S. R. (1988). Toward a policy paradigm of the wildlife sciences. *Renewable Resources Journal*, 7, 7-16.

May, P. J. (2003). Policy design and implementation. In B. G. Peters & J. Pierre (eds.) *Handbook of Public Administration* (pp. 223-233), London, UK: Sage Publications.

Kahlor, L.A., & Stout, P. A. (2010). *Communicating Science: New Agendas in Communication.* New York, NY: Routledge.

Trench, B., & Bucchi, M. (2010). Science communication, an emerging discipline. *Journal of Science Communication*, 9, 1-5.

For Discussion

1. Do you consider that the general public should know more science and if so, why?
2. Consider a science communication initiative with which you are familiar. Who are the 'driving actors' and the 'target actors? What do you consider the goals of the 'driving actors' might be?
3. Consider the 'typologies' of the general public discussed in this chapter. Can you come up with typical examples of each typology?

Project

Research and write a short summary of the 'deficit model' of the public. In the year 2000, this model came under serious attack. Research why this happened and the difference this 2000 critique has made to the current view of science communication. In your opinion, are we still actually operating within a deficit model?

Chapter 2: Engagement with Science: Models of Science Communication.

Further Reading

Lehr, J. L., McCallie, E., Davies, S. R., Caron, B. R., Gammon, B., & Duensing, S. (2007). The value of "dialogue events" as sites of learning: An exploration of research and evaluation frameworks. *International Journal of Science Education, 29*, 1467-1487.

Trench, B. (2008). Towards an analytical framework of science communication models. In D. Cheng, M. Claessens, T. Gascoigne, J. Metcalfe, B. Schiele and S. Shi (Eds). *Communicating science in social contexts*. New York:Springer, pp. 119-138.

Leach. J., Yates, S., & Scanlon, E. (2009). Models of science communication. In R. Holliman, E. Whitelegg, E. Scanlon, S. Smidt and J. Thomas (Eds). *Investigating science communication in the information age*. Oxford: OUP, pp. 128-146

For Discussion

1. Considering Lewenstein's (2003) term "lay expertise", in what areas of science will lay expertise be important, and why?
2. Are there areas of science which can or should be excluded from public input or debate?
3. What contributions can the humanities make to science?
4. "The world of science communication is so big and varied, the "noise" is so frequent and effective that hardly any action can have an effect in a linear way" (Greco, 2004). If this is true, is there any point in trying to communicate about important issues such as climate change? Would it be better to leave it to the experts to decide what to do?

Projects

1. Research and write a short summary of the advantages and disadvantages of the methods of engagement mentioned in this chapter: theatre meetings, interactive meetings, focus groups, citizens' juries, consensus conferences, the Delphi technique, web discussions and written consultations. Which would be appropriate for the following?

   a. Finding out about the public's views on planting street trees in your city or town.

   b. Gathering views for a lobby group, which wants to present the case for more science funding to your Government.

   c. Finding out the nature and level of general public support for a controversial research issue of your choice.

   d. Presenting research findings about a conservation issue of interest to your town (you may choose the issue).

2. "Providing reliable information in an accessible way—in other words, filling the relevant 'knowledge deficit'—is an essential pre-requisite of both healthy dialogue and effective decision making" (Dickson, 2005). Choose an area of science where you think there is a "knowledge deficit" across the science/public interface; identify what you think the deficit is, and explain how you would address the communication problem. The deficit can exist within the world of the general public, or science, or any identified relevant group.

Part II: Challenges in Communicating Science

Chapter 3: Scientists' Engagement with the Public

Further Reading

Bauer, M. W., & Jensen, P. (Eds). Special Issue: Mobilization of scientists for public engagement activities *Public Understanding of Science, 2011; 20*(1). This Special Issue (10 papers) presents evidence of public engagement activities by scientists from several countries.

For Discussion

1. Do you think that scientists' communication with the general public, i.e., people in the street, is as important, more important, or less important than communicating with their scientific peers? Why?

2. Do scientists lose any public or professional status and credibility if they express their own opinions about the implications and value of their research?

Projects

1. Interview a scientist and ask what they think and do to communicate their specialized knowledge with the general public.

2. Survey your friends and family and ask if they can name an individual scientist(s) they value as a source of scientific information, and if so, why?

3. Conduct a role play where at least one scientist and one lobbyist with opposing views are interviewed for television or radio about their views on

a current controversial science-related issue (you pick which one.) The interviewer should question why the audience should believe either of them.

Chapter 4: The Role of Science in Public Policy. What is Knowledge For?

Further Reading

Academic literature

Relevant sources within the academic literature include many of the articles and books discussed in this chapter. Perhaps most relevant are:

McNie, E. (2007). Reconciling the supply of scientific information with user demands: an analysis of the problem and review of the literature. *Environmental Science & Policy*, *10* (1), 17–38. doi:10.1016/j.envsci.2006.10.004

Pielke, R. A. J. (2007). *The Honest Broker: Making Sense of Science in Policy and Politics.* Cambridge: Cambridge University Press.

Sarewitz, D., & Pielke, R. A. J. (2007). The neglected heart of science policy: reconciling supply of and demand for science. *Environmental Science & Policy*, *10*(1), 5–16. doi:10.1016/j.envsci.2006.10.001

Alongside this, a variety of leading scholarly journals has contributed to the discussion on the intersection of science and policy-making via policy-focused editorials and opinion pieces. *Nature*, *Science*, and the *British Medical Journal*—as well as many others—regularly include such writing. Those interested in the dynamics and arguments at the science/policy interface would find much of value in these.

Governmental reports and discussion pieces online

As noted at the beginning of this chapter, governments, policy-makers and institutions have long called for a closer relationship between science and policy-making. Many of these reports and discussion pieces make compelling reading, such as online sites from your own national or local governments. Useful examples from Australia at the time of writing include:

Banks, G. (2009). *Challenges of Evidence - Based Policy - Making. Challenges.* http://www. apsc.gov.au/publications09/evidencebasedpolicy.pdf

Campbell, S., Benita, S., Coates, E., Davies, P., & Penn, G. (2007). *Analysis for policy: Evidence-based policy in practice. London: Government Social Research Unit.* http://bit.ly/n05AAg

Opinion writing

Finally, it is worth pointing to the great swath of opinion writing in both traditional media and online sources discussing or critiquing the interaction between science and policy-making. It would be difficult to provide anything close to an exhaustive survey of this field; in most democratic countries there would be near constant discussion of policy-making and the legitimate role of science in this. Perhaps the best way in to this literature is to follow the opinion-writing in the leading news sources in your country, or to turn as well to the blogs and online sources of the more politically engaged scientists, such as those at *ScienceBlogs.com.*

For Discussion

1. Pick one story from the front of today's newspaper and identify all of the relevant stakeholders and all the possible fields of science that might inform our understanding of the issue. What makes someone a stakeholder? What are the most relevant sciences? What other forms of knowledge are relevant? Which are mentioned in the story, which aren't?

2. Pick one problem discussed on the front of today's newspaper and craft a relevant policy response. How might the problem be solved? What impact might your solution have on the problem, or on other parts of society? What information do you need to craft that policy?

Project

Role playing a policy response to a complex problem: Childhood obesity is on the rise in many parts of the world (Deckelbaum, Williams, & Christine, 2001), yet an adequate understanding of the causes of this problem—and potential solutions—remains somewhat out of our grasp. This task focuses on exploring workable solutions to the problem.

To prepare for the class, you should examine the academic/scientific/governmental literature to find possible causes of the problem. Is it to do with modern diets? Is it exercise? Is it urban design and transport? Is it caused by television and advertising? What other factors might contribute?

Once in class, students will be assigned a role as either a policy-maker or a relevant stakeholder as suggested by the survey of the literature. Policy-makers (perhaps working in competing groups) can present solutions to the stakeholders, while stakeholders can critique the solutions raised. Which solutions are most appropriate? Which might solve the problem best? Which are most palatable for the stakeholders? What does it mean to craft a good policy solution?

Chapter 5: Negotiating Public Resistance to Engagement in Science and Technology

Further Reading

Barnes, B. (2005). The credibility of scientific expertise in a culture of suspicion. *Interdisciplinary Science Reviews, 30* (1), 11-18.

Irwin, A., & Wynne, B. (Eds). (2004). *Misunderstanding Science? The Public Reconstruction of Science and Technology*, Cambridge, UK: Cambridge University Press.

Jasanoff, S. (1997). Civilization and madness: the great BSE scare of 1996. *Public Understanding of Science, 6*, 221-232.

For Discussion

In Table 1, consider each of the categories of science's meaning in turn. Reflect on how you feel about science with respect to each category. What do you absolutely love about science? Is there anything you hate about it? Are you indifferent to or uncertain of what you feel about any of science's meanings? Does any of it put you to sleep? Can you think of any personal experiences that helped shape these beliefs?

Project

Choose a science or technology-based issue that is controversial or new in your community. Investigate the factors that influence people's attitudes to the issue by following these three steps:

a. Review recent media coverage of the issue to identify the people and organisations who are affected by the issue and/or who have opinions about it, such as community groups, businesses, science organisations, politicians, and so on. The more specific you can be, the better. What kinds of opinions do these stakeholders have? Does anyone speaking about the issue make comparisons with other science and technology issues or generalisations about science?

b. Make an opportunity to speak to representatives of these people and organisations. For example, invite different people to come and speak to your group, or meet them individually and interview them about their views. Ask questions that investigate their attitudes to science and technology generally as well as the specific issue at stake. For example, you might want to ask them whether past science controversies or past dealings with scientists and science-based institutions have affected their views.

c. Analyse the findings that your research reveals. Can you identify any patterns in the kinds of experiences people have had in the past with science or technology developments, events, institutions, or individuals? Which categories of science's meaning are invoked by the stakeholders?

Part III: Major Themes in Science Communication

Chapter 6: Communicating the Significance of Risk

Further Reading

Bernstein, P. L. (1996). *Against the Gods: The Remarkable Story of Risk*. New York: John Wiley & Sons.

Heath, R., & O'Hair, H. (Eds.). (2009). *Handbook of Risk and Crisis Communication*. New York: Routledge.

Lundgren, R. E., & McMakin, A. H. (2009). *Risk Communication A Handbook for Communicating Environmental, Safety, and Health Risks*. Hoboken: John Wiley & Sons, Inc.

Sellnow, T., Ulmer, R., Seeger, M., & Littlefield, R. (Eds.). (2010). *Effective Risk Communication: A Message-Centered Approach*. New York: Springer.

Slovic, P. (Ed.). (2010). *The Feeling of Risk: New Perspectives on Risk Perception*. London: EarthScan.

Zinn, J. (2008). *Social Theories of Risk and Uncertainty: An Introduction*. Malden, MA: Blackwell Pub.

For Discussion

1. Culture and Societies
Consider some distinct cultural groups or different societies that you are familiar with and discuss how they may differ from your own and from each other in the manner by which they orient toward various risks.

2. Utopian or Dystopian?
In his work *Against the Gods* Peter Bernstein describes the mastery of risk as the vehicle that propelled society into modernity: "The ability to define what may happen in the future and to choose among alternatives lies at the heart of contemporary societies. Risk management guides us over a vast range of decision-making, from allocating wealth to safeguarding public health, from waging war to planning a family, from paying insurance premiums to wearing a seatbelt, from planting corn to marketing cornflakes" (Bernstein 1996). In a somewhat less utopian vision, sociologist Ulrich Beck describes this condition as the "risk society," in which "dangers are being produced by industry, externalized by economics, individualized by the legal system, legitimized by natural sciences, and made to appear harmless by politics" (Beck 1992).

Bernstein describes risk as a form of liberation and empowerment of people and societies. Beck describes risk more in terms of political and economic dysfunctions that serve to put people in danger. Which point of view do you feel more persuaded by, and why?

3. The Psychometric Model
The approach to understanding risk pioneered by Slovic and colleagues describes two factors, knowledge and dread, that can tell us a lot about why one hazard or another is reacted to as it is. What other factors might you suggest that motivate people to react one way or another to a hazard?

4. Individual Differences
More recent work on risk perception has shifted away from looking at aggregate reaction to a variety of hazards and toward looking at the characteristics of individuals that cause them to react one way or another toward a specific risk. Can you think of any characteristics that individuals might have that would fit this approach?

5. Raising and Lowering Risk Reactions
Anderson and Spitzberg (2009) described nine conditions that can serve to heighten the risk reactions that people have. Can you add some to this list that might increase, or decrease risk reactions?

6. The media
Sharon Dunwoody has commented:

> "When it comes to risk coverage, it seems that the mass media can do nothing right. They are regularly accused of bias, sensationalism, inaccuracy, indifference, and of being simplistic and polarized .... the mass media are—in a word—lousy at conveying appropriate notions of risk to general audiences (p. 75)."
>
> *(Dunwoody 1992)*

Do you agree? Do you think that individuals respond better to risk messages on mass media channels or on inter-personal channels?

Projects

1. Best Practices
The section of best approaches to message formulation provides a wide range of items for creating effective risk communications. Using a broad approach, sketch the main elements of a risk communication campaign that you might undertake in a scientific field that you are familiar with. Consider things like audience factors, channel effects and so forth.

2. Message Creation
Similar to activity 6, a wide variety of suggestions are provided on how to craft an effect risk message. Take a risk from a scientific area that you are familiar with and write a fairly brief statement (less than a page) in which you provide a

neutral and factual risk message to the public. Also craft an outline of what factors you might need to consider if you are delivering this message through the news media.

Chapter 7: Numbers and the Relation between them in Science Communication.

Further Reading

Graham, A. (2006). *Developing Thinking in Statistics.* London: Open University & Paul Chapman Publishing.

Lundgren, R. E., & McMakin, A. H. (2009). *Risk Communication: A Handbook for Communicating Environmental, Safety, and Health Risks* (4th ed.). Piscataway, N.J.: Hoboken, N.J.: IEEE Press; Wiley.

Steen, L. A. (Ed.). (2001). *Mathematics and Democracy: The Case for Quantitative Literacy.* Princeton, NJ: The National Council on Education and the Disciplines.

Tufte, E. R. (1997). *Visual Explanations: Images and Quantities, Evidence and Narrative.* Cheshire, Conn.: Graphics Press.

For Discussion

1. Find some media reports or official reports on a socioscientific controversial issue, e.g. threat and spread of a disease, food safety.

 (i) discuss how quantities and/or graphical representations have been embedded with values that the creator might have intended.

 (ii) discuss the extent to which the reports facilitate awareness, enjoyment, interest, opinion formation and understanding of science.

2. Consider the advancement and widespread uses of technology in/by the public. Compared with what we had two decades ago, tools for gathering, representing, disseminating data have become far more accessible in the workplace and in personal lives. Do you think this is favouring or hampering the development of quantitative literacy among the public? Try to be more specific by gathering examples that may support your view. What other concerns do you have regarding the development of technology and quantitative literacy among the general public?

Projects

1. Compare Nightingale's work with that of Hans Rosling today. How can visualization of data regarding public health enhance the communication of professionals with the general public? Do we benefit from the variety and popularity of digital visualization tools in sharing our understanding and interpretation of quantitative information?

2. According to the World Health Organization weekly report,

> "In 2009, the number of cases of cholera reported to WHO increased by 16% when compared with 2008. A total of 221 226 cases, including 4946 deaths, were reported from 45 countries; the case-fatality rate (CFR) was 2.24%."
>
> *(www.who.int/wer/2010/wer8531.pdf)*

How will you interpret these statistics, graphs and maps in the report?

3. Watch the video clips for: Congressman Roscoe Bartlett at the U.S. House of Representatives (4th April, 2008) (www.youtube.com/watch?v= YyliwrgbLvo) How did Roscoe Bartlett use charts and graphs to explain the Peak Oil crisis?

Chapter 8: Ethics and Accountability in Science and Technology

Further Reading

This chapter only scratches the surface of the immense body of work that has addressed ethics in its many guises and applications: it is necessarily not exhaustive. For the reader interested in going further, the following sources offer a good primer from which to start delving:

Mautner, T. (1997). *Dictionary of philosophy* (2nd ed.). London: Penguin.
Morris, T. (1999). *Philosophy for dummies.* New York: Wiley Publishing, Inc.
Singer, P. (2011). *Practical Ethics* (3rd ed.). New York: Cambridge University Press.
Various. (2011). *Applied Ethics.* Wikipedia. Retrieved from http://en.wikipedia.org/wiki/Applied_ethics.

For Discussion

In small groups, consider the following four scenarios. Each scenario describes a science/technology based initiative that your country is considering introducing. You have been asked to contribute your expert opinion to the debate.

For each scenario discuss:

*   The major ethical concerns that you believe would be associated with deciding whether to go ahead
*   Who the most supportive, and least, supportive people, groups or organisations would be
*   How you would justify your advice to your government using the ethical positions presented in Chapter 7

Scenario 1
Space tourism plans for 25 recreational launches a year from locations around the country starting in 2008 have been proposed. It will initially cost $100,000 for each passenger but prices are predicted to fall as the number of launches increases.

Scenario 2
Xeno-transplantation—Organ donor waiting lists are increasing but donors are decreasing. It has been proposed to begin routinely transplanting pig and monkey organs into people within the next two years.

Scenario 3
An HIV vaccine has been developed—trials have shown a 60% reduction in overall new cases of HIV infection when it is used among high risk groups. It has been proposed to provide the vaccine via prescription to any person who requests it (at the discretion of a general practitioner) by the end of the year.

Scenario 4
A GM wheat strain which can withstand up to ten times the salinity and use 20% of the amount of water required by conventional varieties has been developed. While it hasn't had the long-term testing some other GM crops have had, there is a strong likelihood it will be approved for commercial use within 12 months.

Project

1. In groups of two, four, or six people (keep the numbers even if possible), brainstorm what you consider to be most controversial issues facing your community at the moment. Make sure these are issues that have strong science and technology based elements to them;
2. Decide on one or two issues, preferably ones on which your group disagrees (if you are stuck for ideas, organ donation, the use of natural resources such as forests, and abortion are just three topics that usually get people disagreeing!);
3. Review the differing positions that are presented in supporting and rejecting the issue or issues;
4. Characterise these positions in terms of the more common ethical theories presented in Chapter 7;
5. Split your group into two teams;
6. Organise a debate between the teams in which the teams must try to defend a position they personally *disagree* with or consider to be unethical.

The purpose of this project is to explore what you can learn about your own, and other people's, ethical position by arguing for something with which you disagree. It is often through considering how to defend something we don't support that we find out more about our own beliefs and convictions.

At the end of the debate, write a short piece of reflective prose describing what it felt like to support a position you find ethically disagreeable. What did you learn about the issue, the people who support it, and your position? What did doing this tell you about your relationship to the issue in general, and ethics in particular?

Chapter 9: Beliefs and the Value of Evidence

Further Reading

Long, D. E. (2011). *Evolution and religion in American education: An ethnography.* Dordrecht: Springer.

Meyer, S. C. (2009). *Signature in the cell: DNA and the evidence for Intelligent Design.* New York: HarperCollins.

Miller, J. D., Scott, E. C., & Okamoto, S. (2006). Public acceptance of evolution. *Science, 313*, 765-766.

Reiss, M. J. (2008). Teaching evolution in a creationist environment: an approach based on worldviews, not misconceptions. *School Science Review, 90*(331), 49-56.

Scott, M. (2007). *Rethinking evolution in the museum: Envisioning African origins.* London: Routledge.

For Discussion

1. Does the word 'belief' have one meaning or a range of meanings?
2. Given that science should always be open to the possibility of refutation and change, is it bad science to say that evolution is scientifically not controversial?
3. What if any lessons do BSE and GMOs provide for science communication about evolution?

Projects

1. Would it be better if scientists avoided using the word 'belief' and 'believe' as in, for example, 'I believe in the theory of evolution'?
2. Research so-called 'scientific creationism' (also known as 'creation science'). Do the arguments against the theory of evolution have any scientific validity?
3. Draw up a short list of principles for effective science communication in the light of BSE and GMOs.
4. Obtain two or three school biology textbooks and examine how the topic of evolution is treated. Is it presented as being controversial or not? How, if at all, do the authors deal with recent scientific developments in our understanding of evolution? Are religion or creationism mentioned?

How do you think a creationist would feel about the treatment of evolution? How do you think an atheist would feel about the treatment of evolution?

5. Research the ways in which creationist museums and/or zoos present the topic of evolution. Is this done well or not? Critique what you find using established frameworks for science communication. Draw up a list of recommendations as to how (i) creationist and (ii) national science museums should deal with the topic of evolution.

6. If at all possible, visit a museum that has an evolution display/exhibit. If you are unable to make such a visit, consult the relevant part of the website of a suitable museum (e.g. www.nhm.ac.uk/nature-online/evolution/, http://australianmuseum.net.au/Human-Evolution). How suitable do you consider the materials and messages to be for a diversity of audiences?

7. If you are in the UK: Read the guidance for English maintained (state) schools on dealing with creationism and Intelligent Design (DCSF, 2007). Suggest improvements.

Part IV: Interactions Involved in Science Communication

Chapter 10: Helping Learning in Science Communication

Further Reading

Ainsworth, S. (2008). The educational value of multiple representations when learning complex scientific concepts. In: J. Gilbert, M. Reiner, M. Nahkleh (eds.) *Visualization: Theory and practice in science education.* Dordrecht: Springer (p.191-208).

Hodson, D. (2009). Reading writing and talking for learning. In: D. Hodson, *Teaching and learning about science.* Rotterdam: Sense Publishers (pp. 283-326).

Lakoff, G., Johnson, M. *Metaphors we live by.* Chicago: The University of Chicago Press.

For Discussion

1. What aspects of learning science did you find the most difficult at school? What implications have they had for your learning of science since then?

2. Of the things that you are currently learning, which are best communicated by an assumption of each of the three models of learning?

3. Try to learn something new by interactive discussion with others. What are the attractions/disadvantages of this mode of learning? Suggestions:

   a. A science topic outside your area of expertise but which is well known and understood by some others in your group;

   b. An aspect of lifestyle choice such as an "alternative" therapy, a diet, a nutritional preference, etc. which is well known to some others in your group.

4. Look at a "popular book" about science (for example, one by David Attenborough). What explanatory use is made of the photographs that it contains?
5. How much use is made of gestures in a typical attempted science communication that you experience?
6. In what circumstances is the use of mathematical equations and/or chemical equations helpful in providing science communications?

Projects

1. Find an example in a book, newspaper, TV programme, or on the Internet, of a specific *context* being used to communicate scientific ideas that are new to you. Explain whether and how the context helped you understand the ideas. What factors about the context contributed to this?
2. Read any written science communication, e.g., a chapter in this book. What new words were introduced? Were you able to understand their meaning? What factors assisted or hindered this understanding?
3. Find examples of an analogy being used to explain something scientific in a newspaper or magazine. How successful is that use, in your opinion?
4. Over an extended period (a day or a week), monitor the types of talk that you encounter in all your classes. Is that distribution suitable for the purposes of the classes in which you encountered such talk?
5. Find an example of a concrete/material representation being used in a science communication in any medium. Why was it used? How effective was it, in your opinion?
6. Look at a range of examples of public science communication e.g. in a newspaper, magazine or on the Internet. What types of explanation do they provide?

Chapter 11: Science Communication and Science education

Further Reading

Barnett, J., & Hodson, D. (2001). Pedagogical content knowledge: Toward a fuller understanding of what good science teachers know. *Science Education, 85*(4): 426-453.

van den Berg, E. (2001). Impact of inservice education in elementary science: Participants revisited a year later. *Journal of Science Teacher Education, 12*: 29-45.

Kapon, S., Ganiel, U., & Eylon, B. S. (2010). Explaining the Unexplainable: Translated Scientific Explanations (TSE) in public physics lectures. *International Journal of Science Education, 32*(2): 245-264.

For Discussion

1. Read about Pedagogical Context Knowledge (pp. 436-9 in Barnett & Hodson, 2001) and spend some time to evaluate your own *Knowledge Landscape*. Try to classify the different knowledge types you would use when communicating science to your preferred audience. You might wish to use the *Framework* (see pp. 441-4) to help you with this task.

2. What do you understand by the term "propelling effect" (p.42 in van den Berg, 2001)? Draw up a table and explain the characteristics which you believe are necessary for a *propelling effect* to occur with different groups. You might wish to use the following categories to help you with this task.

   • Target audience
   • Audience conceptual framework
   • Communication objective
   • Scientific content
   • Context of the activity
   • Individual motivation

3. Select a series of public science lectures. You may wish to visit a local university or science centre for this purpose. Alternatively, you may wish to download vodcasts from preferred websites. Using the Translated Scientific Explanations (TSE) explanatory framework (p. 251 in Kapon et al, 2010), identify and select TSE elements from each public lecture. It would be more engaging to work on this activity with your friends as part of a group and compare your findings.

Chapter 12: The Practice of Science Communication in Informal Environments

Further Reading

Pedretti, E. (2004). Perspectives on learning through critical issued-based science center exhibits. *Science Education, 88*(Suppl. 1), S34-S47.

Rennie, L. J., & Stocklmayer, S. M. (2003). The communication of science and technology: Past, present and future agendas. *International Journal of Science Education, 25*, 759-773.

For Discussion

Choose a science-related issue or concept that has current, local media interest. What are the key points about this issue that need to be conveyed to assist public understanding of this issue? When you have made this list, discuss one or more of the following questions.

1. What is the science story that underpins this science issue? What are the central positive and negative arguments that people would need to consider in order to make a decision about a way forward?
2. If you had to design an exhibit or display to communicate these key points to an audience, how would you go about it? Think about: What is the story you wish to tell? How would you engage the audience? How might the key points be represented—using analogies or real artifacts?
3. What are some of the constraints faced by exhibit designers in trying to build an exhibit that informs the local population about this issue?

Project

Choose a relatively complex exhibit from a local science museum or interpretative centre. Develop a list of the science messages you think these exhibits are designed to convey. Interview three or four people who have spent time exploring the exhibit to find out what science-related message(s) they have experienced. Compare their views with your list and see if you can explain the points of difference. How could the communication effectiveness of the exhibit be improved? If it is possible, discuss the exhibit intentions with its developer.

Part V: Communication of Contemporary Issues in Science and Society

Chapter 13: Communicating Global Climate Change: Issues and Dilemmas

Further Reading

Maxwell T., & Boykoff, M. T. (2008). Media and scientific communication: a case of climate change *Geological Society, London, Special Publications*, 305, 11–18 doi:10.1144/ SP305.3.

Rask, M., Worthington, R., & Lammi, M. (Eds) (2011). *Citizen participation in global environmental governance.* London: Earthscan.

Sheppard, S. (2012). *Visualizing climate change. A guide to visual communication of climate change and developing local solutions.* London: Earthscan.

Victor, D. G. (2011). *Global warming gridlock: Creating more effective strategies for protecting the planet.* Cambridge: Cambridge University Press.

For Discussion

1. When students are discussing controversial issues such as climate change should the teacher state their own view or act as a neutral chair? Which approach is more honest?
2. Should museum and science centre exhibitions about climate change primarily aim to encourage visitors to take particular actions or to do no

more than inform people about the scientific consensus? What values should underpin such exhibitions?

3. What communication strategy should an energy company adopt? What would its three main messages be for its customers?

Projects

1. Identify a range (five-ten) of people who you know as friends or colleagues. Ask them to tell you their stories about extreme weather conditions in your country that they have personally experienced in recent years. Next, using the Internet or other sources, identify an equivalent number of reports of people that you do not know from across the world who have experienced extreme weather conditions and collect their stories. Now compare and contrast these stories of local and global experiences. Devise three teaching activities that could be used with the public or school students to communicate information about climate change using the unique resource bank of stories that you have collected. Explain which theories of learning support the design of your activities. [This project is based on one demonstrated at the 2010 World Environmental Education Congress by Professor Bob Stevenson and Associate Professor Hilary Whitehouse from the Cairns Institute, James Cook University, Australia)].

2. Visit a museum or science centre that has an exhibit on a related topic such as the weather or climate change. Examine the text, the installations, and the exhibits in terms of *what* they are attempting to communicate and *how* they are trying to do it. Now devise an evaluation strategy that could be used to test the effectiveness of the exhibition. Take account of the range of visitors such as families, school groups, old/young, scientists/non-scientists, etc. Discuss the use of a variety of data collection instruments—what would be the advantages and disadvantages of each one? How would you carry out your evaluation and what might be the challenges that you would face in carrying it out.

Chapter 14: Science Communication During a Short-Term Crisis: The Case of Severe Acute Respiratory Syndrome (SARS)

Further Reading

Abraham , T. (2004). *Twenty-first century plague: The story of SARS* (Hong Kong: Hong Kong University Press)

Chan, J. C. K., & Wong, V. C. W. T. (eds.) *Challenges of Severe Acute Respiratory Syndrome.* Singapore: Elsevier.

Enserink, M. (2003). SARS in China: China's missed chance. *Science*, 301, 294-296.

Fouchier, R. et al. (2003). Koch's postulates fulfilled for SARS virus. *Nature*, 423, 240.

Guan, Y. et al. (2003). Isolation and characterization of viruses related to the SARS coronavirus from animals in southern China [electronic version]. *Science,* 302(5643), 276-278.

Normile, D. (2004). Viral DNA match spurs China's civet roundup. *Science,* 303(5656), 292.

Pottinger, M., Cherney, E., Naik, G., & Waldholz, M. (2003). How a global effort identified SARS virus in a matter of weeks. *Wall Street Journal* 16 April 2003, A1.

Wong, S. L., Hodson, D., Kwan, J., & Yung, B. H. W. (2008). Turning crisis into opportunity: enhancing student-teachers' understanding of nature of science and scientific inquiry through a case study of the scientific research in severe acute respiratory syndrome. *International Journal of Science Education, 30*(11), 1417-1439.

For Discussion

Discuss the following controversial issues arising from the control of SARS. In debating these issues, critically evaluate the arguments and counter-arguments of different decision alternatives, and their underlying values.

1. Should the government allow public access to information concerning the addresses of SARS patients, down to the street, or apartment/building number?

2. Was it appropriate for the Singapore government to install video cameras to monitor whether SARS patients were observing the home quarantine orders?

3. Should the government impose some restrictions on the mass media in crisis situations to minimize sensationalism to avoid undue panic?

4. During the SARS crisis, the WHO's influenza program manager, Klaus Stohr, appealed to scientists to "share data and set aside Nobel Prize interests or their desire to publish articles in Nature" (Pottinger et al. 2003, p. A1). Nonetheless, competition was still keen in the research on SARS. Abraham (2004, p. 93) commented on the cause underlying this intense competition candidly, "There are few prizes in science for coming in second".

   Discuss how collaboration and competition can be reconciled to gain maximum benefit from each. What are some feasible ways of enhancing collaboration, while recognizing the contribution of individual scientists or research teams? How feasible would it be to use special awards, akin to the Nobel Prize, to recognize collaboration among different institutes to enhance science communication in the long run?

Projects

1. Simulated inquiry

During the SARS epidemic, some people living in Western countries believed that SARS was associated with the Chinese community, its people and its food. If you were a local government health official undertaking an inquiry into whether such a claim is justified or not, how would you plan your investigation? What specific questions would you ask? What scientific evidence do you need to answer these questions?

2. Case study
Choose a health scare, sudden epidemic, or outbreak of a disease which has affected or is affecting your city/state/country, and neighboring regions. Then collect evidence from official documents, news-reports, and web-based sources to help you analyse the following:

1. What roles do scientists, health professionals, government health officials, and lay citizens play in controlling the disease or crisis? To what extent are citizens' actions based on scientific evidence? In what ways does the nature of science communication impact policy-making and citizen action?
2. What factors facilitate or impede the control of the crisis?
3. How and to what extent is the handling of the crisis different between your state/country and neighboring states/countries? What are the possible reasons for these discrepancies, if any?
4. Could the problem be traced to wider social contexts, such as power distribution in society, information flow, health care governance, human rights, or political and economic concerns? Explain your answer.

Chapter 15: Communication Challenges for Sustainability

Further Reading

Keen, M., Brown, V. A., & Dyball, R. (Eds) *Social Learning in Environmental Management*. London: Earthscan, (2005).
MacDonald, C. (2011). Responsibility, meet transparency. *Miller-McCune*, July/Aug., 48-59.
Meppem, T., & Bourke, S. (1999). Different ways of knowing: A communicative turn toward sustainability. *Ecological Economics*, 30, 389-404.
Turner, B. L. II. (2010). Vulnerability and resilience: Coalescing or paralleling approaches for sustainability science? *Global Environmental Change*, 20, 570-576.

For Discussion

Find an article in an academic journal with the word "sustainability" in its title. Then find a news story about "sustainability." Compare their definitions and conceptualizations of the term and what it means. Do you agree with their definitions or apparent definitions? Try to write your own definition of "sustainability."

Projects

1. Calculate your carbon footprint. Search the Internet for several calculators and compare results. Make a change in your behavior and recalculate your footprint. Comment on this result.
2. Visit the website of a major city on your country or a large national company. How do they describe "sustainability" or their environmental commitment? What parts of the definition of *sustain* and *ability* do they utilize? Comment on these findings with reference to this chapter.
3. Choose a common item in your home such as an item of clothing or food. Research and comment upon the production and transportation of that item and its ingredients (e.g. cotton, plastic, or a specific chemical). Investigate some of the tradeoffs or cascading effects that might be associated with the item or a particular ingredient in it. How much of this production process and these trade-offs are visible to the consumer?

Chapter 16: The Value of Indigenous Knowledge Systems in the 21st Century

Further Reading

Fayola, T. (2000). *Africa volume 2: African Cultures and Societies before 1885.* Durham: Carolina Academic press.

Grinker, R. R., & Steiner, C. B. (1997). *Perspectives on Africa.* Cambridge (Massachusetts): Blackwell Publications

Harding, S. (1993). *The racial economy of science.* Bloomington: Indiana University Press.

Higgs, P. (2006). In defence of local knowledge: A theoretical reflection. *Indilinga Africa Journal of Indigenous Knowledge Systems,* 5, 1.

Hoppers, C. A. O. (2002). *Indigenous Knowledge systems and the integration of Knowledge Systems.* Claremont: New Africa Books.

Latour, B. (1991). *We have never been modern.* Cambridge, Massachusetts: Harvard University Press, (Translation in 1993).

Makgoba, W. M. (1999). *African Renaissance: The New Struggle.* Sandton: Mafube/ Tafelberg.

Nabudere, D. W. (2007). Chaikh Anta Diop: The social sciences, humanities and physical sciences and transdisciplinarity. *African Renaissance studies,* 2, 1.

Reagan, T. (2005). *Non-Western educational traditions: Indigenous approaches to educational thought and practice* (3rd Ed.). Hillsdale, N.J.: Erlbaum.

Samovar, L. A., Porter, R. E., & McDaniel, E. R. (2010). *Communication between cultures.* Boston: Wadsworth.

Seleti, Y. (2004). *Africa since 1990.* Cape Town: Ministry of Education, New Africa Books.

For Discussion

1. How relevant is the issue of indigenous knowledge to the people of your own country? Would you include local knowledge on the part of farmers or fishers in this category?
2. Do you think it is possible to integrate Western Science knowledge and Indigenous Science knowledge? What would be the benefits?
3. Do you agree that Latour's four premises of modernity have shaped the modern world?

Project

Investigate the controversies surrounding biodiversity and ownership of indigenous and local knowledge. How might equitable sharing of this knowledge occur?

Chapter 17: Science Communication: The Consequences of Being Human

Further Reading

Dawkins, R. (2006). *The God delusion*. London: Bantam Press.

Gould, S. J. (1996). *Life's grandeur*. London: Jonathan Cape.

Mooney, C., & Kirshenbaum, S. (2009). *Unscientific America: How scientific illiteracy threatens our future*. Philadelphia: Basic Books.

Ridley, M. (2010). *The rational optimist: How prosperity evolves*. London: Harper Collins.

For Discussion

1. Do you agree that there will ever be a point where science communicators are no longer needed?
2. What impact do you consider the Internet may have on human intelligence?
3. How important do you consider new media to be in communicating science, especially the impact of new technologies? Consider both positive and negative aspects.

Projects

1. Repeat Kaufmann's button and strings experiment. What effect does varying the ratio of the number of buttons to the number of strings have on the outcome? What does this mean for evolutionary processes?
2. Set up your own Twitter site to develop a conversation about an aspect of science communication.
3. Use the media (past and present, including electronic) to compile a dossier on a common resurgent diseases such as mumps, measles, rubella or

tuberculosis. Can this be correlated with the rise of antivaccination sentiment?

4. Prepare and carry out an informal survey on the acceptance of evolution in your own community.

ABOUT THE AUTHORS

Chris Bryant AM has a BSc in Zoology from King's College London, an MSc in Biochemistry from University College London, and a PhD in Biochemistry from King's College London. Joining the Australian National University in 1963, he has served as Professor of Zoology, Dean of Science, and Head of the School of Biology. Together with Dr. Mike Gore, he set up the ANU Shell Questacon Science Circus which lead to the establishment of the Australian National Centre for the Public Awareness of Science. Professor Bryant was admitted to The Order of Australia in 1999. email: chrisandanne@grapevine.com.au

Maurice M.W. Cheng is Assistant Professor in the Faculty of Education, The University of Hong Kong. His major research interests are in the teaching and learning of science, with particular focuses on the roles and uses of diagrams, mental visual representations, scientific models, and modelling. His teaching areas include quantitative literacy, science and chemistry education, and learning psychology. He is a registered pharmacist and has taught in secondary school, and he serves on the official curriculum and assessment committees (science and chemistry subjects) of Hong Kong. email: mwcheng@hkucc.hku.hk

Julia B. Corbett is a Professor in the Department of Communication at the University of Utah, USA. She studies science, environmental, and health communication both from a macro-sociological view of social conflict and change, and a micro-social view of attitude and behavior change. She authored one of the first texts in environmental communication, *Communicating Nature: How We Create and Understand Environmental Messages* (2006, Island Press). Recently, her scholarship has shifted to include creative nonfiction essays about human relationships with the natural world. email: corbett.julia@gmail.com

Justin Dillon is Professor of Science and Environmental Education and Head of the Science and Technology Education Group at King's College London. After taking a degree in chemistry he taught in six inner London schools until 1989 when he joined King's. Justin is co-editor of the *International Journal of Science Education* and was President of the European Science Education Research Association from 2007-11. He is one of the coordinators of the Economic Social Research Council's Targeted Initiative on Science and Mathematics Education and is involved in R&D projects with museums, science centres, and botanic gardens in Europe and beyond. email: justin.dillon@kcl.ac.uk

John K. Gilbert is Professor Emeritus of The University of Reading, Visiting Professor of King's College London, Editor-in-Chief of International Journal of Science Education (IJSE)(A), and Co-Editor (with Susan Stocklmayer) of *IJSE(B) Communication and Public Engagement*. Against a background in chemistry, he has progressively worked on conceptual development, models and modelling, visualization, all in respect of the learning of science. Most recently he has begun to link these interests to informal science education for all. email: john.k.gilbert@btinternet.com

Will J. Grant is a researcher and lecturer at the Australian National Centre for the Public Awareness of Science (CPAS) at the Australian National University. Will holds a BA (Hons) and a PhD in Political Science (The University of Queensland). Will's research, teaching, speaking, and writing has focused mostly on the intersection of science, politics and society, and how existing relationships in this intersection are changing in response to new technologies. Current projects include longconversations.net, howtokeepfood.com and society5.net. email: will.grant@anu.edu.au

Marie Hobson is the Senior Audience Advocate at the Science Museum London. Using evidence from research conducted by herself and her team, she advocates on behalf of audiences to ensure that visitors' needs and wants are met and barriers to engagement removed. Marie completed her MA in Museum Studies at the University of Leicester in 2008. email: mchobson84@gmail.com

Rod Lamberts is the Deputy Director of the Australian National Centre for the Public Awareness of Science at the Australian National University. He has been a presence in the international science communication space for 15 years and supervises students across a diverse range of science communication research projects at the Honours, Masters, and PhD levels. Rod has been a pioneer in the development and delivery of science communication courses in Australia since 2000. He is also a science communication consultant for UNESCO in the Pacific, and regular public commentator on science in the media, and science and public policy. email: rod.lamberts@anu.edu.au

Arthur M.S. Lee is a Teaching Consultant at the Faculty of Education in the University of Hong Kong. He was a secondary school mathematics teacher in Hong Kong for 12 years before joining the University. He is interested in understanding mathematics teaching and learning in technology rich environments. His current PhD study focuses on students' conception in geometry through exploratory tasks designed with dynamic geometry tools. Since 2009, he has taken part in design and teaching of new courses with colleagues in the Faculty of Education, focusing on developing quantitative literacy among first year undergraduates. email: amslee@hku.hk

Yeung Chung Lee is an Associate Professor at the Department of Science and Environmental Studies, Hong Kong Institute of Education, Hong Kong SAR, China. He teaches primary science education, biology education and health education from bachelor to doctorate levels. His research focuses on Science-Technology-Society-Environment education, the nature of science, inquiry-based learning, and the use of models and analogies for developing students' conceptual understanding and metacognitive reflection. His recent interest is studying secondary students' decision-making about socio-scientific issues from cross-contextual and cross-cultural perspectives. email: yclee@ied.edu.hk

Ida Ah Chee Mok is Associate Professor and Associate Dean in the Faculty of Education at Hong Kong University. Her research interest includes mathematics teaching and learning and teacher education. Her teaching areas include quantitative literacy and mathematics education. She is co-editor of *Making Connections: Comparing Mathematics Classrooms Around the World* and author of *Learning of Algebra: Inspiration from Students' Understanding of the Distribution Law*. email: iacmok@hku.hk

Masakata Ogawa is Professor of Science Education and Dean of the Graduate School of Mathematics and Science Education at Tokyo University of Science. After a doctorate in plant physiology at Kyoto University, he taught science education at Ibaraki University, Hiroshima University, and Kobe University. His research interests include cultural and policy aspects of science education, science teacher education and science literacy. He received the Award for Distinguished Contributions through Research from the Japan Society for Science Education in 2003. He served as President of the Japan Society for Science Education and President of the East Asian Association for Science Education. email: ogawam@rs.kagu.tus.ac.jp

Lindy A. Orthia is a lecturer in science communication at the Australian National Centre for the Public Awareness of Science at the Australian National University. She trained as a biologist (La Trobe University) and worked for several scientific research organisations before changing disciplines to science communication. Her PhD thesis examined representations of the political, social, economic, and cultural aspects of science in the British television series *Doctor Who*, and science in popular fiction continues to be her primary research interest. email: lindy.orthia@anu.edu.au

Sean Perera is a Research Fellow at the Australian National Centre for the Public Awareness of Science at the Australian National University. Sean's research explores aspects of science education from a science communication perspective and covers teachers' professional development, informal science learning, and communicating science to culturally diverse audiences. He is currently developing programmes to enable participation of non-traditional audiences—such as mothers, grandparents and migrant refugees—with mainstream science communication. After training as a microbiologist and crop scientist (U. Bangalore and U. Peradeniya), Sean completed his PhD in science communication at ANU. email: sean.perera@anu.edu.au

Michael J. Reiss is a biologist and educator. He is Pro-Director: Research and Development and Professor of Science Education at the Institute of Education, University of London, Chief Executive of Science Learning Centre London, Vice President and Honorary Fellow of the British Science Association, Honorary Visiting Professor at the Universities of Birmingham and York and the Royal Veterinary College, Honorary Fellow of the College of Teachers, Docent at the University of Helsinki, Director of the Salters-Nuffield Advanced Biology Project, a member of the Farm Animal Welfare Committee, and an Academician of the Academy of Social Sciences. email: m.reiss@ioe.ac.uk

Léonie J. Rennie is Emeritus Professor of Science and Technology in the Science and Mathematics Centre at Curtin University and Adjunct Research Professor in science communication at the University of Western Australia. Her research focuses on learning science and technology in integrated and out-of-school contexts, and scientific literacy. She has co-authored significant reports for the Australian Government and been involved in nationally funded programs raising the awareness of science in the community. She serves on the editorial boards of several journals. In 2009, she received the Distinguished Contributions to Science Education Through Research Award from the National Association for Research in Science Teaching. email: l.rennie@curtin.edu.au

Suzette D. Searle is a Postdoctoral Research Fellow with the Australian National Centre for the Public Awareness of Science where she focuses on surveying public opinion on science-related topics. Her PhD thesis was "Scientists' communication with the general public: An Australian survey". Before working in science communication for the last 15 years, she was awarded two degrees in forestry and worked for 17 years for CSIRO, the Australian national science research organization. email: suzette.searle@anu.edu.au

Yonah Seleti is the Acting Deputy Director General for Human Capital and Knowledge Systems in the Department of Science and Technology in South Africa. He is also responsible for the National Indigenous Knowledge Systems Office. He has been a leader in the integration of African traditional medicines through the "Farmer to Pharma" initiative, bringing together indigenous knowledge, biodiversity, and biotechnology. He has also taught as visiting professor at Tulane University (New Orleans) and Roskilde University in Denmark. He has been a member of several ministerial committees and is chairperson of the Digital Innovation of South Africa, a digital library of liberation heritage. email: yohan.seleti@dst.gov.za

Susan Stocklmayer AM is Professor of Science Communication and the Director of the Australian National Centre for the Public Awareness of Science, which is also the UNESCO Centre for science communication. Her major research concerns issues related to science learning at the interface between science and the public, and gender and multicultural issues. As part of the University's outreach she has presented science shows, lectures and workshops on all five continents. She is the co-Editor in Chief of the *International Journal of Science Education Part B: Communication and Public Engagement* and was awarded The Order of Australia in 2004. email: sue.stocklmayer@anu.edu.au

Craig Trumbo earned a BA and MS in journalism at Iowa State University and a Ph.D. in Mass Communications at the University of Wisconsin-Madison with a focus on environmental health risk communication. He also holds a certificate degree in Geographic Information Systems from Pennsylvania State University. He is currently Professor in the Department of Journalism and Technical Communication at Colorado State University. Professor Trumbo's research addresses a range of interests located at the intersection of health, risk, and the environment. His areas of university teaching include mass media effects, mass and interpersonal communication theory, communication of science and technology, and risk communication. email: ctrumbo@mac.com

Ka Lok Wong is a Teaching Consultant at the Faculty of Education, The University of Hong Kong, where he received his BSc (mathematics) and Cert. Ed. (mathematics). He also holds a BA (philosophy) from the Catholic University of Leuven, Belgium, and an MA (mathematics education) from the Institute of Education, London. After teaching secondary school mathematics, he became involved in mathematics teacher education. Recently he has also been involved in teacher education for liberal studies where he sees the relevance of quantitative and scientific literacy. email: klwong3@hku.hk

INDEX